T3-BSC-675

GLOBAL CLIMATE CHANGE

HUMAN AND NATURAL INFLUENCES

GLOBAL CLIMATE CHANGE

HUMAN AND NATURAL INFLUENCES

Edited by

S. Fred Singer

An ICUS Book

PARAGON HOUSE
New York

Published in the United States by
ICUS
481 8th Avenue
New York, New York 10001

Distributed by Paragon House Publishers
90 Fifth Avenue
New York, New York 10011

Library of Congress Cataloging-in-Publication Data

Global climate changes : human and natural influences / edited by
S. Fred Singer.
p. cm.
"An ICUS book."
Includes index.
ISBN 0-89226-033-5 : $34.95
1. Climatic changes. 2. Man—Influence on nature. I. Singer, S. Fred (Siegfried Fred), 1924–
QC981.8.C5G66 1989
551.6–dc20 89-32884
CIP

TABLE OF CONTENTS

PART THREE: HYDROSPHERE

PART FOUR: HUMAN INVOLVEMENT

PART FIVE: NATURAL PROCESSES

OVERVIEW

S. Fred Singer

Within the last decade many people have become increasingly concerned with our future on this planet. Aside from its profound philosophical aspects, the subject also attracts interdisciplinary scientific interest and, of course, international concern. This volume, *Global Climate Change: Human and Natural Influences*, is a result of that concern.

Because of the magnitude of our concern, it seemed appropriate to deal with the global environment from a scientific viewpoint and to review the current status of a variety of issues. We focused on the interaction of the environment—atmosphere, land, water and especially ocean—with ecological resources. These resources provide us with a healthful environment in which to live and with food, probably the most basic resource.

This volume deals with three categories of problems. The first arises as the result of inadvertent byproducts of human activities. That is, byproducts are not intended to occur, but their occurrence has important worldwide effects that manifest themselves in chemical changes of the environment, and ultimately in climate changes and perhaps other changes that we do not yet understand.

The second category has to do with modifications of the environment which are purposeful, that is, directed by humans, but with long-range problems that are not desirable. For example, agriculture is certainly a major modification of the land environment. It has many desirable consequences—it provides us with food—but has long-range implications which need to be examined. The provision and transport of water is another important modification. The discharge of wastes into the ocean, the ultimate sink, is something that we need to be concerned about.

Finally, the third category deals with "natural" global environmental issues, i.e., "natural" in the sense that they are not controlled by human intervention but do have important consequences for us. Catastrophic events, of course, are easy to recognize. These are principally volcanic eruptions, earthquakes, and natural disasters which we have not experienced recently, such as impacts of large meteorites or comets. These events can kill populations of living creatures on this earth and are therefore cataclysmic. But we also have gradual natural changes that are difficult to recognize because they are very slow, e.g., changes in the climate. As we will discuss later, there are also many changes in the environment brought about by living creatures other than human beings.

Among the inadvertent modifications of the environment, many have to do with the generation of energy. Certainly the most important one is the generation of carbon dioxide as a result of the burning of the fossil fuels: oil, coal, and natural gas. Carbon dioxide is an innocuous gas: we exhale it continuously, it is naturally present in the atmosphere and is not toxic. We have observed that the concentration of carbon dioxide in the atmosphere has now increased by about 25 percent over its pre-industrial value. More than that, we anticipate that the concentration will continue to increase because we will continue to burn fossil fuels. We anticipate that sometime in the first half of the next century the carbon dioxide concentration in the atmosphere will have doubled.

What is the consequence of such a major change in our natural environment? Here is where differences of opinion emerge. The general view is that the planet will become warmer as a consequence of the increase of carbon dioxide (the enhanced "greenhouse effect"), but there is disagreement about the magnitude of the effect. There are those who argue that the temperature increase will be extreme enough to melt the ice caps and glaciers, raise the sea level, and perhaps inundate coastal cities. Others hold that the warming will be compensated by other effects that cancel most of the temperature increase, so that it will be lost in the "noise" of the natural climatic fluctuations. The majority of scientists believe that there will be modest warming, enough to be noticeable and enough to make a difference in the distribution of agriculture on this planet.

An agreement among the majority of scientists does not mean that that majority is correct. It only means that the majority of scientists agree on something. The minority may well be correct; it could be a minority arguing for a very large increase in temperature (a catastrophic increase, if you like) or a minority arguing that very little will happen. Nevertheless, assuming the consensus is correct, the climate will warm by 2°C by the next century. Our main discussion deals with the consequences to be expected as a result of such a secular change in the climate, and how this will affect living conditions on the planet. The results are not catastrophic. In some cases, they will result in hardship, and in other cases they will result in improvements. Agriculturists believe that as a result of the higher concentration of carbon dioxide we will have larger crops.

The second problem we discuss is that of methane, a rare gas in the atmosphere, whose concentration seems to have been increasing rapidly. The exact cause of the increase is something of a mystery; it is presumed to involve human activities. A large amount of methane comes from swamps which, of course, occur naturally; rice paddies are also a source, as are cattle. Numbers of cattle and other ruminants have increased as people eat more meat. Termites are believed to make a large contribution to atmospheric methane. Finally, energy production is also a source, because natural gas (methane) escapes into the atmosphere in the production of oil, gas, and coal. The effect of all of this methane on the climate should be similar to that of CO_2 and should reinforce the greenhouse effect.

A third important gas that affects the climate is ozone. Scientists are not sure yet how our activities on this planet affect the distribution of ozone. In specific places the concentration of ozone may increase, producing what we call photochemical smog, similar to that in Los Angeles and other cities that have peculiar meteorological problems. On the other hand, the concentration of ozone high up in the stratosphere may decrease as a result of other human activities. Important agents include the Freon compounds widely used as propellants in spray cans and for refrigeration. But if their manufacture and release were controlled, there are still other influences on ozone; for example, an increase in methane can affect the amount of stratospheric water vapor and ozone. Such ozone changes are important to climate, but there are also possible consequences to human health involved, which need to be considered.

Finally, in the burning of fuels, particularly coal, a large amount of sulfur dioxide is generated, the sulfur being a constituent of the coal. Also, atmospheric nitrogen is oxidized. These two constituents acidify the rain. Acid rain is a natural phenomenon, but an increase in both the acidity of the rain and its distribution has been observed. There is little question that the increase is related to human activities.

The real controversy arises about the ecological effects of acid rain. Some believe that in the northeastern United States and Canada, as well as in Scandinavia, acid rain will not only kill fish populations in certain small lakes but will also destroy the soil that maintains the health of forests. Indeed, in some places in the world, particularly in Germany, people have become very alarmed about the death of trees. However, the evidence is ambiguous. We could not discover in our discussion any single mechanism that would adequately explain what is happening to these trees. Indeed, the chemicals in acid rain also act as a fertilizer that can help the growth of trees and crops. The state of knowledge is not yet adequate to decide these questions.

Turning to purposeful modifications of the environment, we discuss human interference with the hydrological cycle, that is, with the distribution of water on the surface of the globe. Water—not just drinking water, but water for agricultural use—is vital to human survival. The greatest use of water in the world today is for irrigation. The greatest threats to human life, now that the great plagues have been suppressed, are famines; these are largely due to periods of low rainfall in particularly sensitive regions. The word often used is "desertification," which means the creation of deserts. Indeed, deserts have expanded in certain areas, particularly in Africa, for reasons that we now understand fairly well. The variability of rainfall in these sensitive regions causes famines, for example in the Sahel region of North Africa, from Senegal in the West to Ethiopia in the East.

The increasing use of water for irrigation is producing worldwide effects because it is putting more water vapor into the atmosphere by increasing evapotranspiration. At the same time, the spread of agriculture is eliminating forests throughout the world, and by substituting grass, maize or other crops, one decreases evapotranspiration. Indeed, as far as one can tell right now, human activities have put less water vapor into the atmosphere and more water into rivers and into the ocean directly.

Irrigation canals were built in the Middle East four to five thousand years ago. The Romans built aqueducts. Today we build pipes with pumps, but it is expensive to move water over long distances. Desalination of sea water is even more expensive. One of the cheapest methods for increasing the supply of water is to use it in a better way, that is, to use less of it for irrigation and to use it more efficiently. This forms an important subject for our discussion. We are able to derive simple rules for improving water use efficiency, which should be of value in many parts of the world. Unfortunately, some of these conservation methods are also expensive.

A cheap method for increasing water supply is to increase the rainfall by cloud seeding. That method seems to work in certain regions of the world. Meteorological science has shown under what conditions such increases in rainfall can be brought about reliably. Statistical methods prove that these projects are indeed working.

A completely new method of increasing the amount of rainfall depends on putting more water vapor into the atmosphere by increasing the heat stored in the top layer of the ocean, and then using this stored heat in the winter to evaporate more water into the atmosphere. Instead of wasting the energy by evaporating moisture during the summer, it is saved for the winter when rainfall is more prevalent. This method may work in the Eastern Mediterranean, where the right conditions exist. Mixing the top ocean layer can increase the amount of heat stored during the summer. The scheme is highly efficient: an input of energy can gain a factor of 20,000 in terms of heat stored. However, even if the active mixing project is not undertaken, the method can be used as a predictive tool to tell how much rain is likely in the coming winter.

We now turn to ocean pollution as an example of human intervention with the environment. As the ocean is the ultimate sink for most human wastes (and parts of the ocean are already being overloaded) there is, of course, concern about its health. The three guidelines that one needs to look at in discharging wastes are toxicity, the persistence of certain wastes, and the accumulation of toxic materials in living things in the ocean, that is, bioaccumulation problems. The substances discharged into the ocean cannot even be listed (there are so many), but roughly speaking they are radioactivity, toxic chemicals of all sorts, toxic metals, sewage, sludge, and oil. A great amount of research is still required to understand how pollutants behave in the ocean, how they move, and what happens to them, so that decisions can be made regarding the best way to reduce damage.

The final set of papers deals with modifications of the global environment that are not controlled or even affected by human activities. Catastrophic modifications are very easy to recognize because they occur suddenly and often unexpectedly. Comets and meteorites are plentiful in the solar system; statistically, one would expect that every so often one will hit the earth. We know that for certain, because we see impact craters on the earth, moon, and other celestial bodies. A body even as small as one km. in diameter may be sufficient to produce a cataclysmic event, by which we mean a major extinction of life on the earth. It has become a widely accepted theory in scientific circles that the extinction of the dinosaurs, 65 million years ago, was due to such an event. In fact, it may be that the transitions between geological ages, which are punctuated by large-scale extinctions and disturbances, occurred as a result of such impacts.

The largest volcanic eruptions on the earth produce effects similar to the impacts of meteorites. The largest volcanic eruption is equivalent in energy to the impact of a meteorite 0.5 km. in diameter. The common feature of these impacts and of all volcanic eruptions is that there is a lot of energy released at one time in one place on the surface of the earth. The amount of energy has to escape somehow; it pushes up through the atmosphere, which is relatively thin and light, so that much of the material from the surface (from the impact crater or the volcano) ends up in the high atmosphere, where it can remain for many years. Such long residence times are not possible in the lower atmosphere because the material will rain out in a matter of days, and will therefore produce only minor climatic influences. If enough material is put into the stratosphere it will essentially "turn off" the sun. As far as the surface of the earth is concerned, the effect is to decrease photosynthetic activity and kill plants, so that many animals will starve.

The climatic effect, of course, depends on many details, such as the size of the particles, their optical properties, and their lifetimes. Little is known about these. Would nuclear exchanges between the superpowers produce similar effects, as has been asserted recently? We are not sure that nuclear exchanges would necessarily produce worldwide cataclysmic effects because they are quite different in character from the impacts of meteorites and from volcanic eruptions, so that one cannot draw a valid analogy.

The final topic deals with more gradual changes in the earth's environment, in particular with James Lovelock's "Gaia" hypothesis. Normally it is thought that the physical environment controls the biosphere. Lovelock's challenging view, that there is a homeostatic mechanism on this planet whereby the biosphere controls the evolution of the earth, has raised a great amount of discussion. We do know that the origin of the oxygen of the earth's atmosphere, on which we all depend, is due to the development of biota early in the history of the earth. (There was no atmospheric oxygen when the earth was formed.) However, the question of whether there exists a purposeful Gaia, a sort of "Mother Earth" controlling in a beneficial way the physical environment, cannot yet be answered with certainty.

The topics discussed in this volume are diverse; they deal with the atmosphere and hydrosphere, with human and with natural influences that change the environment. Yet the topics share common themes, difficult and controversial to describe and predict, but vital to human survival on this planet.

PART ONE:

CLIMATE

ONE

MECHANISMS of CLIMATIC CHANGE

Curt Covey

The purpose of this chapter is to give an overview of the physical mechanisms responsible for climatic change, providing an orientation for the chapters that follow. The stress is on fundamental concepts, and no attempt has been made to present a comprehensive survey of the many different mechanisms proposed to explain various climatic changes, or to provide a complete bibliography. Reference is made, however, to a selection of books and review articles which will provide more details and documentation.

First, we must define the term "climate." The word denotes a summary of the atmospheric environment—including broad-perspective quantities such as regional average temperature—rather than specific events such as a rainstorm occurring at a particular time and place. Although specifying the climate obviously involves taking averages, there is more to it than that. It is often important to know extreme cases: how cold was the century's worst winter? How dry was the worst drought? Formally, a climatic state may be described by a set of statistical properties—averages, standard deviations and higher moments—of atmospheric variables (pressure, temperature, cloudiness, wind velocity, etc.) and related quantities such as ocean temperatures and glacial ice extent (National Academy of Sciences 1975).

Defining the climate statistically can easily lead to the tacit assumption that the climate is stable, that once the statistics are gathered over a few decades for a given region, one may assume that they will not change in the future. Quite the opposite is the case. Geologists have long realized that the climate at certain times in the past, for example during the Ice Ages, was radically different from climate today (Imbrie and Imbrie 1979). More recently, persuasive evidence has been compiled which shows that climatic change takes place on all time scales, from slow geologic evolution to decade-by-decade variations (Lamb 1980). During the past few years, interest in climatic change has been spurred by concern over humanity's impact on climate (Schneider and Londer 1984) and by explorations of Venus and Mars (Pollack 1979) which raise the question of why Earth alone evolved into a habitable world (Walker 1977, Kasting *et al.* 1988).

All of this has focused attention on the question of what determines the climate and causes it to change. As a result the field of climatology has been transformed from a descriptive endeavor to a quantitative science, with theories expressed in terms of numerical models. Given the complexity of the system and the variety of mechanisms that undoubtedly are important, it is not surprising that our knowledge of the subject is incomplete and uncertain. But a framework for thinking about the problem has been firmly established, and some fairly sophisticated models have been developed for simulating climatic change.

Basic Physical Concepts

Global Energy Balance

Many of the important ideas about climatic change can be illustrated by considering the average, over horizontal area and altitude, of the net energy input to the earth-atmosphere system:

$$C\frac{dT_S}{dt} = (S/4)\,(1-\alpha) - F_{IR} \qquad (1)$$

T_S is surface temperature averaged over longitude and latitude and C is effective heat capacity per unit area; so the left side of the equation is the rate at which heat is stored by the earth-atmosphere system (in units of Watts per square meter). Solar radiation, the source of energy for the earth's weather and climate, appears on the right side. S is the "solar constant," the flux of solar energy through a plane perpendicular to the beam at the top of the atmosphere (about 1370 W m-2). Despite the name, the flux is not absolutely constant; it varies slightly since the earth's orbit is not a perfect circle, and recent satellite measurements imply that the sun's emitted power (the "solar luminosity") has varied by a few hundreths of a percent over a period of several years (Willson and Hudson 1988). Averaged over the earth's surface, the solar flux is $\pi R^2 S/4\pi R^2 = S/4$ (where R is the earth's radius). A fraction (the planetary albedo) is reflected back into space and the remainder is absorbed by the earth-atmosphere system. Electromagnetic radiation is also emitted by the Earth and its atmosphere. Almost all of this is in the infrared portion of the spectrum, while solar radiation is nearly all in the visible portion. Therefore one may consider the two types separately. In equation (1) the infrared flux emitted to space by the earth-atmosphere system is denoted by F_{IR}.

In many applications one assumes that the climate is in global equilibrium, in other words, that the left side of equation (1) is zero. In that case the net incoming solar radiation and the outgoing infrared radiation must balance. If, for example, the solar luminosity were to increase, then S would increase and in equilibrium F_{IR} must increase by an equal amount. Since one expects F_{IR} to increase with increasing T_S, we reach the reasonable conclusion that increased solar luminosity leads to increased surface temperature. How much T_S changes for a given change in S is a measure of the sensitivity of the climate to external perturbations. Climatic sensitivity is determined by a host of feedback processes. For example, if T_S increases, then the amount of snow and ice covering Earth's surface ought to decrease. Since snow and ice reflect sunlight very efficiently, this would lead to smaller α and hence to an increase in the absorbed solar energy $S(1 - \alpha)$, reinforcing the original perturbation. This is a case of positive feedback. At the same time, the increase in F_{IR} when T_S increases will oppose the original perturbation (negative feedback.)

It is reasonable to assume that the earth's present climate is close to equilibrium. However, there may well be other equilibrium states even for the same value of external forcing S. For example, a completely frozen Earth could conceivably balance small F_{IR} (due to small T_S) with small $(S/4)(1 - \alpha)$ arising from a value of α near unity. But in fact the geological record shows that liquid water has existed continuously on the earth for nearly four billion years. Even though during the Ice Ages some 30 percent of the earth's land surface was covered by glacial ice, the positive feedback between snow and ice and albedo was not strong enough to drive the earth to the ultimate "ice catastrophe." Nevertheless, the multiplicity of nonlinear feedbacks in the climate system implies that one must always be aware of the possibility of multiple equilibrium states and of transitions among them. Random transitions between equilibrium states would be examples of internally controlled climatic change, in contrast to the externally forced climatic changes which result from changes in parameters that are external to the climate system, for example changes in S.

Although climatic change is often thought of as a change from one equilibrium state to another, the transient response, in which the left side of equation (1) is nonzero, will be important whenever the forcing varies quickly compared with the characteristic time it takes for the system to adjust to a perturbation by coming to a new equilibrium. The characteristic response time depends on the effective heat capacity C and also on the same feedback processes that determine the equilibrium sensitivity; the greater the equilibrium sensitivity, the longer it will take the system to approach to within a given fraction of the new equilibrium.

The concepts discussed above can be demonstrated by use of equation (1) and similar "energy balance" climate models, provided that quantities such as α and F_{IR} can be parameterized in a simple way in terms of surface temperature T_S (North *et al.* 1981). Figure 1-1 provides an example. Here both F_{IR} and $(S/4)(1 - \alpha)$ are plotted as functions of T_S. The former increases monotonically with T_S. The latter is assumed constant at very low T_S, where Earth is completely frozen, and at high T_S, where all snow and ice have melted. Between these two extremes $(S/4)(1 - \alpha)$ increases with T_S because α decreases with T_S.

As drawn, the two curves intersect at three points, labeled A, B, and C, with corresponding temperatures T_A, T_B, and T_C.

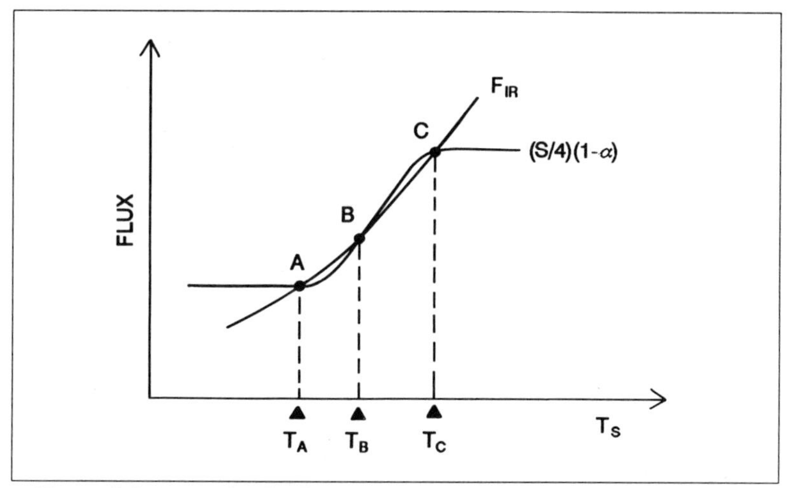

Figure 1-1: A very simple climate model may be constructed by assuming that the average solar flux absorbed by earth and its atmosphere (S/4) (1-α), and the average infrared flux to space, F_{IR}, are both well-defined functions of average surface air temperature T_S. The intersections of the curves then determine possible equilibrium climates.

These represent three equilibrium climates which satisfy equation (1) when the left side is zero. One of these, B, is an unstable equilibrium, analogous to a pencil balanced precisely on its point. The slightest perturbation will cause the system to move away from state B; if T_S increases, moving the climate to the right of point B, then (S/4)(1 - α) will exceed F_{IR}, causing T_S to continue to increase, and if T_S decreases slightly then the opposite will happen and T_S will continue to decrease. Therefore state B is not a physically realizable equilibrium climate. A similar analysis shows that states A and C, in contrast, are in stable equilibrium. State A represents the completely frozen Earth discussed above, and state C may be identified with the present-day climate.

Externally forced climatic changes may be represented in Figure 1-1 by shifting the curves. Increasing solar luminosity, for example, shifts the (S/4)(1 - α) curve upward and causes T_S to increase for both of the physically realizable climatic states. The greenhouse effect, which occurs when substances in the atmosphere trap outgoing infrared radiation, can be represented by lowering the F_{IR} curve to account for the fact that less infrared radiation escapes into space for a given T_S. Note that this also causes increased T_A and T_C.

An analytic treatment of the problem becomes possible if one assumes simple forms for $\alpha(T_S)$ and $F_{IR}(T_S)$. The following example is taken from Schneider and Thompson (1981), who apply it to the CO_2 greenhouse problem. Assuming that the decrease in α with T_S and the increase in F_{IR} with T_S are both linear, that is,

$$F_{IR} = A + BT_S \tag{2}$$

and

$$\alpha = \alpha_0 - \alpha_1 T_S \tag{3}$$

where A, B, a_0, and α_1 are all positive constants, one may rewrite Equation 1 as

$$\frac{dT_S}{dt} + \frac{B - S\alpha_1/4}{C} T_S = \frac{(S/4)(1 - \alpha_0)}{C} - \frac{A}{C} \tag{4}$$

This is a linear, first-order differential equation with constant coefficients. A particular solution may be found by setting the derivative term to zero, obtaining

$$T_S = \frac{(S/4)(1 - \alpha_0) - A}{B - \alpha_1 S/4} \tag{5}$$

This represents an equilibrium climate state. Note that the heat capacity C has dropped out in the equilibrium case. The sensitivity of the equilibrium climate to changes in S may be found by differentiating equation (5) to obtain

$$\frac{\partial T_S}{\partial \ln S} = \frac{(S/4)(1 - \alpha[T_S])}{B - \alpha_1 S/4} \tag{6}$$

Note that the sensitivity is decreased by large negative IR feedback (large B) and increased by large positive ice-albedo feedback (large α_1). Reasonable values to use are B = 2 W m^{-2} K^{-1} (Watts per square meter per degree Kelvin [absolute temperature scale]) and α_1 = 0.0025 K^{-1}. Together with the observed solar constant S = 1370 W m^{-2} and planetary albedo $\alpha = 0.3$, these give $\partial T_S/\partial \ln S \approx 200$ K, which implies that a one percent change in S leads to a 2°K change in T_S. A little numerical experimentation will show that the result is highly sensitive to the assumed values of B and α_1.

The general solution of equation (4) is given by a combination of the particular (equilibrium) solution and a solution to the homogeneous equation formed by setting the right side of equation (4) to zero. The latter is proportional to $e^{-1/\tau}$ where

$$\tau = \frac{C}{B - \alpha_1 S/4} \tag{7}$$

τ measures the response time of the climate system. When the climate is perturbed it will approach a new equilibrium state, and its relative distance to the new state will decrease by a factor 1/e during each time interval τ. If one uses C = 4 x 10^8 J m^{-2} K^{-1} (the value for a layer of water 100 m deep, representative of the uppermost level of the ocean as discussed below), together with the other values given above, one obtains a τ of roughly ten years. Climatic changes which are fast compared with this time scale (for example, the seasonal cycle or a perturbation caused by the sudden release of material into the atmosphere by a volcanic eruption) must be treated as transient problems. Climates which change slowly compared with τ may be assumed to always be near a state of equilibrium (analogous to the quasi-static approximation in thermodynamics).

It is worth noting that the denominator in the right side of equation (6) appears also as the denominator in the right side of equation (7). Thus, as asserted above, the more sensitive the climate is to perturbations the longer it will take the climate to adjust to a new equilibrium.

Interacting Systems

Despite its conceptual usefulness, the idea of global energy balance as expressed in equation (1) or in Figure 1-1 is usually not sufficient to simulate climatic change. Equation (1) is a relationship among globally averaged quantities. As discussed above, it can be solved only if tractable parametric relations among the quantities are assumed. Such parameterizations often cannot represent the complicated behavior of the real climate system. A basic problem is that the averaging process which gives rise to simplified models such as equation (1) masks important interactions among different parts of the climate system. For example, the atmosphere at low latitudes receives more solar energy than does the atmosphere at high latitudes. Some of the energy at low latitudes is transported to high latitudes; this is accomplished by a variety of atmospheric and oceanic circulation systems. The state of the climate depends strongly on the efficiency of this equator-to-pole heat transport, but it is impossible to deal with the phenomenon in terms of globally averaged quantities.

An additional complication arises from the fact that many of the different components of the climate system—the atmosphere, the surface and biosphere, the oceans, ice sheets and sea ice (the cryosphere) and the solid earth, respond with widely different time scales. In terms of equation (1) this means that use of a single effective heat capacity C is invalid. The cycle of the seasons, for example, occurs rapidly enough so that only the upper mixed layer (about the uppermost 100 m) of the oceans becomes involved in heat storage and release. The effective C for this process is much less than the heat capacity that would ultimately come into play if a permanent climatic change took place, involving the entire depth of the oceans. And the thermal response time of the deep oceans (a few thousand years) is less than the time it takes for a continental ice sheet to advance (tens of thousands of years).

The coupling together of various subsystems creates a possibility of climatic oscillations in which energy is transported back and forth between the subsystems. Continental ice sheets, for example, could grow until the underlying bedrock has time to begin sinking under the weight of the ice; the ice sheets could then shrink as they sink to lower and warmer elevations; finally as the weight on the bedrock is removed the elevation could return to colder levels and the cycle could repeat. Such oscillations are another example of internally controlled as opposed to externally forced climatic change.

Models

To examine the multitude of physical mechanisms that could be important for climatic change one performs controlled experiments using models—which are almost always computer programs—designed to simulate the physical processes and their interactions with climate. Ultimately models may prove useful in forecasting future climatic changes, but at the present stage of development climate models are generally regarded as research tools whose purpose is to test theories and to provide guidance in identifying possible future climatic events.

Types of Climate Models

It is impossible for one model to explicitly simulate the full range of physical processes involved in climatic change. A variety of models exist (Schneider and Dickinson 1974). These may be arranged in a hierarchy from simple to complex, but there is also a "horizontal" separation of models which focuses on different components of the climate system—for example, detailed atmospheric models coupled to simplified ocean models versus detailed ocean models coupled to simplified atmospheric models. Most climate models can be put into one of three categories: energy balance models (EBMs), radiative-convective models (RCMs), and general circulation models (GCMs).

Energy balance models (North *et al.* 1981) were mentioned above. One may consider equation (1) and associated parameterizations for $(S/4)(1 - \alpha)$ and F_{IR} to be a very simple EBM. Most EBMs used for climate research are more complex; they generally possess latitude resolution. In such models one considers the earth to be divided into a number of latitude zones, each with its own average surface temperature. Mathematically, the dependent variable T_S becomes a function of latitude as well as time, and equation (1) is replaced by a partial differential equation with $C \partial T_S / \partial t$ appearing on the left side. Such models are called one-dimensional EBMs. There also exist a few higher-dimensional EBMs. In addition, there are EBMs which deal with separate components of the climate system, for example by solving coupled equations for surface-air temperature T_S, ocean temperature T_0, etc.

The distinguishing feature of EBMs is that they parameterize in a simple way the processes of energy transport that are important for climate. These include the absorption of solar energy and emission of infrared radiation to space (the right side of equation (1). In addition, for models with one or more dimensions or with more than one component, one must parameterize energy exchanges among the different parts of the system. For example, EBMs with latitude resolution must parameterize equator-to-pole heat transport. Often this is done by a simple diffusion law which states that heat is transported between adjacent latitude zones at a rate proportional to their temperature difference. Though it is certainly reasonable to expect that heat will generally flow from warmer to cooler areas, it is not clear that a simple parameterization such as diffusion mimics properly the behavior of a complex system (in this case fluid motions in both atmosphere and ocean which transport heat). The issue of parameterization comes up for all climate models and will be discussed further in the next section.

One additional limitation of EBMs deserves mention: since they deal only with energy flows, they can directly calculate only temperatures. Other important climatic variables, most notably atmospheric motions, must either be ignored or parameterized in terms of temperature. Models which do parameterize atmospheric motions are often called statistical-dynamical models (Saltzman 1978).

Radiative-convective models (Ramanathan and Coakley 1978) are another widely used class of climate models. These are characterized by detailed calculation of the process of radiative transfer, that is, the exchange of solar radiation and infrared radiation among different levels of the atmosphere, and among the atmosphere, clouds, other particulate matter in the atmosphere (aerosols), and the surface. RCMs calculate the fluxes of solar and infrared radiation at many levels of the atmosphere, in contrast to EBMs which usually consider only the fluxes at the top of the atmosphere (the right side of equation (1)) and therefore cannot resolve the altitude variation of temperature.

RCMs invariably average out all horizontal differences and are thus one-dimensional models; they predict temperature as a function of height and, unlike EBMs, cannot deal with regional or seasonal variations. A particularly significant problem is the inability to distinguish between land and sea. These have very different values of surface heat capacity C and thus RCMs have a severely limited ability to deal with transient climatic change.

Nevertheless, RCMs are quite useful. Many climatic changes of great interest—for example, the greenhouse effect of human-produced CO_2 and other gases—are driven by radiative processes. RCMs compute these processes in great detail, and since they average over horizontal area, they do not explicitly involve parameterizations for horizontal energy transport; in a sense they are "cleaner" than EBMs. RCMs have been tested extensively in climate simulations not only for the earth but also for other planets. In particular, RCMs have simulated the very high surface temperatures produced on Venus by the greenhouse effect (Tomasko 1983). Successful performance under such radically different conditions gives one a fair bit of confidence in the ability of RCMs to simulate global average equilibrium climatic changes on the earth.

One very important process which RCMs must parameterize is convection, the rising motion which can take place when warm air underlies cool air. Convection transports heat upward and this energy transport must be added to that accomplished by radiative transfer. RCMs generally parameterize convection in a very simple way, called convective adjustment. If the rate at which temperature decreases with altitude exceeds a preset limit, or in other words when $dT/dz < -\Gamma$ where Γ is a positive constant, upward heat transport is assumed to occur until the point is reached at which $dT/dz = -\Gamma$. Theoretical considerations imply that convection does in fact act largely in this way, and the success of RCMs seems to bear this out. There are complications, however, when clouds are present. Clouds are created by the cooling of upward-convecting humid air, and they in turn affect the convective process (and in addition strongly affect radiative transfer processes). As one might imagine from looking at their complicated forms and evolution, clouds are difficult to parameterize reliably.

The most comprehensive climate models are three-dimensional general circulation models (Saltzman 1983). As the name indicates, these are distinguished by their ability to compute the general circulation, or motion, of the total atmosphere or of the oceans. (There exist a few coupled atmosphere-ocean GCMs which do both.) Atmospheric GCMs also compute radiative transfer in much the same way as do RCMs. The circulation is computed from first principles, by numerically integrating equations which express conservation laws for momentum, thermal energy, mass, and water vapor (in the atmosphere) or salinity (in the ocean). Given an initial state of the atmosphere or ocean, a GCM can simulate the subsequent time evolution, including detailed predictions of quantities such as temperatures, winds (or currents), and (for the atmosphere) precipitation as functions of space and time.

GCMs are an outgrowth of the numerical weather prediction models used routinely by meteorological agencies in weather forecasting (Haltiner and Williams 1980). But in climate research these models are used in a fundamentally different way: the simulation is run forward in time far beyond the point where weather forecasts are reliable, and it is understood that although the model will continue to predict specific meteorological events such as storms at particular times and places, it is only the average and other statistical properties that are believable. In fact, the most common way in which GCMs are used is to simulate an equilibrium climatic state by integrating the model until the climate is independent of the initial conditions.

GCMs are one of a small group of computer programs which strain the resources of even the fastest computers. Atmospheric GCMs usually take at least several minutes of mainframe computer time to simulate just a day's evolution of the weather; oceanic GCMs are faster but they require many years of simulated time to reach equilibrium. Computer time is the major constraint on the use of GCMs in climate research; often EBMs and RCMs are used because it is simply not practical to consume the time necessary to simulate a situation with a GCM. This is particularly true of problems involving the coupled response of the atmosphere and ocean to a climatic perturbation. An additional limitation of GCMs comes, ironically, from their detailed representation of the atmosphere or ocean, which results in model output nearly as complex as real data. It is often very difficult to tell from the output what physical mechanism is responsible for a climatic change, or even to tell if the change is statistically significant. Despite these problems, GCMs will certainly be used more frequently for theoretical climate studies as computers become cheaper and more powerful.

GCMs do not, of course, escape the perils of parameterization. The conservation laws which they solve are for all practical purposes exact, but in order to solve them on a computer one must in essence divide the atmosphere or ocean into a large number of boxes and solve for only the average temperature, wind, etc., within each box. With today's computers the boxes can be made no smaller than a few hundred kilometers in the horizontal and a few kilometers in the vertical; otherwise there are too many numbers to handle. Any physical process which is smaller than this (sub-gridscale) must be parameterized by the GCM. These include convection and cloud formation, and in general any other turbulence. GCMs typically parameterize convection and cloudiness in the same way as RCMs, and parameterize turbulence by diffusion laws similar to those used in EBMs to simulate heat transport.

Parameterization and Validation

From the foregoing it should be apparent that all climate models, even GCMs which come closest to first-principles computation, contain a number of adjustable parameters. These come from averaging over spatial and time scales which the model cannot resolve, and also from leaving certain components of the climate system out of the model (for example, purely atmospheric GCMs require sea surface temperatures as a lower boundary condition). Not even an increase in computing speeds of several orders of magnitude will permit models to encompass spatial scales from millimeters to tens of thousands of kilometers, time scales from seconds to millennia, and to treat in detail subsystems as varied as the atmosphere, the oceans, glaciers, and all the components of the biosphere. Consequently there always will be a need for parameterizations, and inevitably some of these parameterizations will be of doubtful validity.

How then can one establish trust in the validity of a climate model? Obviously the ability to replicate a single known climatic state, such as the mean annual present-day climate, is a necessary but far from sufficient condition. Fortunately, the earth undergoes a regular, well-observed series of significant climatic changes: the cycle of the seasons. Figures 1-2 and 1-3, taken from Manabe and Stouffer (1980), show results from an atmospheric GCM which is fairly successful in simulating the seasonal cycle. Contours of surface air temperature (in degrees kelvin) for February and August are shown in Figure 1-2 and Figure 1-3 respectively. In this particular GCM ocean surface temperatures are not assumed; they are calculated from the

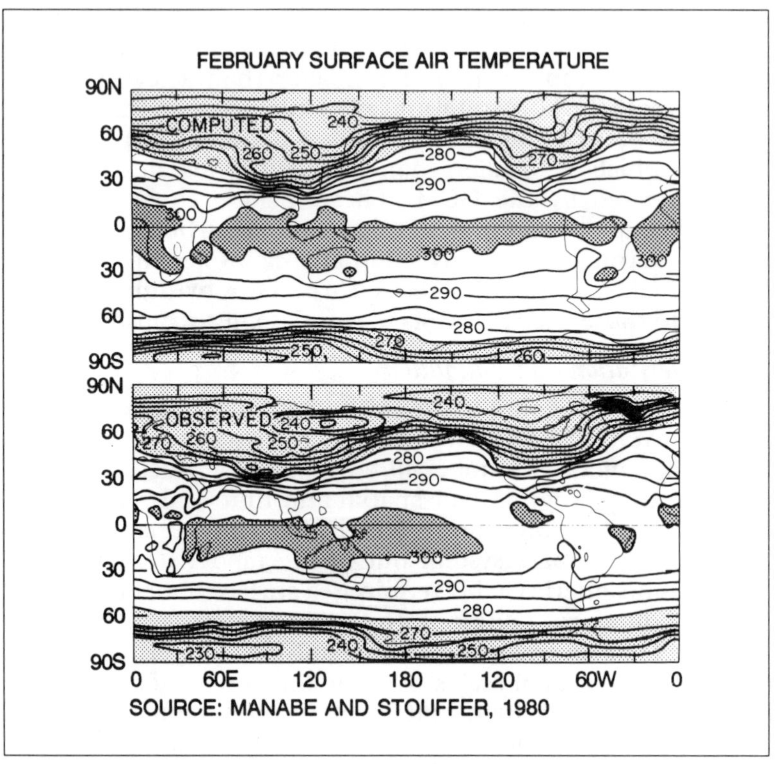

Figure 1-2: Temperature (in degree Kelvin) of near-surface air during February. The top frame shows values computed by a GCM coupled to a static heat reservoir (layer of water 68m. deep) representing the ocean mixed layer; the bottom frame shows observed values. Redrawn from Manabe and Weatherald (1980; published by the American Geophysical Union).

energy exchanges across the sea surface with the oceans represented by a uniform heat reservoir with C equal to that of a column of water 68 m deep. When given the seasonal variation of solar radiation over the different parts of the globe, the model computes a geographical and seasonal distribution of temperatures which agrees remarkably well with observations. Such results relieve some of the anxiety one might have that defects in GCM parameterizations would prevent the models from responding reasonably to external forcing.

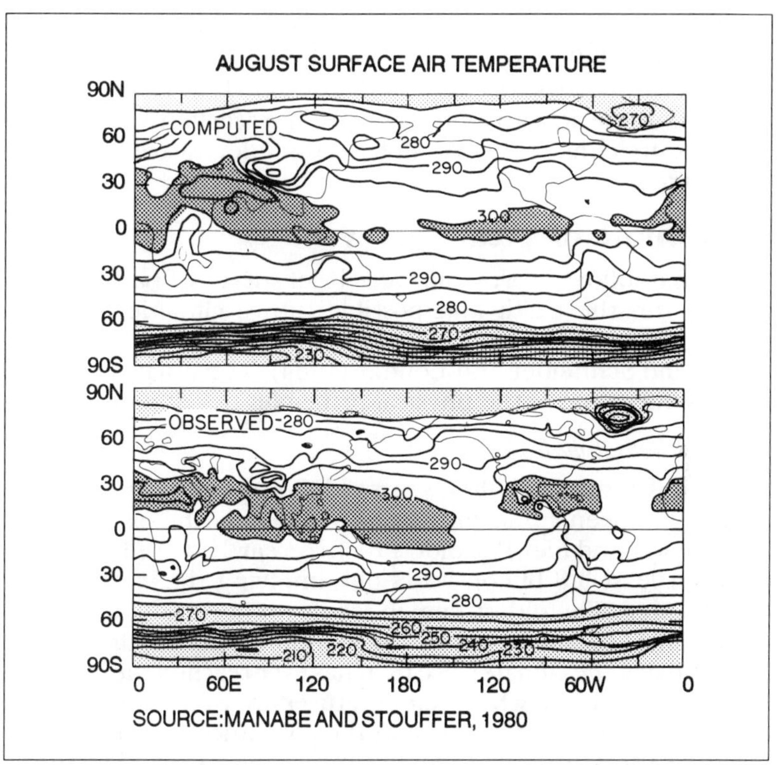

Figure 1-3: Same as Figure 1-2 for August. The GCM reproduces the seasonal variation of temperatures fairly well, providing some degree of confidence in the ability of models to simulate climatic change over longer time periods.

Despite such reassurances there remains the need for careful testing of the parameterizations in any climate model. One simple way to do this is to investigate the sensitivity of model results to variations in uncertain parameters. For example, in the simple EBM discussed above one could vary the coefficients A, B, α_0, and α_1 through a range consistent with data on F_{IR} and α, and examine the resulting changes in T_S. (Such an exercise will reveal that the presence or absence of a completely frozen earth as a viable solution of equation (1) is highly sensitive to the values of the parameters.) Mathematically, this is the same type of sensitivity test that one performs when simulating climatic change (for example one changes S to simulate the effect of changing solar luminosity and examines $\partial T_S / \partial S$). A complete set of parameterization tests must include in addition the sensitivity to changes in uncertain variables of changes in climates—formally, second derivatives such as $\partial^2 T_S / \partial \alpha_1 \, \partial S$.

Brief Climatological Survey

In what follows I present a very brief survey of past climatic changes on the earth and possible future climatic developments. Comprehensive discussions may be found in Frakes (1979) and Schneider and Londer (1984). The purpose here is to illustrate some of the physical principles developed above.

The Earliest Climates

Earth is about 4.6 billion years old. For most of this time the climate has been sufficiently benign to permit the existence of liquid water and life. The oldest rocks on earth are sedimentary rocks, at least 3.8 billion years old, implying the existence of liquid water. The oldest known fossils (of bacteria-like microorganisms) are nearly this old. As mentioned above, there is no evidence that the climate since then has ever reached the extreme "ice catastrophe" in which temperatures are everywhere below the freezing point of water; and the continuous record of life on earth also argues that temperatures have never gone more than a few tens of degrees above the current global average.

The mild climate of earth in the distant past implies a paradox. From some very firm theoretical arguments it is believed that the solar luminosity several billion years ago was significantly less than at present (perhaps 25% less). Simple climate models conclude that Earth should have been locked into the "ice catastrophe." For example the linear sensitivity expressed in equation (6) implies that globally averaged surface temperature would decrease by some 50°K to values well below freezing. Even when one considers the flaws in such simplified models, it seems mysterious that a very strong decrease in S would not lead to a completely frozen earth.

The solution to the paradox may lie in the greenhouse effect. As mentioned earlier, CO_2 and many other gases absorb infrared radiation, decreasing F_{IR} and thus increasing surface temperature. The chemical composition of the early atmosphere was very different from today's atmosphere (Walker 1977). The concentrations of CO_2 and other infrared-absorbing gases may well have been orders of magnitude larger than they are now. This is not to say that the paradox has been fully resolved. The composition of the atmosphere at these times in the very distant past is not known with certainty. Furthermore, the climate modeling efforts relevant to the early Earth have been confined to fairly simple EBMs and RCMs. If feedback processes were simulated more accurately, the need for a large greenhouse effect could decrease. It would be interesting to examine the problem with a GCM.

The Mesozoic Optimum

About half a billion years ago, there was a dramatic increase in the number and kinds of living organisms on earth. As a result there is a detailed fossil record which tells the history of the planet from this time to the present. The period of time from 570 million years ago to the present is divided by geologists into the Paleozoic Era (570 to 230 million years ago), the Mesozoic Era (230 to 65 million years ago), and the Cenozoic Era (65 million years ago to the present).

It appears from the geologic record that for most of the last half-billion years, earth's climate was warmer than it is now, and temperatures were more uniform over the globe. The widespread occurrence of "tropical" plant life (some of which, buried in large quantities, have become today's fossil fuels) and animals such as amphibians implies that warm climates were pervasive. The geological evidence is especially clear for the last period of the Mesozoic Era, the Cretaceous Period (Barron 1983). The global average surface temperature was probably about 5°K warmer than at present, and the temperature difference between the equator and the poles was much smaller. No significant amount of ice was present on land or sea, in contrast to today's climate in which the surface of the Arctic Ocean is frozen and ice sheets up to several kilometers in thickness cover Greenland and Antarctica.

Several factors probably contributed to the warm, equable climate of the Cretaceous Period. Because of continental drift, the geography of the Cretaceous Period was different from today's; a greater area was covered by oceans and shallow seas, which reflect less sunlight than land, so the planetary albedo α was likely smaller. The circulation of the oceans may well have transported more heat from the equator to the poles. There is also evidence that the concentration of CO_2 during the Cretaceous Period was much larger than in more recent times. With relatively abundant geological evidence providing the foundation, modeling of these effects is providing a useful test of the ability of climate models to simulate a significant departure from present-day conditions, and also an increased understanding of the Cretaceous Period itself.

The Cretaceous Period ended 65 million years ago with the disappearance of the dinosaurs together with a majority of the species of life that existed on earth at the time. Many of the species seem to have died out very quickly, perhaps within a single year, at the same time as a large asteroid or comet collided with the earth. The collision would have injected an enormous amount of debris into the atmosphere, inducing a temporary but drastic cooling of the climate as planetary albedo increased and sunlight was cut off from the surface (Pollack *et al.* 1983). Other mass extinctions exist in the geological record and it is possible that these may have been brought about by asteroid or comet collisions; similar though less catastrophic "climate shocks" may arise from volcanic eruptions (Rampino *et al.* 1988) and perhaps even nuclear war (Schneider and Thompson 1988).

The Ice Ages

The Cenozoic Era has been marked by a cooling of the climate. The Antarctic Ice Sheet, which today stores most of the ice on earth, appeared a few tens of millions of years ago. By one or two million years ago, the cycle of the Ice Ages began (Frakes 1979). Periodically, vast continental ice sheets develop in North America and Eurasia from compacted snow which fails to melt in the summer; thus the ice sheets advance toward the equator. Then the ice sheets retreat and ultimately are confined to just the two continents nearest the poles, Greenland and Antarctica. We are now living in such an interglacial period. Advances and retreats have occurred about half a dozen times over the past half million years, where the record is most clear; the cycle of the Ice Ages has a period of roughly 100,000 years. The peak of the most recent Ice Age was about 18,000 years ago. At that time the ice sheets extended over North America as far south as the states of Washington and New York, and over much of northern and central Europe. The global average surface temperature was perhaps 5° cooler than it is today.

The Ice Ages are a good testing ground for theories of externally and internally driven climatic change. The consensus now is that the cycle is brought about by external forcing but amplified by internal interactions (Imbrie and Imbrie 1979). Small oscillations in the eccentricity (non-circularity) of the earth's orbit about the sun, and in the tilt of the earth's axis with respect to the plane of its orbit, occur with periods from tens of thousands to hundreds of thousands of years. These lead to variations in the amount of solar radiation received at different latitudes and seasons. This climatic forcing is called the Milankovitch mechanism, after the astronomer who championed the idea that these variations in received solar radiation (insolation) caused the Ice Ages. The timing of the Ice Ages coincides in many respects with the Milankovitch variations, providing strong circumstantial evidence in favor of the Milankovitch theory. But by themselves the changes in insolation appear too small to drive the large observed variations in climate between peak glacial and interglacial periods, so some sort of amplification mechanism appears necessary. There are a variety of possibilities. The ice sheet/bedrock interaction described earlier could provide an internal resonance with Milankovitch forcing. Changes in deep oceanic circulation, which is very sensitive to conditions at high latitudes, could conceivably act to reinforce the cooling or warming phases of the cycle. Finally, recent geologic evidence indicates that the concentration of CO_2 in the earth's atmosphere has varied in synchrony with Ice Age temperature cycles (Barnola *et al.* 1987); this brings up the possibility of a positive feedback loop involving the CO_2 greenhouse effect and climate during the Ice Ages.

Present and Future Climates

The present interglacial period appears to have peaked out about 6,000 years ago, with temperatures significantly warmer then than they are today. Simple extrapolation would indicate that the earth is steadily but surely sinking toward another Ice Age. But a new actor has appeared on the scene: humankind. Human numbers and technological prowess have now reached the point where humanity's impact on the climate may be significant enough to tip the balance between warming and cooling. As will be described later in this book, CO_2 and other infrared-active gases produced by deforestation and by the burning of fossil fuels could very well produce, within a generation or two, the warmest temperatures on earth since the Mesozoic Era (Ramanathan 1988). Of course natural events, such as volcanic eruptions that inject significant amounts of aerosols into the atmosphere, will continue to be significant.

Space prohibits discussion of the multitude of factors which may be influencing the climate today and in the near future. Suffice it to say that a reliable forecast of the climate for the next several generations is beyond our current understanding either of the climate system or of human behavior; nevertheless the best bet, barring natural or human-made catastrophes, is an overall warming trend from the greenhouse effect.

Conclusion

In the 19th century, modern earth science was established by refuting the literal biblical interpretation which advocated a short planetary history marked by divine creation and by the catastrophe of the Great Flood. This picture was replaced by one of very slow evolution. Climatic change was recognized, but generally considered to be important only over very long periods of time. The prevailing opinion was that the earth's environment is reliably benign—at least in the sense that the nuisances Nature has to offer can usually be overcome by human efforts. Such a viewpoint was perfectly consistent with the ideology of inevitable progress which dominated the thinking of the technologically advanced societies through the turn of the century (Berggren and Van Couvering 1984).

In the wake of the two world wars, stunning and often frightening technological developments, and extraterrestrial exploration which underscores the uniqueness of the earth as an abode of life, the general viewpoint of humanity and its environment has become far less certain. It has been recognized not only that climatic change can significantly affect humans over a reasonably short period of time, but also that humanity today can itself inadvertently alter the climate. These considerations form a large part of the discussions in this book.

In this chapter I have summarized what are to me the most important concepts regarding mechanisms of climatic change. Climatic change is based ultimately on well-known physical conservation laws, such as conservation of energy, applied to the climate system which consists of the atmosphere, the oceans, and many other components. In principle these may be written down as a complete system of mathematical equations and solved on a computer. But in practice the climate system is far too complex to be given a complete and rigorous treatment, and this is likely to remain true even with the advances expected in computing power over the next few decades. Climate models become workable only if they focus their attention on a subset of the vast range of components and spatial and time scales important to climatic change; the other factors enter through boundary conditions and parameterizations. The challenge is to use the results of such models, together with observational data and, above all, physical insight, to explain climatic variations of the past and to map out possibilities for the future.

Acknowledgements

Most of my education in the fundamentals of climate dynamics occurred during scientific visits in the National Center for Atmospheric Research in Boulder, Colorado. My sincere thanks go to NCAR for its hospitality and to NCAR scientists, particularly Starley L. Thompson and Stephen H. Schneider, for their stimulating conversations and collaborations. (NCAR is sponsored by the National Science Foundation; any opinions or conclusions expressed herein do not necessarily reflect those of the National Science Foundation.) I also thank Dr. Thompson for a thoughtful reading of this manuscript.

References

Barnola, J.M., D. Raynaud, Y.S. Korotkevich and C. Lorius. 1988. "Vostok ice core provides 160,000-year record of atmospheric CO_2," *Nature, 239:*408–414. Barron, E.J. 1983. "A warm, equable Cretaceous: the nature of the problem," Earth Science Reviews 19: 305–338.

Berggren, W.A. and J.A. Van Couvering, eds. 1984. Catastrophes and Earth History: The New Uniformitarianism (Part I: The concept of catastrophe as a natural agent). Princeton University Press, Princeton, New Jersey.

Frakes, L.A. 1979. Climates Throughout Geologic Time. Elsevier, Amsterdam.

Haltiner, G.J. and R.T. Williams. 1980. Numerical Prediction and Dynamic Meteorology, 2nd ed. John Wiley and Sons, New York.

Imbrie, J. and K.P. Imbrie. 1979. Ice Ages: Solving the Mystery. Enslow Press, Short Hills, New Jersey.

Kasting, J.F., O.B. Toon and J.B. Pollack, 1988. "How Climate Evolved in the Terrestrial Planets." Scientific American, February 1988, 90–97.

Lamb, H. 1980. Climate: Past, Present, and Future, Vol. 2. Methuen Press, London.

Manabe, S. and R.J. Stouffer. 1980. "Sensitivity of a global model to an increase in CO_2 in the atmosphere." *Journal of Geophysical Research* 85:5529–5554.

National Academy of Sciences. 1975. Understanding Climatic Change: A Program for Action. National Academy of Sciences, Washington, D.C.

North, G.R., R.F. Cahalan and J.A. Coakley Jr. 1981. "Energy Balance climate models." *Reviews of Geophysics and Space Physics* 19:91–122.

Pollack, J.B. 1979. "Climatic change on the terrestrial planets." *Icarus* 37:479–553.

Pollack, J.B., O.B. Toon, T.P. Ackerman and C.P. McKay. 1983. "Environmental effects of an impact-generated dust cloud: Implications for the Cretaceous-Tertiary extinctions," *Science* 219:287–289.

Ramanathan, V. and J.A. Coakley Jr. 1978. "Climate modeling through radiative-convective models." *Reviews of Geophysics and Space Physics* 16:465–489.

Ramanathan, V. 1988. "The Greenhouse Theory of Climate Change: A Test by an Inadvertent Global Experiment." *Science* 240:293–299.

Saltzman, B. 1978. "A survey of statistical-dynamical models of the terrestrial climate," *Advances in Geophysics* 20:183–304.

Saltzman, B., ed. 1983. Theory of Climate (Part I: History and application of general circulation models). *Advances in Geophysics* 25.

Schneider, S.H., and R.E. Dickinson. 1974. "Climate Modeling." *Reviews of Geophysics and Space Physics* 12:447–493.

Schneider, S.H., and R. Londer. 1984. The Coevolution of Climate and Life, Sierra Club Books, San Francisco.

Schneider, S.H., and S.L. Thompson. 1981. "Atmospheric CO_2 and climate: importance of the transient response." *Journal of Geophysical Research* 86:3135–3147

Schneider, S.H. and S.L. Thompson. 1988. "Simulating the Climatic Effects of Nuclear War." *Nature* 333:221–227.

Self, S. and R.B. Stothers. 1988. "Volcanic Winters," Rampino, M.R., *Annual Review of Earth and Planetary Sciences.* 16:73–99.

Tomasko, M.G. 1983. "The thermal balance of the lower atmosphere of Venus," in Venus, D.M. Hunten *et al.*, eds, University of Arizona Press, Tucson.

Walker, J.C.G. 1977. Evolution of the Atmosphere. Macmillan Press, New York.

Willson, R.C. and H.S. Hudson. 1988. "Solar luminosity variations in solar cycle 21." *Nature* 332:810–812.

PART TWO:

ATMOSPHERE

TWO

CARBON DIOXIDE and CLIMATE CHANGES

IMPLICATIONS FOR MANKIND'S FUTURE

William W. Kellogg

As early as the turn of this century the possibility that people could modify the climate of the earth was hypothesized, but this startling notion attracted little public attention for the next half-century. The suggestion was that we are taking carbon out of the earth in the form of fossil fuels and burning it, thereby adding carbon dioxide to the atmosphere. An increase in atmospheric carbon dioxide warms the lower atmosphere, as will be explained below.

In the past two decades we have learned a great deal about the global system that determines the earth's climate, and the question of the influence of increasing carbon dioxide has received special attention—both in scientific circles and in the media. Our best guess is that the next several decades may witness a continuing gradual global warming, and that patterns of prevailing winds, temperatures, and rainfall could shift dramatically. To be sure, there are many uncertainties about just how these shifts will take place, but some of the features of the future climate that may be in store for us are beginning to emerge more distinctly.

It is therefore not too early to begin to ask the next set of questions concerning climate change. These have to do with the impact that climate change will have on human activities that are sensitive to weather and climate—as indeed most of our activities are. And this is only the beginning, for we must then be concerned with the implications that such climatic impacts will have for the future of our economic and societal systems. We must somehow seek to trace these implications into the very structure of our complex global society—at least, to the extent to which we are able.

This hierarchy of problems leading from the expected physical changes to possible social effects is shown schematically in Figure 2-1. Notice that there are closed loops or "feedbacks" here, since at each stage there are reverse influences on society and on the natural environment in which we live. For example, as climate change is perceived by people and their governments there will inevitably be actions to modify our way of doing things in order to cope with the change.

It is an unfortunate fact, however, that our knowledge of how our societal systems function is even more incomplete than our knowledge of the climate system. The tools that are at hand to assess and predict the outcome of a climate change in human terms are woefully blunt. Most economists and social scientists are pessimistic about our ability to make such predictions, except perhaps in the broadest terms. No doubt we can improve those tools, but we have a long way to go.

In the meantime, starting immediately we are obliged to provide some guidance to those who are in a position to make provisions for the changes ahead. We cannot ignore that challenge. What can we say to our policy makers in government, industry, agriculture, forestry, and development of natural resources when they ask for advice? We cannot tell them: Go away and do not bother us until we have done more homework.

Instead, scientists of many disciplines who have been studying the subject must carefully explain in nontechnical language what they think about the implications of future climate change—and where our uncertainties lie as well—and then

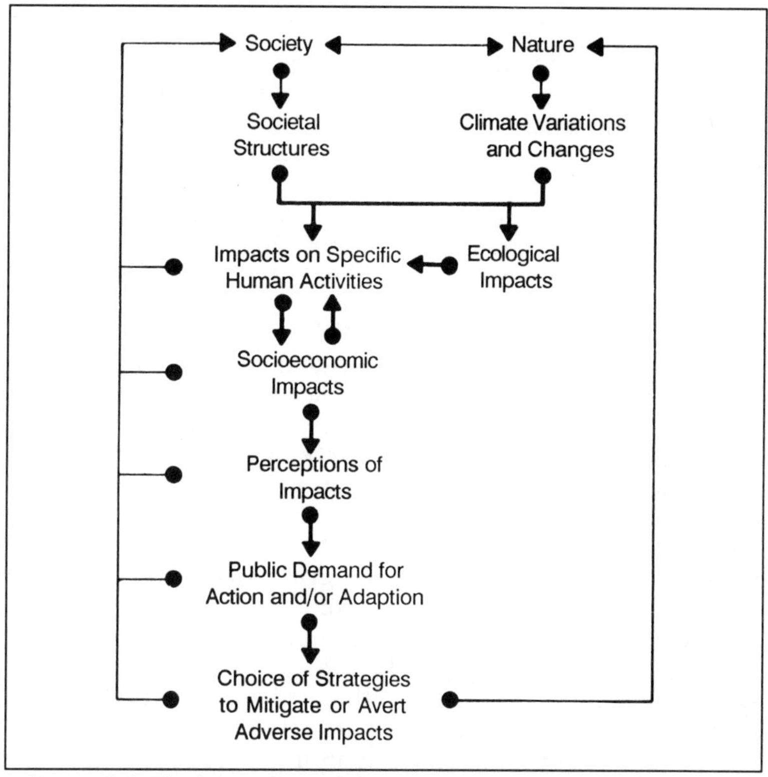

Figure 2-1: The interconnected factors that are involved in climate in past studies. Note that the impacts are determined by both "nature" and "society"— climate changes interact with existing societal structures to affect people, and the activities of people affect the climate.

proceed to give sensible advice about the adaptive measures that seem to make the most sense.

That is the subject of this chapter. It is based to some extent on the book *Climate Change and Society* by this author and Robert Schware (1981), and on our article "Society, Science, and Climate Change" (1982). In addition, there have been a number of national and international meetings devoted to the impacts of climate change, and the Villach/Bellagic Workshop Report (1987) is particularly noteworthy. Further details may be sought in the References at the end of the chapter.

Development of Climate Scenarios

Research on the influences of increasing carbon dioxide on the climate system has been pursued vigorously by a number of organizations and individuals, and has been competently reviewed elsewhere. If we conceive the ultimate objective of this research to be a self-consistent and credible picture of the climate of a future warmer earth, and also an estimate of the rate at which the change will occur, then the effort can be characterized by several components.

Estimating the Timetable for the Human Influences

There are a variety of large-scale changes that humans have brought about on the surface of the earth and in the atmosphere, and all must have had some influence on climate, at least regionally. Notable are the modifications of the face of the earth by agriculture, deforestation, and urban-industrial development; the introduction of smoke and chemicals into the atmosphere; and, most important from the climatic point of view, the production of atmospheric carbon dioxide from burning fossil fuels. Carbon dioxide is a long-lived and infrared-absorbing gas that traps some of the outgoing infrared radiation from the surface that would otherwise escape to space; the resulting warming of the surface is often referred to as the greenhouse effect. (There are other persistent infrared-absorbing trace gases that we are producing, such as methane, the chlorofluoromethanes, carbon monoxide, ozone, and nitrous oxide, all of which can contribute further to the greenhouse effect of carbon dioxide.)

The rate of increase of carbon dioxide depends on the rate at which we burn fossil fuels. Other possible anthropogenic sources of carbon dioxide exist (such as deforestation in the tropics). The operation of natural sinks for this trace gas also affects the rate of increase. The oceans are undoubtedly the major sink.

The dominant factor that will determine the build-up of carbon dioxide in the decades ahead is the rate of burning of fossil fuels. Estimates of this future rate vary widely among the experts. It is not a geophysical problem (though geophysical factors enter into the calculation in an important way), but rather a problem of guessing what the global energy demands will be, the technological developments that will lead to alternatives to fossil fuel, and the world economic order. There are no crystal balls that we can use to answer this kind of question.

A not unreasonable *upper limit* of rate of increase of carbon dioxide can be based on the assumption of a continued 2% per year rate of increase of fossil fuel use, leading to a doubling of its pre-industrial (natural) concentration of about 270 parts per million by volume (ppmv) around the middle of the next century. Similarly, a *lower limit* would be based on the assumption of a decreasing rate of growth in the use of fossil fuels to the point of no further growth in 50 years; this would still lead to an increase of concentration of carbon dioxide from the present 35 ppmv to over 400 ppmv in the next 50 years, and perhaps a doubling late in the next century. The most likely course is somewhere in between these two limits, and this gives us an idea of the time scale for the change to occur. There may, however, be delays of a decade or two in the resulting temperature change due to the thermal inertia of the oceans, and the other greenhouse gases will accelerate the change.

Figure 2-2 is a nomogram for estimating the date at which we may reach the doubling of carbon dioxide from the pre-industrial concentration. It depends on the rate of increase of fossil fuel use in both the developing and developed world, assuming that approximately the same fraction of the added carbon dioxide remains airborne as in the past.

If we combine a kind of best guess about the rate of future fossil fuel use (about 1% per year increase) with the theoretical estimate of the greenhouse effects for a given carbon dioxide increase (see next section), we can arrive at a rate of change of global mean temperature with time, shown in Figure 2-3. It also shows the rates of change for the upper limit ("high") and lower limit ("low") just discussed, and an estimate of the larger changes in the Arctic.

Describing Regional and Seasonal Changes

Our models of the climate system give relatively little information about the regional distributions of temperature change and shifts of precipitation patterns, yet it is the regional effects that are of most importance to planners. We are convinced that the temperature difference between equator and pole will

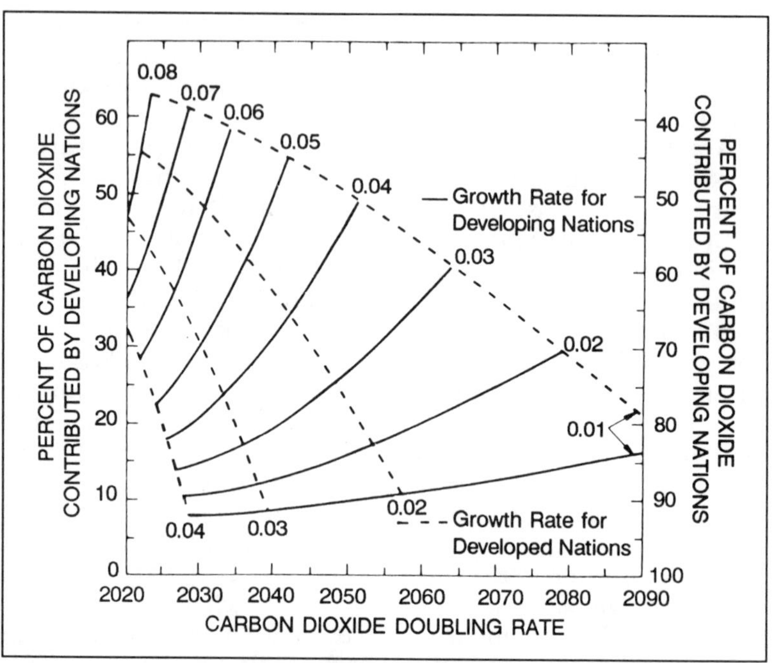

Figure 2-2: A nomogram relating the annual rates of increase of fossil fuel use in both developing and developed nations to the date at which carbon dioxide will be twice its preindustrial value of about 270 ppmv. It was prepared by Edward Friedman of the Mitre Corporation. (Source: Kellogg and Schware 1981; 1982.)

become less with an increase in carbon dioxide, and that this will undoubtedly change the large-scale circulation patterns that determine the weather of each place on the earth, but the actual patterns of change are still unclear. This situation has led us to look at a variety of ways to estimate the regional and seasonal changes, and the effort has been termed "development of climate scenarios."

Climate scenarios, or depictions of conditions on a warmer earth, should *not* be taken as predictions, but rather likely examples of what *could* occur. They should be based on physical reasoning and a knowledge of what has happened in the past, and should be as self-consistent as possible. Some aspects of a given climate scenario may be identified as quite probable, while others must admittedly be more conjectural.

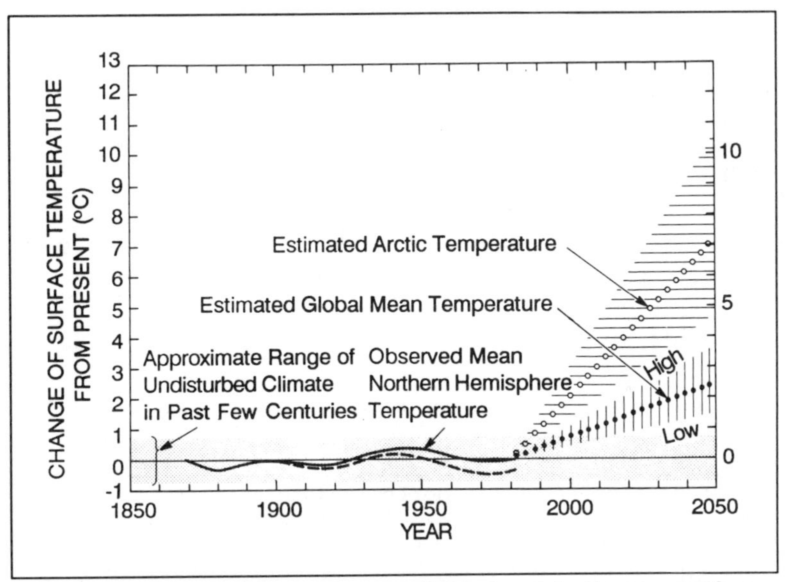

Figure 2-3: Past and future changes of global and Arctic mean surface temperature. The dashed line shows what it might have been without the addition of carbon dioxide. The stippled area shows the approximate range within which the global mean temperature has varied during the past 1,000 years or more. See text for an explanation of the estimation of future trends. (Source: Kellogg 1979; Kellogg and Schware 1982.)

Such climate scenarios are essential first steps in studies of how various economic sectors (especially agriculture, ranching and forestry) would be impacted by a carbon dioxide-induced climate change, in studies of likely societal responses, and in making plans for the future. They must, however, always be accompanied by caveats to the effect that they are illustrative examples and not yet firm predictions—though there is always the danger that they will be misused in spite of the caveats. It is this danger that has caused most climatologists to shy away from climate scenario development until very recently. Nevertheless, in Figure 2-4 we present a climate scenario depicting the possible changes in *soil moisture* that could occur throughout the world with a warming due to increase of carbon dioxide.

It will be noted that in the scenario shown in Figure 2-4 we call for a general decrease of soil moisture in the middle of the continents at mid-latitudes, and for a general increase in the

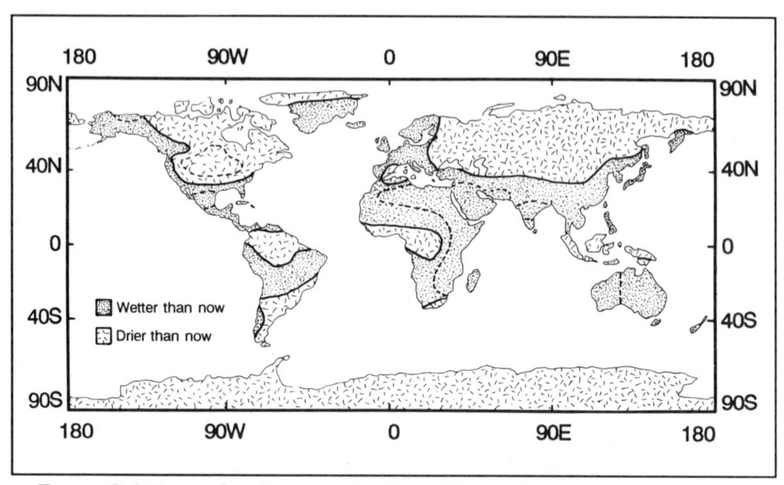

Figure 2-4: Example of a scenario of possible tendencies of soil moisture on a warming Earth. It is based on paleoclimatic reconstructions of warm periods in the distant past, comparisons between recent warm and cold years or seasons in the northern hemisphere, and several climate model experiments. Where there seems to be general agreement on the tendency between two or more of these sources we have indicated the area of agreement with a dashed line and a label. (Source: Kellogg and Schware 1981; 1982.)

semi-arid and arid regions of the subtropics—areas where the majority of the less-developed countries exist. That rainfall and soil moisture are highly important factors in the formation of both natural ecological systems and agricultural or pastoral productivity is unquestionable. Figures 2-5 and 2-6 illustrate this point. It is therefore evident that such shifts of soil moisture as are illustrated in Figure 2-4 could have significant impacts on where the major food crops are grown, and this is suggested in Figures 2-7 and 2-8. It will be noted that the countries that currently account for most of the production of wheat, maize, and barley (the United States and the Soviet Union) are just the ones that could suffer a trend towards drier average conditions. We referred to food (grain) production here, and in the next section impacts on other economic or social sectors will be outlined as well.

As pointed out above, the time scale for this transition to a warmer earth will depend to a large extent on the rate at which

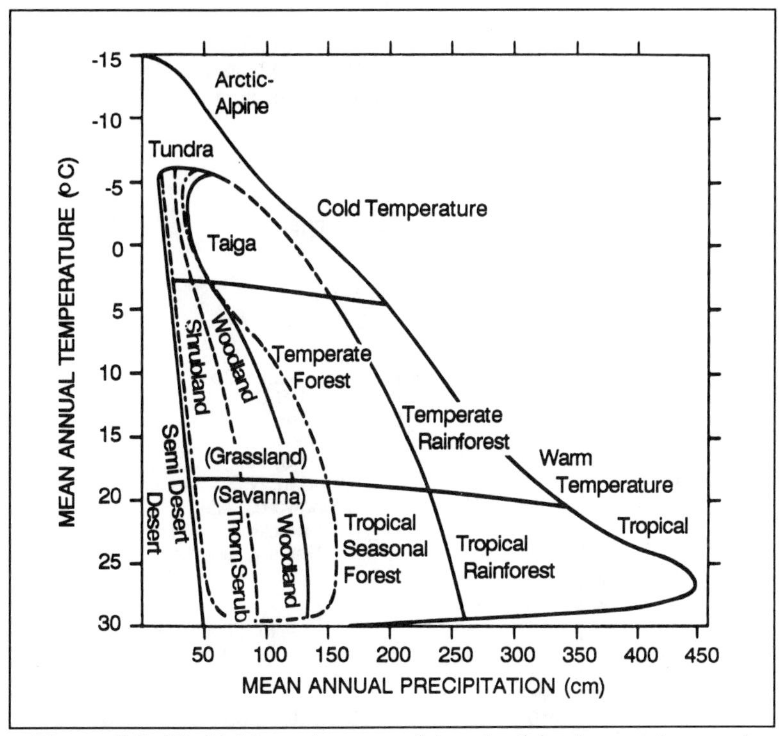

Figure 2-5: World biome types as determined by temperature and precipitation. The boundaries between types are approximate, as there are other factors involved such as soil condition. The dot-dash line encloses a wide range of environments in which either grassland or one of the types dominated by woody plants may form the prevailing vegetation. (Source: R.H. Whittaker: *Communities and Ecosystems*, New York: Macmillan, 1975.)

mankind uses fossil fuel—and also on some geophysical factors such as the thermal inertia of the oceans that could slow the rate of change of temperature. We may be talking about 50 years or over 100 years for a doubling of carbon dioxide, but all evidence now seems to suggest that in any case we will definitely begin to notice the effects of climate change well before the turn of the century—in fact, it is undoubtedly already taking place, though hard to identify unambiguously as being due to increasing carbon dioxide.

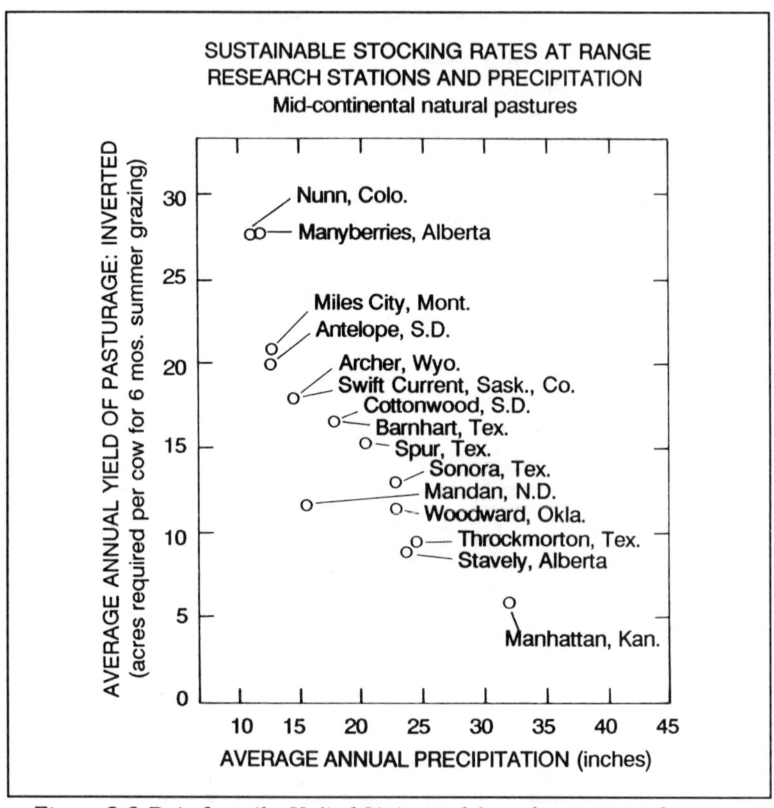

Figure 2-6: Data from the United States and Canada on acres of pasture land required per cow vs. average annual precipitation. It shows that precipitation rather than temperature is the major factor. (Source: *Climate Impact Assessment Program CIAP*, 1975: Department of Transportation, Washington, D.C.)

Impacts of Climatic Change

The World Climate Program, established in 1979 by the World Meteorological Organization (WMO), has as one of its goals defining climatic variability and change in terms of what they mean for societies. In the past few decades short-term weather anomalies and slowly developing changes of climate have triggered severe economic, social, and political dislocations. Developed and developing countries alike find themselves increasingly vulnerable to "abnormal weather" of the sort that struck a number of countries in 1972 and again in 1982, and 1988. Furthermore, there is a growing realization that a carbon dioxide-induced climatic change could have much larger impacts than any of these short-term and more or less random events.

Studying such climatic impacts is already an established practice. What appears to be new about climatic impact studies is the attempt to take into account many factors, all dealing specifically with the carbon dioxide/climate problem. Here are some of the potential impacts on several climate sensitive activities.

Energy Supply and Demand

The largest single use of energy in the temperate latitudes of the industrialized world is for space heating and cooling; this is markedly influenced by temperature variations. Energy is also needed in many areas for pumping water for irrigation; during dry years more is needed. There are other sectors, such as transportation and tourism, the energy demands of which are affected by climate. However, in practice it is often hard to separate the effects of climate variations from the other economic and social factors involved.

Energy supply is also affected by climate, as in the case of the cold winter of 1976-1977 in the eastern United States, when fuel shortages accounted for an estimated one million unemployed by February 1977. However, part of this shortage was due to the fact that the utility and pipeline companies neglected to plan ahead and store fuel reserves.

There is a current move to replace centralized fossil fuel and nuclear power sources with renewable sources such as hydroelectric, solar, wind, and biomass. Though this may be desirable, it will tend to make our energy supplies more sensitive to the vagaries of the climate, depending as they do on rainfall, sunshine, and steady winds.

Energy demand in the tropics would probably change little if global temperatures rose due to an increase in carbon dioxide. However, at middle latitudes there would be a decrease of demand for heating in the winter and an increase in demand for air conditioning in summer, shifting the kind of power from heating fuel to electricity. The pattern of energy

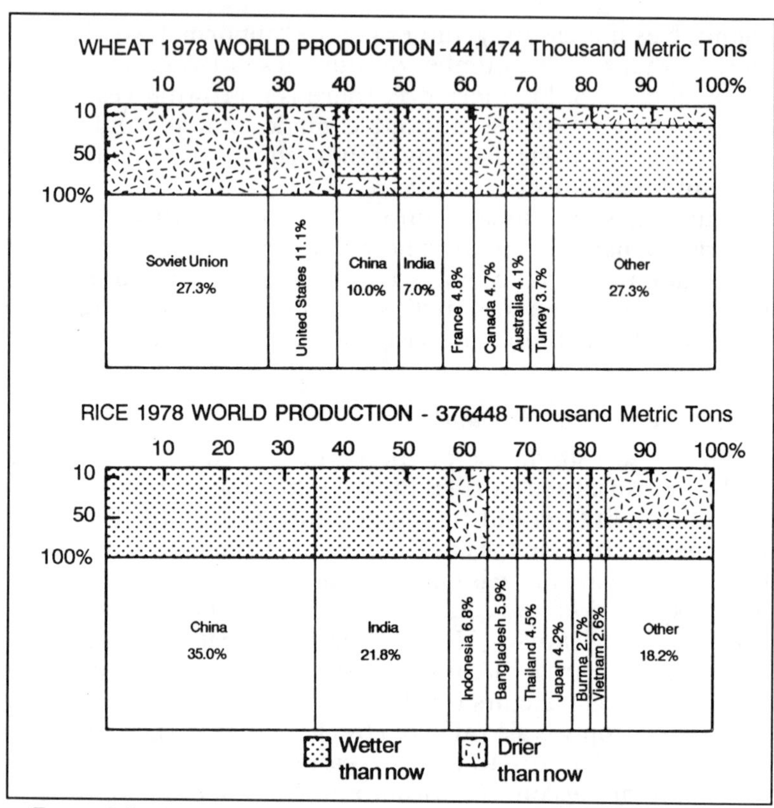

Figure 2-7: Distribution between countries of world production of wheat and rice in 1978, combined with a scenario of possible changes of soil moisture on a warmer earth (see Figure 2-4). World food production data compiled by the Food and Agriculture Organization, Rome. (Source: *FAO Production Yearbook*, 1979: Statistics Series No. 22, Vol. 32.)

demands will not shift uniformly poleward, since, as we have emphasized, the regional changes of temperature and rainfall will be complex.

World Food Production

Some aspects of the projected climatic changes due to carbon dioxide could be beneficial for crops and other ecosystems, depending on where they are growing. More carbon dioxide is known to enhance photosynthesis and plant growth, and we

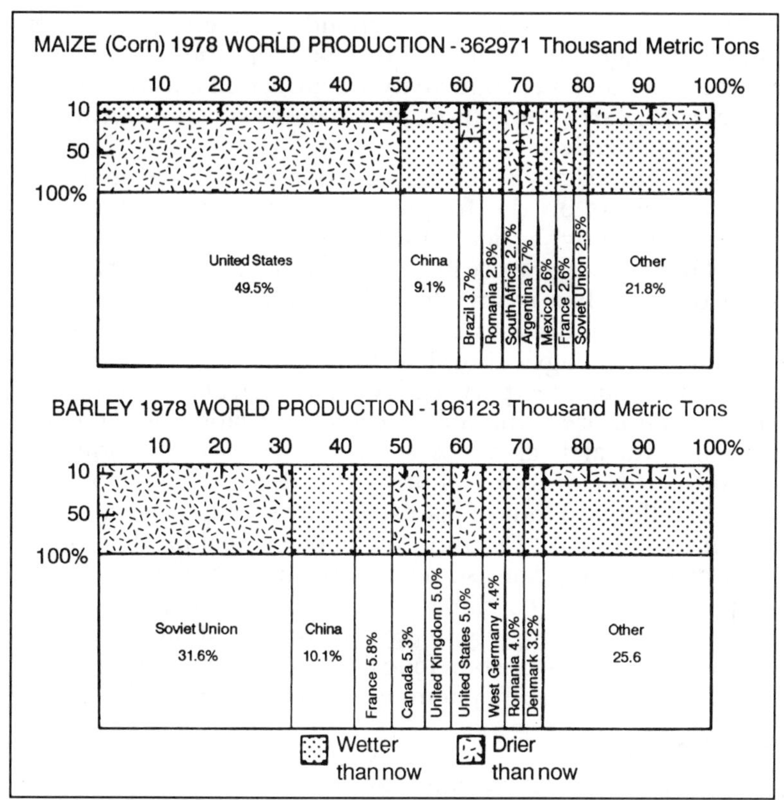

Figure 2-8: Same as Figure 2-7, showing production of maize (corn) and barley.

may expect an increase in the average growing rate of plants from this cause alone of about 5% by the turn of the century. It is estimated that the effect in this century has already been 10% or more.

A 1° rise in average summertime temperatures at middle latitudes increases the average frost-free growing season by roughly 10 days, an obvious advantage. The increase of precipitation in the semi-arid regions of the subtropics called for in our scenario could also be advantageous. Figures 2-7 and 2-8, however, illustrate that the major producers of grain (except for rice) may be adversely affected by a hotter and drier climate.

Of course, every food crop responds differently to a given climatic change. For several major food crops the relationships between climate and productivity are fairly well known. But for other crops, particularly those grown in the subtropics, further study is still needed.

Equally important to food production may be the changing climate's effects on the frequency and severity of pest outbreaks. Currently, agricultural losses due to pests are about 25%. Thus, a temperature increase may make pest control even more difficult than it already is. The same may be true for a number of plant diseases, such as stripe rust on winter wheat.

Even though agricultural technology has given us hardy crop strains, many experts believe that agricultural monocultures are far more vulnerable to climatic variations and change than the natural diverse ecosystems they replaced. It should be noted that 95% of human nutrition is derived from no more than 30 different kinds of plants; just three crops—wheat, rice, and maize—account for over 75% of our cereal production.

In principle we can use new crop varieties that are better adapted to a changed climate. But plant geneticists are concerned about the rapidly decreasing stocks of vigorous wild strains and disease-resistant food crops; these are being pushed out as their natural ecosystems are damaged or eliminated.

Global Ecology

While agricultural systems may depend on a few specialized plant species, natural ecosystems—or "biomes"—are usually characterized by a great diversity of plants and animals. These organisms interact and live together in an intricate balance—that is, until this balance is disturbed by humans or by climatic change. Few biomes, if any, remain untouched by human influence. Among the major biomes that are still close to their original states are the remote tropical forests that have not yet been exploited, and the unpopulated tundra areas of the Arctic.

The kind of biome that will thrive in a given region is determined by temperature, precipitation, soil type, and availability of sunlight, among other variables. The first two are generally the most important, as illustrated in Figures 2-5 and 2-6. A slow climatic change will affect temperature and precipitation, forcing the biomes to shift, as some species in each region die out and others succeed them. This has occurred many times throughout geological history. For example, during the warm Altithermal period, some 5000 years ago, the spruce forests of central Canada extended 300 to 400 kilometers farther north than they do now. At that time the Sahara was not a desert but a semi-arid grassland that supported grazing animals and nomadic people.

Whether ecosystems can adapt successfully to a climatic change will depend on how fast the change occurs. While the life span of individual trees and some other plants is many decades, the response of an entire ecosystem occurs over several plant lifetimes.

A special biome exists in the Arctic tundra where few trees grow and permafrost (permanently frozen ground) inhibits drainage, slowing the decay of dead plants. The result is the accumulation of deep layers of water-saturated or frozen organic matter, called peat. If there is a general warming trend—and recall that the warming is expected to be greatest in the Arctic, as indicated in Figure 2-3—permafrost will gradually retreat northward, trees will encroach on what had been tundra, and the peat bogs will thaw to a greater depth in summer and dry out in their upper layers. As a result, the upper layers of organic matter will oxidize once they become exposed to the air for the first time. This will probably release carbon dioxide and also some methane and add further to the warming, an example of a positive feedback.

Hence, a gradual climate change will cause the distribution of biomes to shift as each seeks to adapt and achieve a new equilibrium. These shifts are fairly predictable, provided that natural processes alone are involved. However, in many parts of the world human intervention will have a larger ecological impact than the shifting climate in the next 50 to 100 years.

Water Resources

We depend on a reliable supply of fresh water for survival. A climate change will surely shift patterns of precipitation, and this will directly affect the water resources of every region. Areas with marginal water resources will be the hardest hit if there is a decrease in rainfall; our tentative scenario indicates that such areas may be in the midwestern United States and the Soviet Union, as shown in Figure 2-4.

Similarly, many developing countries in the semi-arid parts of the subtropics are now expected to experience a general increase of precipitation and soil moisture if a warming occurs. Changes in precipitation and soil moisture are key elements in food and forest production. Hence, the climatic warming could perhaps adversely affect the two superpowers and help a great many developing countries.

Many of the dams, aqueducts, pumping stations, reservoirs, and water distribution networks around the world have been designed to cope with seasonal and year-to-year fluctuations of water supply and demand. The lifetime of such facilities is typically 50 to 100 years or more, so the anticipated climatic change will occur while they are still in place. Whether they will still be adequate is the question that must be answered.

In any case, any measures taken now to guarantee more reliable water supplies will be advantageous regardless of longer-term climatic changes. Even without a future shift of precipitation patterns they will probably be cost-effective.

Fisheries

Up until the early 1970s rivers, lakes, and coastal areas of the world were generally viewed as vast, almost limitless resources of food. As late as the 1960s and early 1970s scientists had reported that fish catches could be greatly increased, and would provide a large supplement to food production on land. However, after 1972, global fish catches declined from their peak of 26.5 million tons to 18.5 million tons in 1973, according to the United Nations Food and Agriculture Organization. There has recently been a partial recovery, but not to the levels attained in the early 1970s.

Decreases in fish landings can be caused by a number of factors, either singly or in combination. These factors include overfishing, poor fishery management practices, and oceanographic and climatic fluctuations or changes. In 1972 and 1973 climatic fluctuations resulted in shifting ocean currents, sea surface temperatures, and wind patterns that may have been partly responsible for the general reduction in fish landings, and specifically in Peruvian coastal waters. The anchovy population there depends on an upwelling of nutrient-rich cold, deep water to the surface. There are other coastal zones that depend on upwelling to provide nutrients to the fish population; all are subject to similar year-to-year fluctuations of currents and water temperatures.

Another example of the influence of climate on fish catches is the West Greenland cod supply. Up to about 1950 North Atlantic temperatures increased, and during this period the catch also increased, reaching a peak of some 450,000 tons in the early 1960s. However, since then there has been a cooling trend, and in recent years catches of cod have been so low that they have been banned off Greenland.

While there seems to be good reason to believe that winds, temperatures, and ocean currents all play a part in determining favorable environments for fish, the marine ecosystem is still poorly understood. It is difficult to separate the climatic impact from the human influence, such as overfishing. In any case, it appears that the sea is not as bountiful a source of food as once believed.

Health, Comfort, and Disease

Human beings can survive environments from extremely hot (50°C) to extremely cold (-60°C). However, most people thrive best in a temperature range known as the "comfort zone," ranging a few degrees above or below the optimum annual average of about 10°C.

The prevalence of most diseases is also affected by climate, demonstrated by the seasonal outbreaks of certain illnesses in temperate regions and the limitations of other diseases to certain climatic zones in the tropics. Some human diseases depend on insects, snails, or other carriers for their spread, carriers which are subject to temperature, moisture, and other climatic constraints.

But diseases are not solely linked to climate, since they are also affected by the condition of water supplies, food sanitation, and refuse disposal. Not surprisingly, most of the serious diseases are disproportionately concentrated in the poor and developing countries of the world, nations which tend to be in the tropics or subtropics. In short, poverty provides favorable conditions for the spread of disease, which itself helps to perpetuate poverty.

In the course of a global warming a number of diseases that are now mostly confined to the tropics might spread to more temperate regions. These diseases include schistosomiasis, bacillary dysentery, hookworms, yaws, and malaria. However, they could be controlled by hygienic and other measures, so their spread is by no means inevitable.

Population Settlements

Many population shifts have been partly due to an adverse climate. The emigration to the United States and England during the Irish Potato Famine of 1845 to 1851, and the abandonment of farms in the Great Plains of Nebraska, Kansas, Texas, and Oklahoma during the dry periods of the 1890s, the 1910s and the 1930s, are examples. Of course, the latter was triggered not only by abnormally low rainfall and high temperatures in the Great Plains, but also by poor grazing and ploughing practices and economic depression. The drought of the 1950s in the same region did not cause as much disruption.

The developing world, which will probably contain more than three-quarters of the world population by the year 2000, is especially vulnerable to the pressures of climatic variations and change. The Sahelian disaster of 1968–1973, when over 100,000 North African nomads died, and the Ethiopian famine of 1984–1985, illustrate this point.

If a climatic warming causes sea levels to rise, and a 1 m or more rise is a definite possibility in the next century, the densely populated coastal regions would face serious consequences. It is obvious that a sea-level rise would affect all coastal areas of the world.

It must be emphasized that glaciologists do not agree on the time scale for such a rise, in which a primary cause could be the shrinking or breakup of part of the great ice sheets of Greenland or Antarctica. The West Antarctic ice sheet has attracted the most attention, since it sits largely on bedrock below sea level, but whether it would disintegrate in a matter of centuries or millennia following a major warming is still being debated. If it were to disintegrate and slip into the sea it would raise the global sea level some 5 to 7 m.

Thus, while it is clearly too early to make evacuation plans, we can begin to prepare for this contingency by applying what we know about land use and flooding in areas where water levels are already rising. For example, land is subsiding or shifting in Venice, Italy; Long Beach, California; and Galveston, Texas. This creates an apparent sea level rise for non-climatic reasons.

Tourism and Recreation

An increasing number of communities, and even entire countries, are becoming dependent on the income from tourism. It is an important economic sector that is also remarkably sensitive to climatic variations and change. The best example of this is a ski resort without adequate snowfall.

Health through recreation is becoming more popular throughout the world—that is, where people can afford the luxury of recreation. Climatic change will clearly affect the development of vacation facilities.

Policy Decisions and Measures Dealing with the Carbon Dioxide/Climate Problem

Long Range Strategies

While there are many long-range strategies that can increase our resilience to climatic variability and change—or perhaps delay the change—there are also certain obstacles to accepting and implementing them. Among these obstacles are the following:

Incomplete information: Since climatologists still cannot predict with certainty future temperature and precipitation changes, either regionally or seasonally, it is not surprising that policy makers and the public are reluctant to commit resources to long-range measures in order to mitigate a situation that might never occur.

Poor planning and information distribution: Climatological information, though available in archives, has not been properly consulted in a number of well-documented cases. This seems to have been due to a lack of awareness on the part of the public and its planners of the opportunity and need for taking such information into account.

Discounting the future: When weighing risks and benefits there is the tendency to opt for near-term gratification rather than longer-term rewards. This tendency is especially strong when dealing with benefits to be reaped by future generations. What is their future worth to us today? Residents of flood plains, for instance, have denied the possibility of another flood occurring, at least, in their lifetime. This irrational discounting of the future often serves to their disadvantage, and governments as well as individuals can fall into this trap. The same is true for carbon dioxide. Are we willing to accept the future risk of added carbon dioxide in the atmosphere for the present benefit of burning fossil fuels?

When societies have perceived the need to adopt some kind of long-range strategy there are, in principle, four possible alternative courses of action. These are shown diagrammatically in Figure 2-9. They range from the most aggressive on the left to the most passive on the right. The next two sections deal with the likelihood of the adoption of the more aggressive courses of action and the kinds of steps that could be taken to adapt to the change, once the inevitability of the change is recognized.

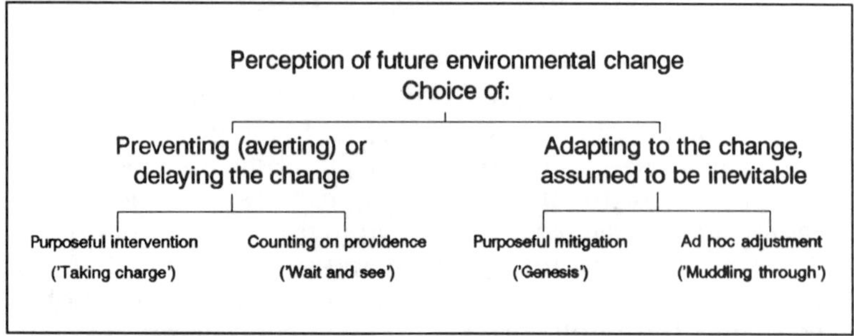

Figure 2-9: The choice of strategies when faced with the prospect of a future environmental (climate) change. Basically the alternatives are to take action to prevent the change or to accept it as inevitable and learn to live with it. The latter strategy is in effect a decision to do nothing at all, at least for the time being.

Averting the Change

A common first reaction to the thought that a climate change could be caused by burning fossil fuels is to call for a cutback in their use. This is in principle a reasonable course to take, but it seems we do not live in a reasonable world. Some of the reasons for skepticism with regard to such a cutback of fossil fuel use are the following:

1. Fossil fuel is a major economic commodity, and a great majority of countries have developed the costly infrastructures needed to use it. The powerful vested interests and the large capital outlays involved will both dictate against the abandonment of such a convenient and economical energy resource.

2. Since the buildup of carbon dioxide is a global phenomenon and the cause is the worldwide use of fossil fuels, any course of action to reduce their use would have to be agreed upon and implemented internationally. There is no international mechanism now in place that could arrive at such a decision—much less enforce it. (See the section on *International Legal Mechanisms* below.)

3. The climate change in store for the world will very likely be perceived as advantageous by some countries, at least in its early stages. They would be reluctant to make a sacrifice of a convenient energy source for a negative reward.

4. The scientific community is admittedly uncertain about many of the features of a future warmer earth, and there are even some who do not agree with the consensus that the change will occur. These well-publicized debates among scientists suggest to the public and its leaders that no major decisions should be taken until the prediction of climate change can be made with more certainty. Scientists on both sides of the issue may have to wait until the earth has demonstrably proven their prediction to be right.

These are but some of the arguments against relying on international action to reduce fossil fuel burning. There may be economic and technical incentives to encourage a changeover, but they will probably be slow to operate. Thus, it seems prudent to proceed on the assumption that atmospheric carbon dioxide will continue to increase, that a climate change may well occur, and that people should take the necessary steps to mitigate potential impacts.

Mitigating the Effects

Regardless of whether carbon dioxide builds up and results in a future climatic change, problems caused by weather and climate variability occur now and will continue to do so. Thus, we should strive to reduce the vulnerability of human settlements and activities. As it turns out, a variety of strategies to mitigate the impacts of future climatic change would also be wise steps to take for more immediate problems. Therefore these strategies should be implemented in any case.

We have identified three classes of long-range strategies with short-term benefits: those that increase resilience (or decrease vulnerability) to climatic change; those that help to slow the increase of carbon dioxide; and those that lead to making better choices.

Strategies that Increase Resilience to Climate Change

Protect arable soil: Agriculture and animal husbandry depend on soil that can grow plants. This resource is being poorly managed and lost through erosion and salinization. This is one of the world's most pressing environmental problems, and it threatens our ability to produce enough food.

Improve water management: Societies will continue to depend on water supply systems that can ameliorate floods and provide water during periods of drought.

Apply agrotechnology: New agricultural techniques led to the "Green Revolution," and they may help us provide food for growing populations in the face of adverse climatic events and longer-term climatic change.

Improve coastal land use policies: Coastal communities need to make better use of climatic information to mitigate the effects of floods, hurricanes, and typhoons. A future sea level rise is one more factor that should be a part of their planning, even though its time scale is still uncertain.

Maintain global food reserves: Future shifts of temperature and precipitation patterns will affect some for the better, some for the worse. Thus, maintaining global food reserves to help the losers is a sound policy, both now and in the future.

Provide disaster relief: Catastrophes will still require emergency aid from international relief organizations to alleviate their impact.

Strategies that Help Slow the Increase of Carbon Dioxide

Conserve energy: There are strong incentives to make energy conservation the basis for a sound energy policy, since it reduces the demand for all fuels, including those from fossil sources.

Use renewable energy resources: By the same token, use of renewable energy sources, such as solar energy, biomass conversion, and hydroelectric power will reduce the demand for fossil fuels.

Increased use of nuclear energy: While there is considerable public opposition to nuclear energy on the grounds of its possible health and ecological hazards, it also has some advantages, one of which is the fact that it does not generate carbon dioxide.

Reforest: Replanting trees in areas where deforestation has taken place, as in many parts of the tropics, not only replaces a valuable economic resource and prevents soil erosion, but also takes some carbon dioxide out of the atmosphere as the trees grow.

Strategies that Lead to Improved Choices

Employ environmental monitoring and warning systems: International organizations have already established global environmental monitoring and warning systems, notably the United Nations Environment Program's Global Environmental Monitoring System (GEMS).

Provide improved climate data for direct use: The newly established World Climate Data Program and World Climate Applications Program, part of the WMO's World Climate Program, should improve the availability and application of climate data for planning and operations. There are many countries that lack the expertise or the mini-computers needed to make use of our knowledge of climate and its influence on human activities.

Inform and educate the public: Dissemination of the latest results of climate studies should raise the general level of public awareness about the carbon dioxide/climate problem. This will lead, one can hope, to better choices by political leaders and a greater public acceptance of measures needed to cope with climatic variations and change.

Transfer appropriate technology: The transfer of technology to developing countries has been taking place for a long time, but it has not always been appropriate. Technology tailored to developing countries' needs will help in agriculture, water resources, the development of export products, and land use.

Clearly, nations need to take vigorous initiatives to cope with our variable and changing environment. Whether societies will have the resources needed to adapt to climatic changes will be determined by several indices, most of them economic: gross national product (GNP), an indicator of the economic resources available to build new facilities or move people; ratio of investment to GNP (or gross rate of investment), an indicator of the rate of turnover of new facilities and capital stock (a faster turnover implies more flexibility); the flexibility and diversity of the capital stock, which, though it may increase costs and lower output somewhat, offers insurance against changing conditions; and the ability to foresee changing conditions and adapt to them quickly, which comes with public information and education.

A first conclusion of the above—hardly an original conclusion—is that the wealthier, generally industrialized countries will be more likely to be able to cope with a worldwide environmental change than the developing countries. This "first conclusion" may be misleading, however, when we consider the distinct possibility, illustrated in Figure 2-4, that the developing countries in the semi-arid and sub-tropics may actually benefit from the climate change as they receive more rainfall. The major powers in the temperate latitudes may not fare so well if they become both hotter and drier. We will return to this point later on.

International Institutions and Legal Mechanisms

History of Organizations for Environmental Cooperation

Formal recognition of the need for international collection and exchanges of meteorological and oceanographic observations occurred in 1873 with the establishment of the International Meteorological Organization (IMO). In 1951 the IMO became the WMO, one of the United Nations' specialized intergovernmental agencies; it now has its headquarters in Geneva, Switzerland.

Until recently the WMO devoted most of its effort to improving weather forecasting and exchanging worldwide meteorological observations twice a day through the World Weather Watch (WWW) system. However, more attention has been paid to climate in the past few years. In February, 1979, the WMO held the first World Climate Conference in Geneva to approve a draft proposal for a new World Climate Program. This program was formally implemented at the beginning of 1980.

During the 1970s there was a growing concern for the environment and its finite natural resources. Following the United Nations Conference on the Human Environment in Stockholm in June 1972, the United Nations Environment Program (UNEP) was established, with headquarters in Nairobi, Kenya. Its charter includes dissemination of "information on major environmental problems and the efforts being made to respond to them, in order to identify gaps, set objectives, and establish priorities." To meet these goals, one step UNEP has taken was organizing the Global Environmental Monitoring System (GEMS). Among GEMS' functions is "assessing global atmospheric pollution and its impact on climate."

The International Council of Scientific Unions (ICSU), a non-governmental body, founded the Scientific Committee on Problems of the Environment (SCOPE) in 1969. This has expanded the opportunity for international collaboration. One of SCOPE's activities is a long-term study of biogeochemical cycles of carbon, nitrogen, phosphorus, and sulfur which would be affected by climate change; and it has recently organized a study of methodologies for assessing climatic impacts.

Another organization conducting research on global problems is the International Institute for Applied Systems Analysis (IIASA), located in Laxenburg, Austria. Presently IIASA's membership consists of 17 national scientific academies, and its research has included studies of world energy and food problems and their climatic implications.

There are also several regional associations of nations that are becoming more active in this field. The Council of European Communities, for example, has adopted a European Climate Program for which the WMO and UNEP share responsibility, similar to the United States National Climate Program.

International Legal Mechanisms

We should not expect too much of international law when it comes to resolving cases that deal with injuries or damages caused by carbon dioxide-induced climate change. Ultimately, countries are free to comply with or to ignore international law. In any case, there is currently no mechanism by which to set carbon dioxide standards and establish universally applied control measures, nor a policing body to enforce the so-called "global right to a clean atmosphere."

It would make sense, however, to encourage new international mechanisms that could, among other things: examine the likely effects of a carbon dioxide-induced warming on national and international activities; recommend to governments, as far as practical, measures to deter excessive use of fossil fuel or large-scale deforestation; and exchange information on climate change and its impacts. The Climate Impact

Studies Program, a component of the World Climate Program, may be in a position to fulfill some of these functions.

There are some existing international mechanisms that could become applicable to carbon dioxide related issues in the next few decades. They include:

International agreements: There are several types of treaties and conventions being negotiated that apply principles of international law to established areas of transnational significance, such as marine pollution, nuclear weapons tests, exploitation of the Antarctic, and the use of outer space. Nations engaging in these agreements have taken preliminary steps toward clarifying the procedures through which regulations can be implemented. A recent example is the Montreal Protocol, concluded in 1987, designed to control the worldwide production and emission of chlorofluorocarbons.

Regional organizations: There seems to be a growing acceptance of regional organizations such as the European Economic Community, the Organization of Economic Cooperation and Development, the Council for Mutual Economic Aid, the Association of Southeast Asian Nations, and the Latin America Free Trade Association. These groups are set up to deal with specifically supranational problems. They could be valuable in implementing carbon dioxide control strategies, and in compensating member countries for damages resulting from climate damage.

International commissions: The findings and recommendations of international commissions such as "conciliation commissions," "arbitration tribunals" and "commissions of inquiry" can sometimes influence international behavior. For carbon dioxide problems such commissions may be useful ad hoc devices to either arbitrate disputes or gather expert opinions on controversial legal, political, and scientific issues.

International conferences: Conducting international negotiations through special conferences is becoming an accepted practice among countries. The United Nations has encouraged this by convening such meetings as the Conferences on the Environment, the Law of the Sea, the Law of Treaties, and Science and Technology for Development. The value of recommendations made at international conferences is that they tend to call attention to matters that *should* be dealt with by the member states. To date, the World Climate Conference has been the largest international conference to investigate the impacts of climate change.

We have only dealt with a few of the possible international legal mechanisms for handling the political and economic consequences of a global carbon dioxide-induced climatic change. Special mention should also be made of diplomatic negotiations, the Permanent Court of Arbitration, and the International Court of Justice. Each of these may be used to resolve the possible carbon dioxide-related damage and compensation disputes. It seems unlikely, however, that any of them can be effective in achieving worldwide action to reduce fossil use, as pointed out in the section designated *Averting the Change*.

Conclusions

These are a few of the more important points that emerge regarding societal impacts of a carbon dioxide-induced climatic change.

It should be clear by now that, while the general features of a global carbon dioxide-induced warming are beginning to become clearer as a result of research efforts undertaken during the past decade or more, there is still an urgent need for the development of climate scenarios describing the expected changes on a *regional* and *seasonal* scale. These climate scenarios need not be considered definite predictions of the future, but they must be possible and internally consistent pictures of what could happen. The process of climate scenario development must be a continuing one, moving hand in hand with the improvement of theoretical climate models and better information about climates of the past. In fact, this process is an inescapable first step in any studies of the societal impacts of climate change, studies which must start with the specifics in various regions of the world.

While we may never be able to predict future climatic change in adequate detail or exactly how individual societies will respond to it, we can clarify the relative merits of alternative long-term policies or strategies. We believe the most useful strategies are those that can mitigate the adverse impacts of the changes and make specific activities—such as agriculture and human settlements—less vulnerable to climatic impacts, from whatever cause.

Having made an initial and by no means exhaustive examination of the various useful strategies, we can see that each one will also help cope with short-term weather and climate variability—devastating events such as droughts, floods, heat waves, and cold spells that cause hardship every year. Some strategies will also help alleviate chronic environmental problems such as the loss of arable soil and tropical forests. International and national measures to preserve these precious natural assets are long overdue. Perhaps the stimulus of a potential impending climate change will help to spur such actions.

Climatic impact studies will help to estimate the effects of climatic changes, which should better enable us to plan agricultural production, land use, and human settlements, and to develop water resources and marketing strategies. However, new and improved methodologies of climate impact assessment are needed as well as more scientists interested in working on impact studies.

Apart from the research needed to improve our ability to carry out impact studies, it is imperative that we implement international programs to cope with carbon dioxide-induced changes. Some international mechanisms already exist that might help us deal with this unprecedented set of global problems. All of them should be explored and their use encouraged. Most importantly, nations should be made aware of the problem of climatic change and of the strategies that could mitigate its effects.

For some countries the change may be favorable, while for others it may be adverse. But we must emphasize that the prospect of a global climatic change is just one of a number of global environmental and societal problems. In the first half of the next century the world could well have twice as many people, consume three times as much food, and burn four times as much energy. Thus, impacts of climatic change must be superimposed on this backdrop. Our future problems will be serious even without a shifting climate.

References

Climate Impact Assessment Program CIAP, 1975. Department of Transportation, Washington, D.C.

FAO Production Yearbook, 1979: Statistics series No. 22, Vol. 32.

Kellogg, W.W. 1979. "Influences of mankind on climate." *Ann. Rev. Earth Planet Sci.* 7:63–92.

Kellogg, W.W., and R. Schware. 1981. *Climate Change and Society: Consequences of Increasing Atmospheric Carbon Dioxide* (Report of Aspen Institute for Humanistic Studies, sponsored by United States Department of Energy), Westview Press, Boulder, Colorado (178 pp.).

Kellogg, W.W., and R. Schware. 1982. "Society, Science and Climate Change." *Foreign Affairs* 60: 1076–1109.

Kellogg, W.W. 1987. "Mankind's Impact on Climate: The Evolution of an Awareness." *Climatic Change* 10:113–136.

Villach/Bellagio Workshop, 1987. "Developing Policies for Responding to Climate Change." Report of workshops held in Villach, Austria, and Bellagio, Italy. World Meteorological Organization (WMD/TD ho. 225), Geneva, Switzerland.

WMO. 1979. *Proceedings of the World Climate Conference*. WMO No. 537, World Meteorological Organization, Geneva.

WMO. 1980. *Outline Plan and Basis for the World Climate Programme 1980–1983*. WMO No. 540, World Meteorological Organization, Geneva.

Whittaker, R.H. 1975. *Communities and Ecosystems*. Macmillan, New York.

THREE

RESPONSE to KELLOGG'S PAPER

Hugh W. Ellsaesser

Kellogg has presented the consensus view of the CO_2 (carbon dioxide) climate problem and then outlined the cascade of deleterious effects this *may* produce throughout all aspects of human activity and the global ecology. He did depart a bit from the script made standard by the Climate Impact Assessment Program CIAP for the SST. He devoted many of his pages to possible beneficial consequences; he mentioned caveats and uncertainties several times; and he even admitted that "there are some who do not even agree with the consensus that the change will indeed occur."

I am one who does not agree with the consensus and there are at least three points I would like to make:

1) I would like to have you step back and take a critical look at the process by which the now fashionable "Interdisciplinary Science through Workshops of a Committee of Experts" operates; how it transforms preliminary "guestimates" into a consensus supported by a constituency with a vested interest in confirming and perpetuating the problem rather than solving it.

2) I would like to have you visualize the hazard from the receiving end of the hazards cascade and contrast this with the description starting from the origin as given by Kellogg.

3) I would like to present a few of the reasons why I have refused to accept the consensus view.

Development of the Consensus View

Following a particularly *ad hominem* article in either the technical or popular literature, an individual or a group is able to initiate organized action. In preliminary planning sessions the basic hazard or problem and its cascade of possible ramifications down through the global environment to the human economy and social practices are sketched out. Experts in the various disciplines involved are recruited and then one or a series of workshops are scheduled. In order to facilitate and speed the impact assessment process, specialists in the primary impact area—in our case, climate modeling—are asked to draw up preliminary estimates of the basic impacts in terms of quantitative changes in temperature, precipitation, wind, solar radiation, sea level, etc. (independent of the uncertain time of occurrence). These are invariably extracted from the most recent publications (unchallenged) and are passed down the line to the ecologists, biologists, agronomists, economists and social scientists as input for assessing impacts in their areas of interest.

In the process a number (or set of numbers) of little concern to anyone has suddenly acquired a large constituency among the recruited specialists, their employees and employing institutions. The basic scientific problem lies in a discipline foreign to most of those recruited to work on the impact assessment and is generally both outside their areas of expertise and of little intrinsic interest. Thus a large constituency has been established that is more concerned with confirming the hazard and prolonging the impact assessment process than in solving or increasing our understanding of the specific problem which gave rise to the impact assessment effort.

As the work proceeds, each group of researchers writes their report and these are successively meshed and condensed into a final one-paragraph statement of findings—and almost invariably, an eventual single headline. No matter how carefully the original reports are written, at each step in the meshing and condensation, the caveats get dropped, shortened or weakened while the hazards or impacts of dramatic interest acquire successively more dominant roles, i.e., they are made to appear both more severe and more certain. Remaining uncertainties which cannot be resolved or ignored are covered by parameterizations and error bars.

I witnessed this process firsthand during CIAP. Early on I recognized the adoption of "The One-Way Filter" approach (Ellsaesser 1974a) in which only those pathways of the effects cascade leading to deleterious effects are pursued and described while any possible benefits are either studiously ignored or deemphasized.

CIAP's evaluation of the supersonic transport (SST) was comparable to a debate on whether we should continue to propagate our species, in which it was contrived to have the discussants consider only the problem of dealing with bodily excreta. "Harmful ultraviolet" became a single inseparable word—no statement or report was complete without a reference to its causing an increase in the incidence of skin cancer. Rickets, which may be alleviated by increased exposure to ultraviolet, might as well have been an unknown phenomenon. It may be of interest to note that the original concern that aircraft exhaust would cause destruction of ozone in the stratosphere is now shifting to a concern for ozone generation in the troposphere, since this now appears to offer the only saleable argument for continuing research on this subject. However, no one, as far as I am aware, has yet suggested that increase of ozone in the troposphere might cause a decrease in the incidence of skin cancer.

Since my experience with CIAP I have preferred not to get too involved in "big science" and to dare to try to be right rather than merely to reflect the prevailing consensus. It has been my experience that the greater the concern, as indicated by media attention to a research program, the more difficult it appears for the participating scientists to maintain and exercise objectivity.

If there is a parameter in another discipline, such as oceanic thermal lag, impacting the answer, I do not want to have to accept the opinion of the committee's oceanographic expert; I want to be able to investigate for myself and determine whether it is any more firmly established than other parameters impacting on the problem.

As a final note in this section, I would like to recall our scientific "batting average" in the past. History is replete with widely accepted theories later abandoned. Let me cite only two. All students of climate surely remember that 10 to 15 years ago it was agreed that the atmosphere was not responding to increasing CO_2 because the warming effect was being overwhelmed by the greater cooling power of increasing anthropogenic particles (Ellsaesser 1974b–1975). Going further back, John Mayow demonstrated in 1668 that both candles and animals expired after a short time in small confined vessels due to the exhaustion of "the igneoaerial particles of the air." This near-discovery of oxygen was "discredited" by Stahl's phlogiston theory which held the consensus for more than three-quarters of a century until oxygen was "rediscovered" by Lavoisier in 1775. However, Lavoisier decided that not lack of oxygen but excess of CO_2 was the cause of discomfort and occasional disaster in overcrowded closed rooms. Another three-quarters of a century passed before Max von Pettenkofer clearly demonstrated that the CO_2 content of badly ventilated rooms remained far below that at which harmful effects appeared in laboratory experiments.

Attention then focused on toxic organic effluvia—termed morbific matter, kenotoxin or anthropotoxin—detectable by odor, by darkening of sulfuric acid and potassium permanganate, but otherwise then unmeasurable. Assuming this unmeasurable toxin was proportional to the level of CO_2, Pettenkofer, in 1863, derived his ventilation standard of 30 cubic feet of fresh air per minute per occupant—soon frozen into law for schools and public buildings. In 1905, Flugge finally exorcised the anthropotoxin theory by experiments in which one subject inside a cubicle inhaled fresh outside air while another located outside the cubicle inhaled air from inside the cubicle—thus demonstrating that it was the cooling power (determined by temperature, humidity and relative movement) of the *air acting on the skin* which first impacted comfort and ultimately survivability.

Thus, of the five air constituents modified by human occupation of a confined space: oxygen, CO_2, organic effluvia, heat and moisture; we required over three quarters of a century each to eliminate the first three (1668–1905) as being of minor importance. In the process we invented at least two nonexistent substances—phlogiston and anthropotoxin. If these were isolated cases, we could laugh them off. However, the historical record, up to and including today, is replete with similar cases demonstrating the human fallibilities of scientists and their predilection for finding what they, at the time, expect to find, rather than what turns out to be supported by the evidence.

The Impact of Climatic Change as Viewed by the Impactee

As reported by Kellogg, the primary impact now expected from doubled CO_2 is a rise in planetary mean temperature of 3 ± 1.5°C, a rise that can be expected to occur in the time range from mid to late 21st century. I ask you to visualize yourself as a farmer, a natural gas distributor, or a construction engineer. You are engaged in producing and delivering a product for sale in a competitive market. There is a premium for delivering when no one else can and a penalty for not being able to deliver on demand. On a daily and annual basis you contend with the seasonal temperature cycle and with its day-to-day and interannual variability. The annual range of surface temperature can reach 120°F (67°C); day-to-day variability in Chicago is such that the daily temperature forecast was on at least one occasion missed by over 100°F (56°C). I ask you, to what extent are you going to modify your business practices when you are told that the mean temperature, not *will*, but *may* rise 3°C in the next 50 to 100 years? Over the past century the global mean surface temperature appears to have increased ≈ 0.5°C—had this been predicted in advance, would anything have been done differently?

The Climatic Effect of Carbon Dioxide

The consensus that doubled CO_2 will lead to a global temperature increase of 3°C is actually quite narrowly based—it exists primarily among climate modelers and their associates. While climate models come in a wide assortment, the two considered most reliable are the radiative-convective models and the general circulation models (GCMs), both of which are quite complex and detailed. The optimum attributes for developing such models are burning ambition and an uncluttered mind—which helps explain why most such models have been developed by graduate students and post-doctoral students. I do not believe that most of us would agree that such people are in general the ones who best understand how the atmosphere as a whole works.

Among non-climate modelers—which includes a good portion of the meteorological profession—there are many who doubt the consensus view on climatic effects of CO_2 but their opinions rarely appear in print. The ones who did get their doubts in print (Newell and Dopplick 1979; Idso 1980) have been severely attacked by the establishment (Crane 1981; Kandel 1981; NAS 1982, Schneider *et al.*, 1980; Watts 1980, 1982). My own reasons for doubting the consensus view are as follows:

Climate modelers have so far concerned themselves mainly with two climatic feedback processes, both of which are claimed to amplify any CO_2 warming: the so-called ice-albedo feedback, and the water vapor feedback. There are at least four reasons for believing that the ice-albedo feedback is currently overestimated—if not actually of the wrong sign:

1) Very little sunlight is received to be reflected in those latitudes and seasons in which seasonal snow and ice cover occur.

2) Planetary albedo is also strongly influenced by solar zenith angle. Once this is allowed for, there is relatively little difference in high latitude planetary albedo between ice-in and ice-out (Lian and Cess 1977). The poor satellite data available in these areas suggest a change of no more than 0.1 to 0.25 albedo units at 50–70°N and 60–80°S (Campbell and Vonder Haar 1980; Stephens *et al*. 1981).

3) Ice and snow, on the other hand, do have a very strong ice-insulation *negative* feedback. An ice cover reduces the wintertime loss of latent and sensible heat to the atmosphere and outer space from open water bodies by orders of magnitude. Also, snow-covered land and sea ice can reduce their radiational loss of energy by cooling to radiating temperatures well below those reached without an insulating snow blanket. Reduced winter loss of heat to space represents a warming for the planet.

4) Polar ice and snow behave quite differently in a seasonal than in an annual mean model in which the sun shines all the time.

That the negative ice-insulation feedback is the one that predominates is suggested by the observations that successive summer and winter Antarctic ice cover anomalies tend to have the opposite sign (Zwally *et al.* 1983) and that Arctic ice cover shows a maximum negative auto-correlation of sea ice cover, i.e., sea ice leading to additional sea ice.

The important point to remember about the water vapor feedback is that it does not become effective until *after* the surface temperature has been increased by some other mechanism. Over the tropical oceans, doubled CO_2 will give a minimal increase in IR (infrared) flux to the surface due to the fact that the lower atmosphere there is already opaque to IR radiation due to the continuum absorption by water vapor. In addition, as temperatures rise, an increasing fraction of any enhanced flux to the surface will be used for evaporation as opposed to temperature rise. Thus, the effect of doubled CO_2 on tropical ocean temperatures should be minimal. From a study of maximum temperatures attained by plant foliage and the surface waters of tropical oceans, Priestley (1966) deduced that there should be a "rather sharply defined upper limit to which air temperature will rise above a well watered surface." Using monthly averages of daily maximum temperatures he identified the limiting temperature for land stations which had not exhausted their soil moisture as about 92°F (33°C). Priestley and Taylor (1972) found that the Bowen ratio (ratio of sensible to latent heat flux from the surface to the atmosphere) decreased monotonically with increasing temperature, suggesting that sensible heat flux became negative (flowed from the atmosphere to the surface) at temperatures above about 32°C. A reversal of the heat transfer between plant leaves and air in the vicinity of 33°C was reported earlier by Linacre (1964). A review of Bowen ratio data by Brutsaert (1982) placed the apparent sign reversal near 30°C. Newell and Dopplick (1979) sought to illustrate the physics of this situation, but since they held the surface atmospheric conditions fixed at 27°C and 65% relative humidity, their computation of a requirement for 30 $W/m^2 °C$ to warm the surface water of the tropical oceans has been "discredited" (NAS, 1982). However, their lack of realism was not unusual for the introduction of a new idea, and in any case, in no way negates the requirement for substantial increments in radiative flux to warm tropical ocean waters in contact with the atmosphere to temperatures above their current level of 26°to 30°C. From observational data the change in net energy flux into the ocean with change in ocean surface temperature has been estimated at 40 $W/m^2 °C$ in the tropical Pacific (Niiler 1981) and in the Eastern Mediterranean (Assaf—herein).

This point is supported by the paleoclimatologists. Matthews and Poore (1980) proposed the working hypothesis that the surface temperatures of the tropical oceans are tied to the solar constant and have maintained their present values back even through the Cretaceous Period—the period of warmest terrestrial climate yet documented (Barron *et al.*, 1981). In their reconstruction of the climate of 18,000 YBP (years before present), CLIMAP (1976, 1981) found very little change in tropical ocean surface temperatures other than that attributed to stronger Chile and Benguela ocean currents and stronger equatorial upwelling. Additional support is provided by the pattern of seasonal change in ocean surface temperatures. Newell's map (1979) shows vanishingly small changes near the equator, bands of maximum changes at 35–45°N and S, and decreasing changes poleward of the midlatitude maxima. While they had no data to show it, the seasonal *water* temperature change must vanish at the most poleward summer boundaries of the sea ice. Seasonal temperature changes and gradients near the equator remain small despite both the considerable changes in solar flux due to the seasonal change in zenith angle and the 7% change in solar flux due to ellipticity of the earth's orbit.

Broecker (1982) presents maps of the difference between current ocean surface temperatures and those for 18,000 YBP as reconstructed by CLIMAP (1981). These indicate that the pattern of ocean surface temperature change for a change in terrestrial climate (glacial to interglacial) is analogous to that for the seasonal changes, i.e., vanishingly small at the equator, latitudinal bands of maximum change in mid-latitudes, and decreasing changes toward the poles. While we do not know the cause of the climate change which occurred over the past 18,000 years, the change in mean global temperature was at least as large as is currently anticipated for a doubling of CO_2.

Doubling CO_2 alone will increase the IR flux to the surface by 1–1.5 W/m^2 averaged over the globe. If continuum water vapor absorption is considered, doubling CO_2 will produce hardly any increase in IR flux to the surface in tropical latitudes where the moist surface layer is already essentially opaque to IR radiation (Kiehl and Ramanathan 1982). Yet most doubled CO_2 model experiments appear to compute a tropical ocean warming of about 2°C. I have yet to hear a plausible explanation of this apparent gross discrepancy.

Analyses of polar glaciers (Neftel *et al.* 1982; Oeschger *et al.* 1982) and oceanographic data (Chen 1980) have recently been added to inventories of the pre-industrial biosphere as reasons for lowering the range of estimates for the pre-industrial level of atmospheric CO_2—this range now extends from 250 to 290 ppm. With today's level of 340 ppm and a model-computed doubled CO_2 warming of 3.0°C, we should, assuming equilibrium, have already had a warming of 0.68° to 1.33°C. Increases in other anthropogenic greenhouse gases such as methane, nitrous oxide and chlorofluorocarbons have presumably contributed an additional warming of approximately one-half this magnitude (Lacis *et al.* 1981; Craig and Chou 1982). If, to warm the ocean surface, it is necessary to warm the total depth of the ocean, there is of course no problem. However, if we examine the oceanic temperature profiles where there are significant seasonal variations in surface temperature, we find a negative correlation between the amount of surface warming and the depth to which it penetrates (Newell 1979). In the regions of large seasonal temperature change and even in the Gulf Stream we find that as the surface layer warms, the water at depths of 100 to several hundreds of meters actually cools (Newell 1979; Niiler and Richardson 1973). These data do not suggest that the whole or even a significant fraction of the oceanic layer has to be warmed before the mixed layer temperature can rise. They rather suggest that any general warming of the ocean surface will be reflected in an upward extension of the thermocline to a warmer but shallower mixed layer. The more the horizontal pattern of ocean surface warming conforms to the pattern of seasonal temperature change described above, the more plausible this vertical pattern of temperature change becomes. And if this vertical pattern of warming occurs, the ocean thermal lag will be less than a decade. These three conditions (i.e., the lowering of the estimates of the pre-industrial level of CO_2, the shortening of the possible oceanic thermal lag, and the inclusion of other greenhouse gases) are all making it increasingly difficult to explain why the model-predicted CO_2 warming has not yet been unequivocally detected. The situation is made even more untenable by the significant biospheric source of CO_2 which must have been released prior to 1900 and thus should have produced its warming effect by now for any credible oceanic thermal lag.

However, my strongest reason for doubting current climate model estimates of the CO_2 warming are the gross differences I see between how the atmosphere works and how it is modeled to work. Water vapor feedback in particular is usually visualized as originally modeled by Manabe and Wetherald (1967). Under the assumption of a fixed relative humidity profile and a fixed tropospheric lapse rate, every increase in surface temperature instantaneously increases the temperature and absolute humidity throughout the tropospheric column. Clearly this leads to a significant positive water vapor feedback. However, both seasonal and latitudinal variations reveal a negative correlation between temperature and relative humidity.

Water vapor in the real atmosphere is cycled mainly by convection which is organized into deep cells by orography, thunderstorms, convergence lines including both wave cyclones and tropical cyclones, ITCZs, monsoons, and the Hadley circulation. Each of these on its own space scale sweeps the warm moist air from the boundary layer into narrow ascending columns where it is dried by condensation and precipitation of the water vapor; the latent heat is carried above the climatological moist layer and then spread by horizontal divergence. This diverging air is frequently accompanied by a cirrus layer which both reflects incoming solar radiation and provides a black-body emitter above approximately half of the atmospheric mass and above a much larger fraction of our principal greenhouse gas—water vapor. At the same time each convective cell is surrounded by a region in which subsidence, compensating for the air lifted within the cell, dries the surrounding troposphere.

Any objective analysis of tropical convection leads to the conclusion that it is a highly effective nonlinear planetary temperature regulator. Consider the tropical cyclone for a moment. These are very sensitive to both surface temperature and tropospheric lapse rate and can only form over the low friction surface provided by the ocean. They rarely if ever form at ocean temperatures below 27°C. Each one individually removes a tremendous amount of heat and moisture from the vicinity of the planetary surface in the forms of precipitation, latent heat carried to and spread horizontally in the upper troposphere, solar radiation reflected back to space by the cloud shield, and even sensible heat removed from the oceanic mixed layer by mixing it down into the thermocline by wind-induced turbulence. In addition, most if not all of the air carried aloft is compensated by subsidence which dries the surrounding areas. It is inconceivable that a tropical cyclone does not temporarily reduce the precipitable water vapor and thus its greenhouse effect for hundreds of kilometers on either side of its path. Most models appear to predict that increased CO_2 will intensify the Hadley circulation. This could as well lead, through increased subsidence and drying of the tropospheric air column in the subtropics, to a negative rather than a positive water vapor feedback throughout most of the tropics, or nearly half of the global surface.

Lindzen *et al.* (1982) has demonstrated that replacing the conventional convective adjustment in a radiative-convective model with a physically-based, cumulus-type parameterization of convection reduced the model sensitivity to doubled CO_2 by up to 80% for tropical temperatures. The sensitivity continued to decrease for even higher, unrealistic temperatures in such a way that they concluded that "cumulus convection effectively inhibits any possiblity of a run-away greenhouse on earth." It is noteworthy that Lindzen *et al.* (1982) obtained these results without considering the drying of air columns surrounding convection cells due to the subsidence which compensates the convective updrafts. Idso (1982) concluded from a completely different approach that a run-away greenhouse was not possible under terrestrial conditions. Some of the more sophisticated models used to perform the doubled CO_2 experiment presumably contain the physics to handle the Newell and Dopplick (1979) problem correctly. However, the fact that they continue to predict equatorial surface warmings of about 2°C represents an enigma which merits priority investigation.

Current climate models overestimate the effect of ice-albedo feedback in high latitudes and overestimate ocean warming and water vapor feedback in tropical regions and as a consequence are in general overestimating the climatic effect of doubled CO_2 by at least 2- to 3-fold. This upper bound of 1° to 1.5°C for the effect of doubled CO_2 is derived from the magnification effect usually attributed to the two feedbacks cited and from the fraction of the currently predicted warming allowed by available temperature records. While I find reasons to expect the effect of doubled CO_2 to be even smaller than this upper bound, they cannot be quantified without extensive additional model tests.

Robinson (1983) has reminded us that both cloud coverage and the radiative effect of clouds differ significantly between day and night and pointed out that a model which does not explicitly attack this problem "does not provide safe ground for an assessment of the carbon dioxide climate problem."

Acknowledgement

This work was performed under the auspices of the CO_2 Research Division, Office of Basic Energy Sciences, United States Department of Energy by the Lawrence Livermore National Laboratory under Contract No. W-7405-Eng-48.

References

Barron, E.J., S.L. Thompson and S.H. Schneider. 1981. "An ice-free Cretaceous? Results from climate model simulations." *Science* 212: 501–508.

Broecker, W.S. 1982. "Ocean chemistry during glacial time." *Geochim. Cosmochim Acta* 46: 1689–1705.

Brutsaert, W. 1982. *Evaporation into the Atmosphere*, D. Reidel, Dordrecht, Holland.

Campbell, G.G. and T.H. Vonder Haar. 1980. "Climatology of radiation budget measurements from satellites." *Atmos.* Science Paper 322, Colorado State University, Ft. Collins, CO.

Chen, C-T. A. 1980. "On the distribution of anthropogenic CO_2 in the Atlantic and Southern Oceans." *Deep-Sea Res.* 29: 563–580.

CLIMAP Project Members. 1981. "Seasonal reconstructions of the earth's surface at the last glacial maximum." *Geol. Soc. Amer. Map and Chart* Series No. 36.

CLIMAP Project Members. 1976. "The surface of the ice-age earth." *Science* 191: 1131–1137.

Craig, H. and C.C. Chou. 1982. "Methane: The record in polar ice cores." *Geophys. Res. Lett.* 9: 1221–1224.

Crane, A.J. 1981. "Comments on recent doubts about the CO_2 greenhouse effect." *J. Appl. Meteor.* 20: 1547–1549.

Ellsaesser, H.W. 1975. "The upward trend in airborne particulates that isn't." pp. 235–269 in *The Changing Global Environment*, S.F. Singer, ed., D. Reidel, Dordrecht-Holland.

Ellsaesser, H.W. 1974a. "The dangers of one-way filters." *Bull. Am. Meteor. Soc.* 55: 1362–1363.

Ellsaesser, H.W. 1974b. "Has man, through increasing emissions of particulates, changed the climate?" pp. 41–53, Proceedings Atmosphere-Surface Exchange of Particulate and Gaseous Pollutants (1974), URDA 38 Symposium Series, CONF-740921, NTIS Springfield, VA 1976.

Idso, S.B. 1980. "The climatological significance of a doubling of earth's atmospheric carbon dioxide concentration." *Science* 207: 1462–1463.

Idso, S.B. 1982. *Carbon Dioxide: Friend or Foe?* Tempe, AZ, IBR Press.

Kandel, R.S. 1981. "Surface temperature sensitivity to increased atmospheric CO_2." *Nature* 293: 634-636.

Kiehl, J.T. and V. Ramanathan. 1982. "Radiative heating due to increased CO_2: The role of H_2O continuum absorption in the 12–18mm region." *J. Atmos. Sci.* 39: 2923-2926.

Lacis, A., J. Hansen, P. Lee, T. Mitchell and S. Lebedeff. 1981. "Greenhouse effect of trace gases 1970–1980." *Geophys. Res. Lett* 8: 1035–1038.

Lian, M.S. and R.D. Cess. 1977. "Energy balance climate models: A reappraisal of ice-albedo feedback." *J. Atmos. Sci.* 34: 1058–1062.

Linacre, E.T. 1964. "A note on a feature of leaf and air temperature." *Agr. Meteorol.* 1: 66–72.

Lindzen, R.A., A.Y. Hou and B.F. Farrell. 1982. "The role of convective model choice in calculating the climate impact of doubling CO_2." *J. Atmos. Sci.* 39: 1189–1205.

Manabe, S. and R.T. Wetherald. 1967. "Thermal equilibrium of the atmosphere with a given distribution of relative humidity." *J. Atmos. Sci.* 24: 241–259.

Matthews, R.K. and R.Z. Poore. 1980. "Tertiary $\delta^{18}0$ record and glacioeustatic sea-level fluctuations." *Geology* 8: 501–504.

NAS. 1982. *Carbon Dioxide and Climate: A Second Assessment.* National Academy Press, Washington, D.C., 72 pp.

Neftel, A., H. Oeschger, J. Schwander, B. Stouffer and R. Zumbrunn. 1982. "Ice core sample measurements give atmospheric CO_2 during the past 40,000 yr." *Nature* 295: 220–223.

Newell, R.E. 1979. "Climate and the ocean." *Am. Scientist* 67: 405–416.

Newell, R.E. and T.G. Dopplick. 1979. "Questions concerning the possible influence of anthropogenic CO_2 on atmospheric temperature." *J. Appl. Met.* 18: 822–825.

Niiler, P.P., ed. 1981. "Tropical Pacific upper ocean heat and mass budget, a research program outline." Hawaii Institute of Geophysics, Honolulu.

Niiler, P.P. and W.S. Richardson. 1973. "Seasonal variability of the Florida current." *J. Mar. Res.* 31: 144–167.

Oeschger, H., B. Stauffer, A. Neftel, J. Schwander and R. Zumbrunn. 1982. "Atmospheric CO_2 content in the past deduced from ice-core analyses." *Annals of Glaciology* 3: 227–232.

Priestley, C.H.B. 1966. "The limitation of temperature by evaporation in hot climates." *Agr. Meteorol.* 3: 241–246.

Priestley, C.H.B., and R.J. Taylor. 1972. "On the assessment of surface heat flux and evaporation using large-scale parameters." *Mon. Wea. Rev.* 100: 81–92.

Ramanathan, V. 1981. "The role of ocean-atmosphere interactions in the CO_2 climate problem." *J. Atmos. Sci.* 38: 918–930.

Robinson, G.D. 1983. "Carbon Dioxide: Friend or Foe" Sherwood B. Idso, ed. Tempe, Az, IBR Press, 92 pp., *Bound-Layer Meterol* 26: 207.

Schneider, S.H., W.W. Kellogg, V. Ramanathan and C.B. Leary. 1980. "Carbon dioxide and climate." *Science* 210: 6–7.

Stephens, G.L., G.G. Campbell and T.H. Vonder Haar. 1981. "Earth radiation budgets." *J. Geophys. Res.* 86: 9739–9760.

Watts, R.G. 1980. "Discussion of questions concerning the possible influences of anthropogenic CO_2 on atmospheric temperature." *J. Appl. Meteor.* 19: 494–495.

Watts, R.G. 1982. "Further discussion of questions concerning the possible influence of anthropogenic CO_2 on atmospheric temperatures." *J. Appl. Meteor.* 21: 243–247.

Weisenstein, D.K. 1978. "A study of arctic sea ice variation and its relationship to global climate." B.S. thesis, Dept. of Phys., MIT, Cambridge, MA.

Zwally, H.J., C.L. Parkinson and J.C. Comiso. 1983. "Variability of Antarctic sea ice and changes in carbon dioxide." *Science* 220: 1005–1012.

FOUR

CLIMATE from a MODELING POINT of VIEW

IN REPLY TO COMMENTS BY HUGH W. ELLSAESSER

Andrew A. Lacis

Substantial progress has been achieved over the past 20 years in expanding our understanding of climate and in improving our ability to model numerically the principal processes of climate with general circulation models. Our present knowledge gives us confidence to predict the forthcoming increase in global mean surface temperature by several degrees over the next century caused by increasing levels of atmospheric CO_2 and other greenhouse gases due to human activity. However, much remains to be learned before we have the modeling capability in hand to make reliable forecasts for the seasonal and regional changes in temperature and precipitation patterns that are expected to occur as the global temperature slowly increases.

Climate is a very complex system of interacting processes spanning many disciplines of research. The early energy balance models which were introduced by Budyko (1969) and

Sellers (1969) and the 1-D radiative-convective equilibrium models developed by Manabe and Wetherald (1967) are necessarily very crude approximations of the real world. Nevertheless, these climate models are very useful tools for illustrating the nature of the climatic response to external forcing and for demonstrating the important role of positive feedback effects which amplify the magnitude of the initial forcing. For example, in these models an increase in solar radiation will not only warm the atmosphere directly, but will also melt some of the existing snow/ice cover and cause the warmer atmosphere to retain more water vapor. This reduction of the snow/ice cover decreases the planetary albedo, causing additional solar radiation to be absorbed to warm the atmosphere and ground further. Likewise, the added water vapor increases the thermal blanketing efficiency of the atmosphere thereby contributing to a warmer surface temperature. Together, the water vapor and snow/ice feedbacks effectively double the temperature change that would be produced in the absence of feedbacks.

Because of their approximate nature, however, results computed with these simple climate models are open to criticism of the kind directed by Ellsaesser (this volume). Indeed, one major reason for the development of comprehensive 3-D climate models was the pressing need to obtain adequate answers to important questions that the simple models were unable to address.

Thus, 3-D general circulation models, a number of which have been specifically adapted for climate applications (e.g., Hansen *et al.* 1983; Washington and Meehl, 1984; Manabe and Wetherald, 1987), are now state-of-the-art in climate modeling. These models compute atmospheric temperatures, winds, clouds, and precipitation, as well as ground hydrology, snow cover and depth, ocean ice extent, and ocean temperature by solving a set of simultaneous equations which conserve energy, mass, momentum, and moisture. In these models the atmosphere is allowed to "run" as in the real world, i.e., the sun is programmed to rise and set daily throughout its seasonal cycle while temperatures, winds, clouds, precipitation, and other physical quantities are statistically sampled every few hours of simulated time to compile monthly, seasonal, and annual climatologies of model-generated climate. An important difference between the state-of-the-art models and the simpler 1-D models is the ability to compute climate from basic physics rather than rely on incomplete empirical parameterizations. As a result, radiative and dynamic interactions, seasonal dependence, solar zenith angle effects, and water vapor interactions with atmospheric temperature gradients (processes that were cited by Ellsaesser as not having received adequate consideration) are automatically included in 3-D model computations.

To be sure, model complexity is not necessarily a measure of modeling reliability. The credibility of a particular climate model is established by comparing model performance with observations. The means for doing this are the extensive diagnostic climatologies generated by 3-D models which can be compared with meteorological observations of our current climate. The ability of a climate model to reproduce the current climate norm for daily and seasonal temperature variations, seasonal and regional precipitation patterns, zonal and meridianal winds, etc. is an essential prerequisite for establishing some degree of confidence in climate change computations. A further important test to determine climate sensitivity and response time is provided by comparing model simulations with the observed 0.5°C rise in global surface temperature over the past century and with the more extreme climate conditions that existed during the last Ice Age. The results of these climate simulations show that (1) the climate system operates with strong positive feedbacks which magnify climate forcing perturbations by a factor of the $f = 3–4$ yielding a global warming of 3–5°C for doubled CO_2, and (2) the response time to reach thermal equilibrium is very long (~100 years) because of the large ocean heat capacity and varies as the square of the feedback factor f.

Climate feedbacks arise naturally from the interactive physical processes that are included in 3-D climate models. The most important feedback arises from the increased ability of a warmer atmosphere to hold more water vapor which increases the thermal blanketing efficiency of the atmosphere and thus strengthens the greenhouse effect. A warmer climate also causes more snow and ice to melt which reduces the surface albedo and permits more solar radiation to be absorbed at the ground. Likewise, in response to a warmer climate, cloud change in 3-D climate models result in a decrease of low clouds and an increase in high clouds which further enhances the greenhouse effect. Thus, the principal climate feedbacks are positive, i.e., they magnify the initial climate forcing perturbation. Of these, the cloud feedback is the most uncertain because of the greater complexity and difficulty in modeling cloud formation processes.

Climate modeling has become in many respects a "big science" and can no longer be considered a one-man operation. For example, the effort to develop a state-of-the-art 3-D climate model may require a 100 man-year investment of top-level expertise in numerical analysis, atmospheric dynamics, cloud physics, radiative transfer, meteorology, oceanography, and ground hydrology, not to mention the need for a large dedicated computer and supporting staff. Climate has also become a focal point for interactive studies with other disciplines. Of particular importance are the synergistic ventures between climate and the fields of atmospheric chemistry, oceanography, paleogeology, and plant biology—activities which promote new ideas and expand our present knowledge of climate and of physical processes in the other disciplines.

Climate modeling has also attracted considerable interest in the political and economic spheres as water resources planning commissions along with energy, agriculture, and forestry industries begin to examine different options and possible consequences of shifting climate trends. The questions that are now being asked are both specific and practical, and typically they cannot yet be answered with full confidence by our present climate models. How will regional precipitation and temperature patterns change? Will the frequency and severity of storms increase? By how much and how soon will the sea level rise? Other questions relate more directly to the modeling effort and deal with such topics as model numerics, subgrid processes, convective transports, cloud feedback effects, ocean-atmosphere interactions, and their effect on climate sensitivity and response time.

To obtain answers to the questions that are being asked requires a continued effort in climate model development with particular emphasis on cloud and ocean physics. Also essential are improved remote sensing and geophysical measurements to help refine the input data for climate models and to help verify model performance against more stringent observational tests. Research in climate is a major effort on an international scale for which substantial financial resources have been committed. All this effort, however, is but a small investment considering that potential economic benefits to be gained from reliable climate modeling predictions are measured in the billions of dollars.

References

Budyko, M.I. 1969. *Tellus* 21: 611.

Ellsaesser, H.W. 1989. "Response to W.W. Kellogg's Paper." This volume.

Hansen, J., G. Russell, D. Rind, P. Stone, A. Lacis, S. Lebedeff, R. Ruedy, and L. Travis. 1983. *Mon. Weather Rev.* 111: 609.

Hansen, F., A. Lacis, D. Rind, G. Russell, P. Stone, I. Fung, R. Ruedy, and J. Lerner. 1984. *Climate Processes and Climate Sensitivity.* AGU, Washington, D.C.: 130.

Manabe, S., and R.T. Wetherald. 1967. *J. Atmos. Sci.* 32: 3.

Manabe, S., and R.T. Wetherald. 1987. *J. Atmos. Sci.* 44: 1211.

Sellers, W.D. 1969. *J. Appl. Meteor.* 8: 392.

Washington, W.M., and G.A. Meehl. 1984. *J. Geophys. Res.* 89: 9475.

FIVE

WHERE DO WE STAND with the CO_2 GREENHOUSE EFFECT PROBLEM?

Helmut E. Landsberg

Even though there is mention of a consensus opinion on the effect of the atmospheric increase in CO_2 and related so-called greenhouse gases in the atmosphere, very few facts are really agreed upon. Even if learned panels of the National Academy of Sciences-National Research Council have voiced the conclusion that doubling of atmospheric CO_2 would lead to a global temperature rise of around 3°C, Academy panels have been known to revise their estimates when new information becomes available. Kellogg's chapter predicates its impact assessment on this rise. Ellsaesser raises a number of questions with respect to the assumed temperature rise. There are other doubters.

The established fact is that since 1958 the carbon dioxide in the atmosphere has risen from 315 parts per million by volume (ppmv) to 340 ppmv by 1984. There is also little argument that this rise can be attributed to increased use of the fossil fuels by a rising world population, motorization, and industrialization. From there on we face only uncertainties. These are brought about by extrapolations into and predictions of the future. One of the major uncertainties is the question of future use of fossil fuels. Earlier exponential growth rate predictions, of which the most notable are those of the International Institute for Applied Systems Analysis (IIASA), have overestimated this growth rate and have been seriously questioned. The next uncertainty is the partitioning of the CO_2 among various carbon reservoirs: biota, the oceans, the atmosphere. In the biological world, carbon is incorporated into plants and into animal shells. There is not even an inventory of the living biomass, let alone an adequate estimate of annual conversion from CO_2. There are only inadequately documented guesses of deforestation rates, but the increases due to increased growth rates are not known. The oceanic biomass is also largely unknown. The uncertainty has grown with recent discoveries of brown algae and other biota at great depth. The uptake and storage of CO_2 in the ocean is quite inadequately understood. Oceanic circulation models are just in their infancy. Their inadequacy has a major impact on the atmospheric climate models. These models have largely figured in the projection of an atmospheric temperature rise, but various models do not agree with each other. Aside from the already mentioned deficiency of the role of assumptions about the oceanic effect on the atmosphere, these models lack any biotic component. Probably the most critical element not adequately represented in the numerical models is the development of cloudiness on an earth with rising temperatures. This applies not only to total cloud cover and cloud types, but also to cloud types and cloud heights. These govern the earth's albedo. A relatively small increase in albedo can have a material effect on temperature rises. One may further remember that the 15% rise in atmospheric CO_2 concentrations since the end of World War II should have had a small but measurable effect on the temperatures of a little under 1/2°C. Yet at least the northern hemisphere temperatures have cooled or stayed at least steady over the last four decades. The greenhouse effect is a far more complex problem than many scientists are willing to admit, the growing literature attests to that. Some very good recent sources are Clark (1982); Liss and Crane (1983); and Schneider (1984).

References

Clark, W.C., ed. 1982. *Carbon Dioxide Review 1982*, Oxford University Press, New York, 469 pp.

Ellsaesser, H.W., 1989. "Response to Kellogg's paper." This volume.

Kellogg, W.W., 1989. "Carbon dioxide and climate changes: Implications for humankind's future." This volume.

Liss, P.S., and A.J. Crane. 1983. *Man-made Carbon Dioxide and Climatic Change—A Review of Scientific Problems*, Geo Books, Norwich, 127 pp.

Schneider, S.H., ed. 1984. *Climatic Change 6* (4): 317–415.

SIX

METHANE in the ATMOSPHERE

Ralph J. Cicerone

The study of atmospheric methane (CH_4) is motivated both by scientific and by practical concerns. Early research on atmospheric methane emphasized its physical properties, the physical chemistry of its atmospheric behavior and physical (optical) methods of measurement were employed. More recently, a broader and more scientifically unified approach has characterized the research: the importance of biological processes is apparent to all, and the roles of methane in atmospheric chemistry and the earth's climate are now appreciated. In this chapter, I will review salient facts about atmospheric methane: its abundance, spatial distribution, sources, and removal processes (sinks), and its roles in atmospheric chemistry and climate, along with evidence for, and possible causes of, the recently measured global increase in atmospheric methane levels. Finally, I will discuss the possible environmental consequences of a continued increase in atmospheric methane levels.

Atmospheric Distribution of Methane

Knowledge of the atmospheric distribution of a chemical species such as methane is important both for descriptive, qualitative understanding and also for more quantitative studies of its behavior and effects. Fortunately, a great deal of the necessary information is available at this writing. From the date of its first measurement in the atmosphere (1948) until about 1965, methane was detected and measured by infrared absorption techniques. These optical experiments were necessarily limited to long line-of-sight determinations (kilometers). The advent and adoption of high-sensitivity modern gas chromatographic techniques has allowed post-1965 experiments to measure methane accurately in very small air samples (1-100 cm^3) taken from ground stations and from airborne platforms. Reliable measurements have shown that there is more methane in the earth's northern hemisphere than in the southern hemisphere; representative values of CH_4 concentrations in 1987 are 1.75 parts per million (ppm) of air at northern mid-latitudes and 1.65 ppm at southern mid-latitudes (Blake and Rowland, 1988). In the vertical, methane is well mixed throughout the troposphere, i.e., its concentration (fraction of total air volume) is essentially constant from the ground to the tropopause along a vertical line. Above the tropopause in the stratosphere, measurements show that CH_4 concentrations decrease with increasing altitude (see, e.g., Ehhalt and Heidt 1973; WMO 1986). At northern mid-latitudes, the concentration decreases from its ground-level value to about 1.0 ppm at 25 km altitude while larger decreases with altitude are observed at high latitudes and smaller decreases at low latitudes. A few data points are available for altitudes over 40 km; by 50 km the CH_4 concentration at mid-latitudes decreases to about 0.3 ppm. Dynamically induced seasonal and interannual variations of stratospheric methane concentrations have been observed with satellite-borne remote sensing instruments (WMO, 1986).

It is reassuring that the overall picture of methane distributions, more in the northern hemisphere's troposphere than in the southern, constant vertical distribution through the troposphere and decreasing concentrations in the stratosphere is understood and explained by our knowledge of the atmospheric chemistry of methane. For reasons noted below, it has become important to obtain a more complete quantitative understanding of methane's atmospheric distribution and its recently observed seasonal cycles and secular trend.

Roles in Atmospheric Chemistry and Climate

Gaseous methane plays many important roles in atmospheric chemistry and climate. An understanding of these roles, combined with the fact (described below) that the earth's biological processes are principally responsible for the existence of atmospheric methane, suggests that the earth's physical environment and its biosphere are closely coupled if not components of an integrated system. Major roles include: acting as a source of atmospheric carbon monoxide, hydrogen and stratospheric water vapor; as an inhibitor of chlorine's destruction of stratospheric ozone; as a regulator of tropospheric hydroxyl radical concentration, serving as a tracer of large-scale atmospheric motions; as a principal source of upper atmospheric hydrogen, some of which escapes to space; as a potentially potent climatic greenhouse gas; and possibly as part of an overall regulatory system that serves to stabilize atmospheric oxygen levels.

How can any gas such as methane, in concentrations as low as 1.6 ppm exert significant influence on the global atmosphere or climate as suggested above? First, even at this low concentration, methane is the eighth most abundant gas in the global atmosphere after N_2, O_2, Ar, H_2O, CO_2, Ne, and He. Second, the chemical and radiative energy-balance effects of a given trace or minor species are often grossly out of proportion to their concentrations in air. Enhanced chemical effects arise through distinct chemical reaction properties of certain species toward others; there are many examples of atmospheric conversion or decomposition processes that are initiated or completed by only one distinct chemical. A second way in which enhanced chemical effects materialize is through the phenomenon of catalysis. In catalytic processes chain reactions occur; these reactions comprise an overall process in which a substance is produced, converted or destroyed, and the principal agent, the catalyst itself, is not consumed in the process. Catalysis can be very efficient—there are industrial and atmospheric catalytic reactions that convert 10^5 or 10^6 times as much material as is present in the driving catalyst. Through catalysis a trace substance present at less than one part per billion can control the behavior of far more concentrated substances. When methane is decomposed in the atmosphere, some of the products of this decomposition become important players in atmospheric catalytic processes.

Independently, trace atmospheric gases, even those in extremely small concentrations, can affect atmospheric radiation and the earth's climate in important ways. Two examples are: the nearly complete shielding of the earth's surface from harmful ultraviolet sunlight by stratospheric ozone (present there at 0.5-10 ppm); and the blocking of the important atmospheric "window" in the infrared 7-13 μm wavelength interval. The earth's radiation energy balance and climate are open to strong influence by changes in the numbers of polyatomic gases in air whose infrared absorption spectra display strong features in this window region. Thus, chlorofluorocarbons growing to parts-per-billion concentrations can cause significant global warmings.

To see how atmospheric methane can be involved in the diverse roles outlined above, it is necessary to review briefly our knowledge of its atmospheric chemistry. First there are no *in situ* atmospheric sources of methane. This is so because our atmosphere is oxidizing; both kinetic and equilibrium considerations favor the formation of CO_2 and CO in candidate

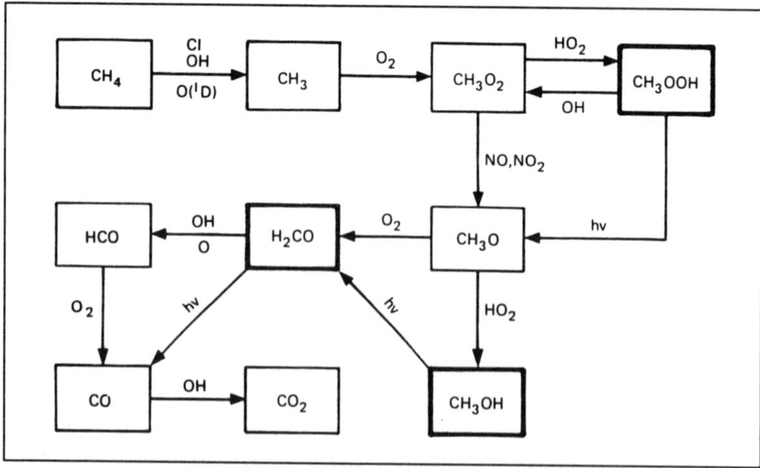

Figure 6-1: Atmospheric oxidation (destruction) pathways for gaseous methane (CH_4). Attack by OH in the troposphere is the dominant pathway. Principal end products are CO_2, CO, H_2O and H_2, and formaldehyde (CH_2O) is a key intermediate.

processes, not methane. Instead, the atmospheric chemistry of methane is that of decomposition. Elements of the atmospheric oxidation pathways of methane appear in Figure 6-1. Because CH_4 is a strongly bonded molecule, there are very few atmospheric reactants that can attack and dissociate it at ambient temperatures. In Figure 6-1, three specific attacking species are shown, Cl atoms, gaseous OH radicals, and $O(^1D)$ (electronically excited oxygen atoms). All available knowledge states that these are the only relevant processes in the atmosphere below 50 km altitude. Further, in the troposphere,

$$OH + CH_4 \rightarrow H_2O + CH_3$$

is the only process active. Only 5–10% of atmospheric methane survives the upward journey through the troposphere to the stratosphere where

$$Cl + CH_4 \rightarrow HCl + CH_3$$

$$O(^1D) + CH_4 \rightarrow OH + CH_3$$

$$\rightarrow H_2 + CH_2O$$

essentially complete the process of methane destruction (oxidation). Thus, 85–90% of the CH_4 that enters the atmosphere is removed in the troposphere. At very high altitudes (over 50 km) Lyman-alpha radiation from the sun becomes a fourth methane-destroying process. While details of Figure 6-1 (including several uncertainties) are not relevant here, it is important to see that the final products are CO_2 and CO and that intermediate steps lead to creation of atmospheric formaldehyde (CH_2O) and other small organic species. Also, not shown in Figure 6-1 is that H_2 and H_2O are formed as methane is decomposed. Although most methane is decomposed in the troposphere, the amounts of H_2O and H_2 so created in the stratosphere are significant. In the present-day atmosphere about half the water vapor in the upper stratosphere is due to *in situ* oxidation of methane (Scholz *et al.* 1970). Singer (1971) has noted that changes in atmospheric methane levels are thus capable of affecting the global stratosphere. More discussion on global environmental effects of methane changes is presented toward the end of this chapter.

Global Trends and Cycles

Evidence for Global Atmospheric Increase

Since 1980 compelling evidence has materialized to prove that atmospheric methane concentrations are increasing with time around the globe. Although such a trend has been suspected or even predicted in some quarters (Singer 1971), the contemporary rate of increase is larger than had been envisioned; the fact that major methane sources are biological and in some sense natural appeared to preclude rapid, sustained methane changes.

The data showing increases of atmospheric methane are very direct, based as they are on reliable methods of modern gas chromatography. Figure 6-2 shows one such data set from Blake and Rowland (1988) for the period 1978 through late 1987.

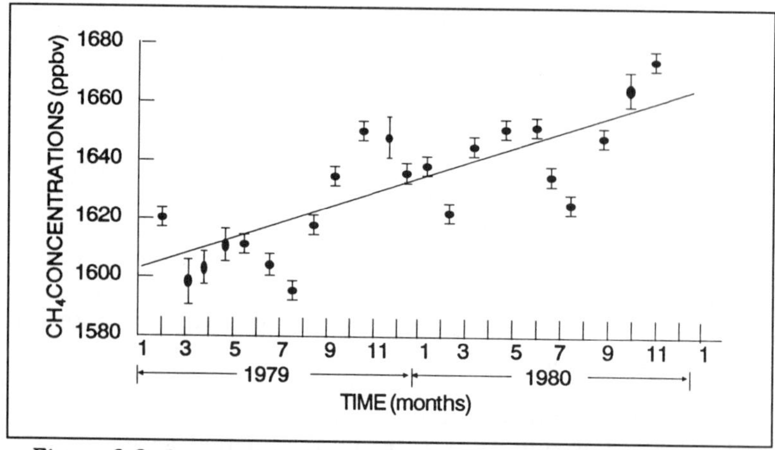

Figure 6–2: Average monthly dry air concentration of CH_4 in ppbv obtained from nearly continuous measurements made at Cape Meares, Oregon, between January 1979 and October 1980. Error bars are the 90% confidence limits of the average, and the least squares estimate of the changing concentration is represented by the solid line (after Rasmussen and Khalil 1981).

Each of the points in Figure 6-2 represents the global average of the methane amounts measured in air samples gathered nearly simultaneously from sites at various latitudes all over the world. Sampling was performed at approximately three-month intervals and in computing global averages, methane concentrations were weighted by the area that they represent in the global network. The straight line that is fit to the data in Figure 6-2 shows an annual increase of 0.016 ppm per year, or about 1% per year over this ten-year period. Additional ground-level measurements made by Blake *et al.* (1982) are summarized in Figure 6-3. Each data point represents only one air sample but because of the great care per sample that is feasible with small sample lots the resulting data base appears very significant.

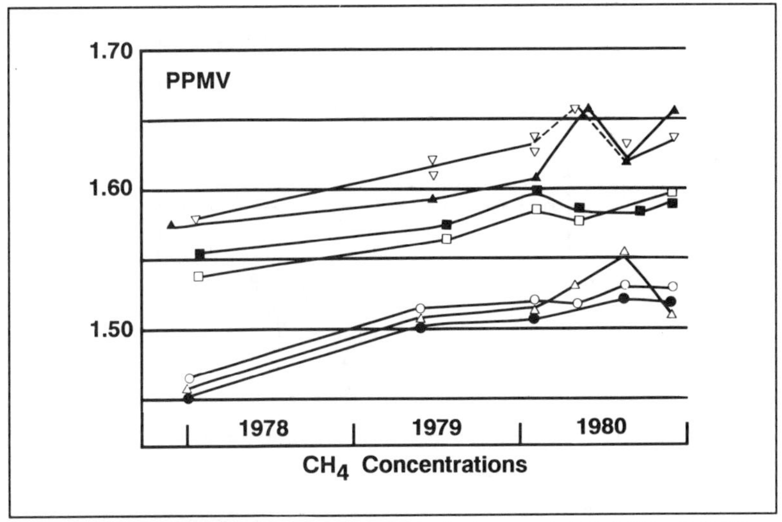

Figure 6–3: Methane concentrations between 1977–1980. ▼67°N; ∇56°N; ▲45°N; ■18°N; □13°N; Δ23°S; ● 42°S; O 53°S (after Blake *et al.* 1982).

Figure 6-3 shows that between November 1977, and November 1980, methane increased both in the northern and southern hemispheres, at rates similar to that shown in Figure 6-2. There is also no doubt that methane concentrations are increasing throughout the troposphere, not just at ground level. Figure 6-4 shows data from Cape Grim, Australia from ground level and at three altitude levels up to 10 km (Fraser *et al.* 1984). Based on the quality of the analytical procedures (absolute calibrations, etc.) and on the general agreement between independent laboratories, one must conclude that atmospheric

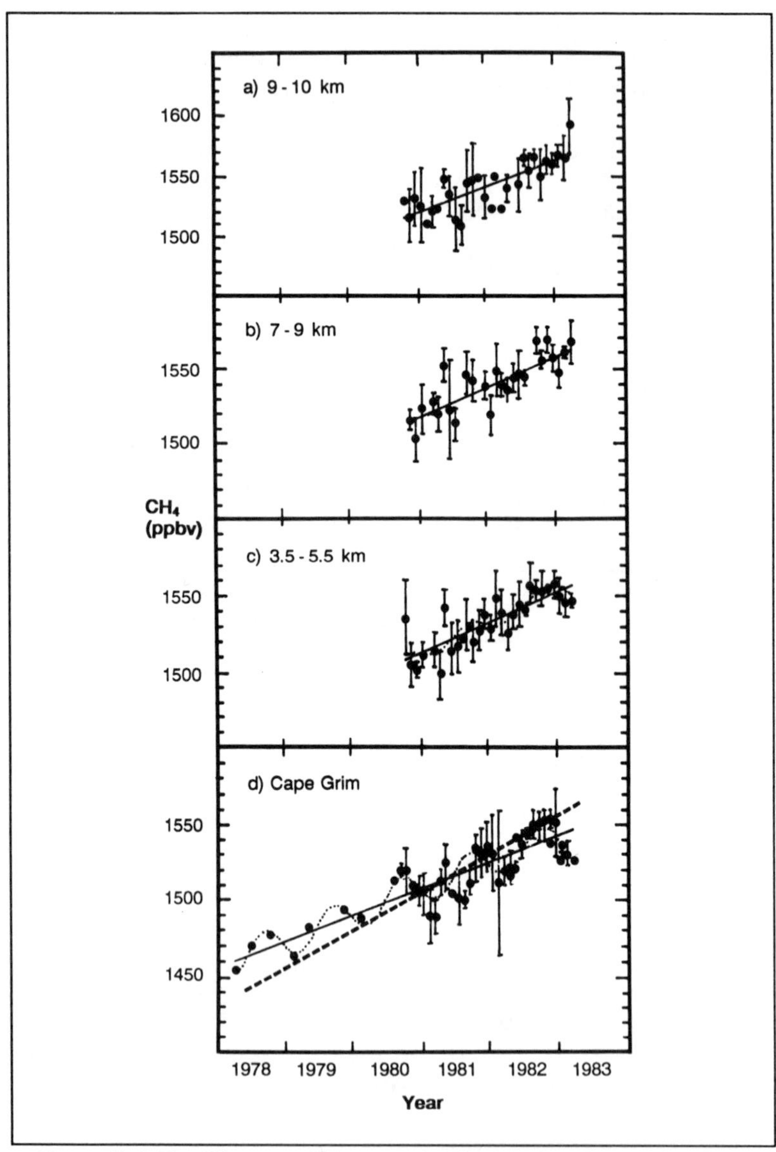

Figure 6–4: Monthly mean CH_4 concentrations (ppbv) observed at Cape Grim, Tasmania (d) and at three altitudes over south-eastern Australia (a-c.) Error bars are one standard deviation (s.d.) from the mean (after Fraser *et al.* 1984).

methane has been increasing at least since 1978. In an authoritative review of many data bases, Ehhalt *et al.* (1983) have concluded that atmospheric methane concentrations increased by about 0.5% per year between 1965 and 1975 and that the rate of increase apparently increased to 1–2% per year by 1978. Further evidence of increasing methane concentrations comes from Rinsland *et al.* (1985) who reanalyzed some solar absorption spectra taken at Jungfraujoch in 1951. From these data Rinsland *et al.* deduced an atmospheric methane concentration of 1.14(±0.08) ppm in April, 1951 and with similar instrumentation and techniques they found a value of 1.58(±0.09) ppm for Kitt Peak, Arizona in February of 1981.

A different kind of data, equally dramatic, indicates that atmospheric methane levels have been increasing for the past 400 years or so. These data, shown in Figure 6-5, were obtained by analyzing the gas contents in dated polar-ice cores (Craig

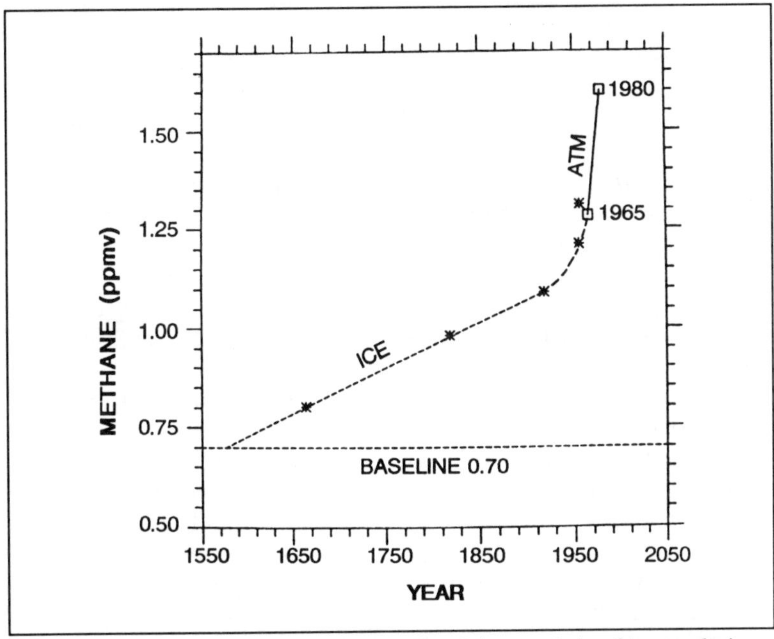

Figure 6–5: Methane mixing ratio in Dye 3 ice vs. date of trapped air, after age correction for firn closure. The present atmospheric ratio (1980) and the estimated 1965 value (squares) are connected by the solid line (Rasmussen and Khalil 1981; after Craig and Chou 1982).

and Chou 1982). While it is possible to interpret these data differently, not as a methane increase in air (Ehhalt *et al.* 1983), the present writer finds the Craig and Chou evidence to be strong; see also Stauffer *et al.* (1985) and Pearman *et al.* (1986). One can only speculate on why methane has increased by over 100% in the last few centuries.

Before proceeding to discuss sources of atmospheric CH_4, the size of several quantities should be noted. First, with about 1.7ppm (by volume) of CH_4 there are about 5.0×10^{15}g of CH_4 in the earth's atmosphere. A 1% increase thus amounts to 50 million metric tons of CH_4. Efforts to explain global changes, described below, focus on these quantities.

Observations on the Nature of Methane Sources

One of the most interesting scientific questions about atmospheric methane, and one with some practical importance, is: What are its dominant sources? Table 6-1 lists a few of the most pertinent facts about contemporary atmospheric methane. The residence time, t, of a gas in air is defined as the quotient $t = B/Q$ or B/L where B is the steady state total atmospheric burden and Q and L are the steady state source and sink (input and removal) rates, respectively. Obviously, in steady state, $Q = L$. Available information on atmospheric OH concentrations allows us to estimate L; the burden (5000×10^{12}g) and L estimates lead to 8 years $< t <$ 12 years. This estimate has been made from data on atmospheric distributions of methyl chloroform ($C_2H_3Cl_3$), a synthetic compound whose only sink is OH, as for CH_4 (Prinn *et al.* 1987). Thus, it appears that the total annual sources of atmospheric methane amount to (415 to 625) x 10^{12}g. Further, measurements of radiocarbon (^{14}C) amounts in atmospheric methane have shown that 80 or 90% of the carbon in atmospheric methane has been incorporated in living material recently, as opposed to fossil or primordial material. Most of these ^{14}C measurements were made before 1960, prior to widespread atmospheric bomb testing. Newer data, coupled with analysis by an atmospheric model, suggest that only 70% of all the methane is of recent biogenic origin (Lowe *et al.* 1988). Such data tell us that postulated large inputs from releases of primordial methane from earth faults cannot occur, or occur very infrequently at best.

Instead, it is clear that the dominant sources are biogenic—^{14}C from the atmosphere is released back to the atmosphere (as CH_4, CO_2, CO, etc.) after being cycled through living material in times very short compared to the half life (5700 years) of ^{14}C. Biological mechanisms of methanogenesis have been researched widely by microbiologists. There are many capable specific pathways and microbes (Oremland, 1988).

Table 6-1: Atmospheric CH_4 1988

Concentration:	Northern Hemisphere 1.75 ppm
	Southern Hemisphere 1.65 ppm
Global Burden:	5000×10^{12}g
Residence Time:	8 - 12 years
Annual Source (sink):	$(400 \text{ to } 625) \times 10^{12}$g
^{14}C Data (pre 1960):	Imply source is over 80% biogenic
Rate of Increase (global):	(1 ± 1)% per year

Summary of available information on total of global atmospheric methane sources. From the total atmospheric content and estimates of residence times, one can deduce total source strengths (see text).

Sources of Atmospheric Methane

Biological Sources in Nature

Let us focus on biological sources of methane, given that ^{14}C data show that only 10-15% of atmospheric methane at most can be abiogenic. Table 6-2 lists estimates of the annual methane sources attributable to individual biological sources. These table entries have been drawn from several original sources and the author's own estimates and some guesswork. Information sources of special importance include Ehhalt (1974), Cicerone *et al.* (1983), Zimmerman *et al.* (1982), and Khalil and Rasmussen (1983), and private communications from W. Seiler, and Cicerone and Oremland (1988). United Nations data were also used to update estimates of cattle populations and of rice-agriculture patterns.

There is much food for discussion in Table 6-2. For example, one immediately wonders how accurate the individual estimates and ranges are and, if the total of these sources exceed the total consistent with other data (see section on *Observations on the*

Nature of Methane Sources), which sources are overestimated? For our present purposes, it is most important to deduce which of these sources (and nonbiological ones: see section on *Abiogenic Sources*) are under human control and which are wholly natural biological processes. Of those entries in Table 6-2 in the latter category, emissions from swamps and marshes appear to be the largest. Considering the importance of the topic, too little experimental research has been performed to measure these fluxes, the relative amounts of release due to bubbles exiting the water, and seasonal dependences. Draining of wetlands and recovery of the underlying land areas for other uses should be quantified and the effects assessed.

Table 6-2: Biological Sources of Methane

Source	Annual Production (10^{12}g CH_4/year)
Enteric fermentation of animals (cows, sheep, horses, goats, buffalo, elephants)	90-130
Rice Paddy emissions	70-300
Swamps, marshes	100-200
Biomass burning and wildfires	40-75
Termite emissions	25-125
Freshwater lakes	1-25
Tundra	15-35
Oceans	1-17
Landfills	30-70
TOTAL:	372-1000

Estimates of individual biological sources of atmospheric methane. In cases where changes are known to occur with time, e.g., animal populations and areas of rice agriculture, estimates are for 1980 A.D. The upper figure, 1000 x 10^{12}g CH_4 per year, is probably too high to be consistent with other information on total annual CH_4 sources (see text).

The entry for biomass burning in Table 6-2 includes contributions from wildfires, a recurring happening in nature. Again, actual field data from such fires are sparse. Perhaps the newest and most interesting category of methane source, termite emissions, is also a major natural source. Because the digestive system of termites depends on symbiotic bacteria (in higher termites) and on protozoa (in lower termites) the release of volatile species such as CO_2, CH_4, and H_2 into the global atmosphere represents a fascinating blending and interdependence of microbiological and global environmental processes. If global termite emissions of CH_4 turn out to be the higher end of the range shown in Table 6-2, it will become more important to determine if our land-clearing operations affect termite activity regionally. Finally, although there are very few published data on methane in oceanic surface waters, an extensive data base of R. Weiss appears to show small fluxes into the air as originally estimated by Ehhalt and indicated in Table 6-2, while for tundra new data from Whalen and Reeburgh (1988) are used in Table 6-2.

Anthropogenic Biologic Sources

Several of the sources listed in Table 6-2 are under our direct control. The most accurate figures there are for enteric fermentation of animals, where the populations of domesticated cattle (responsible for most of the emissions) are known within about 15%, and the phenomenon of methane emission has been studied in many individual animals. In the recent past cattle populations have approximately tracked human populations; cattle populations increased worldwide by over 40% from 1955 (800 million) to 1980 (1,250 million). Continued increases are likely (Singer 1971) although the rates are hard to predict, as they depend on grain supplies, human appetites for meat and dairy products, and obviously, the economy in general.

Emissions of methane from rice paddies have been seen to be the largest potential human-controlled source of atmospheric methane since the early work of Koyama (1963) in Japan and Ehhalt's use of Koyama's data in constructing methane budgets. Koyama's laboratory cultures of rice-paddy soils, his measurements of methane evolution from these soils and his extrapolations for other soils led to estimates of 190×10^{12}g CH_4 per year entering the global atmosphere annually from rice paddies. In 1974 Ehhalt raised this estimate to 280×10^{12}g by noting that areas under rice and multiple cropping of rice had increased worldwide. The figure of 300×10^{12}g per year in Table 6-2 is based on further increases of rice acreage by 1980.

Rice-paddy soils are an ideal environment for methanogenesis—organic substrate, moisture, and anoxic conditions are present. Thus, in principle, Koyama's results are correct—there is great potential for global rice agriculture as a methane source. Yet, there might be complicating factors that reduce our ability to estimate this source accurately. These include soil organic contents, variations in agricultural practice such as field drying and plowback of stubble, and fertilizer type and rate of usage. (Holzapfel-Pschorn and Seiler, 1986) Also, it is not clear that all methane produced in rice-paddy soils escapes to the atmosphere. Field studies have now shown that the principal means of escape is through the rice plants themselves, not through bubble breaking or molecular diffusion through the water covering the soil. A strong variation of methane flux with time during the growing season has been observed in California (see Figure 6-6). This type of variation was reproduced in later studies by Cicerone, Shetter and Delwiche, and by Seiler and colleagues in Europe. Methane fluxes in these later studies led to the lower end of the range listed in Table 6-2.

Finally, it should be noted that some fraction of the total biomass burned annually, and the resultant methane emissions, is directly attributable to human activities. This fraction is apparently large, much more than 90% (Seiler and Crutzen 1980), and is associated mostly with agriculture. Altogether, from examination of our available information on human influence on atmospheric methane, it seems clear that Singer's (1971) proposition was well founded—human beings do control 50% or more of the input of atmospheric CH_4.

Abiogenic Sources

Because ^{14}C measurements in atmospheric methane have found about 80% as much ^{14}C as in live organic-material standards, it has been deduced that atmospheric methane is 80% biogenic. Using the arguments of the discussion surrounding Table 6-1, we would deduce annual abiogenic methane sources of about (80 to 200) x 10^{12}g CH_4. Contributions from several "dead" carbon sources have been identified; see, e.g., Ehhalt (1974) and Sheppard *et al.* (1982). These include fugitive emissions from coal mining and from oil and mineral exploration, venting and transmission loss of natural gas, automobile exhaust, volcanic emissions, and postulated escape from the earth's mantle through fault zones (Gold 1979). Attempts to total such emissions are fraught with difficulties, but typically

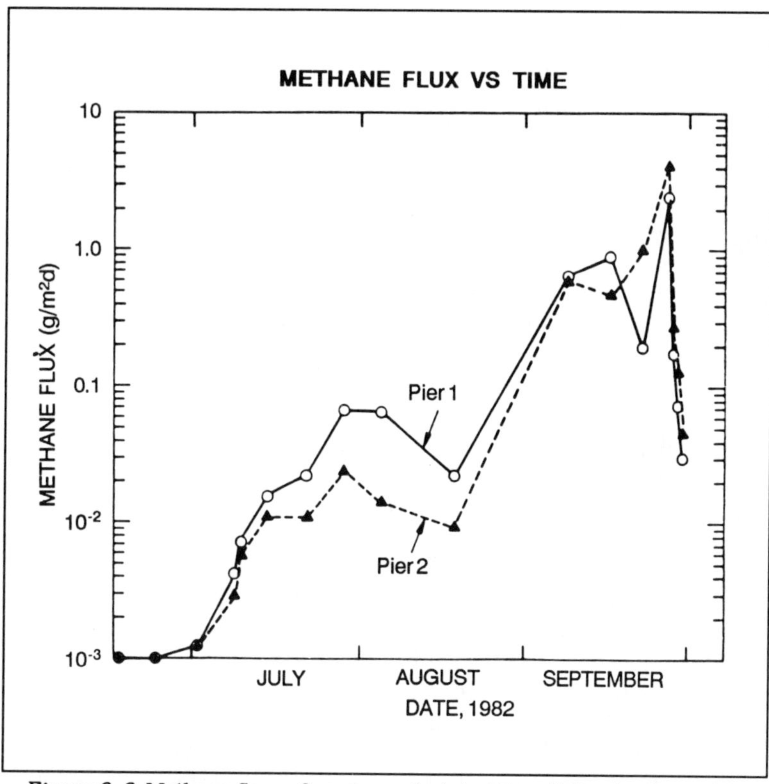

Figure 6–6: Methane fluxes from two subplots, P1 and P2, of a rice paddy near Davis, California, during the 1982 growing season. Fertilization rates and dates and information on irrigation are in the text. Individual data points and related data are shown in Table 1. Note logarithmic scale (after Cicerone *et al.* 1983).

they have yielded totals smaller than seem to be mandated by the ^{14}C data (Cicerone and Oremland, 1988). It is possible that some of the ^{14}C methane analyses have been biased by samples taken too near urban pollution (richer in ^{12}C compared to ^{14}C than clean background air). While others have been biased by $^{14}CH_4$ from certain nuclear reactors (Levin *et al.* 1980). Until this question is settled, we must at least be aware that the possible emissions of fossil methane should increase in the future as more natural gas is transferred, more mining occurs, etc.

Projections of Future Methane Levels

Atmospheric Methane, 1990–2020 A.D.

Continued increases of atmospheric methane concentrations are likely given the observed increases of the years up to 1987. To be able to predict and understand the rate of future increases requires detailed, mechanistic knowledge of the relative importance of the various sources (described in the previous section) and of the time evolution, if any, of the principal atmospheric sink, OH. As noted, we do not have a sufficient data base and understanding for this kind of a prediction so a more empirical, less mechanistically based projection is attempted below.

Before proceeding to estimate future methane levels, we should at least indicate the kinds of information that will be needed for improved estimates. First, it will be necessary to obtain accurate population counts for ruminating animals. The United Nations Food and Agricultural Organization (FAO) has provided estimates of this kind in recent years. The recent data show perhaps a 1.5% increase per year in cattle populations. Another quantity that has been estimated by FAO personnel is the area in rice cultivation. This quantity has increased at 1–2% per year since about 1950 but perhaps as importantly, the incidence of multiple cropping (made possible by irrigation) has also grown so that total global rice yields have grown much faster than the cultivated area. Further, it seems necessary to attempt to quantify trends in amounts of biomass burned for all purposes, a quantity that is probably increasing secularly. Also, the possibility that global methane emissions from world wetlands and marshes are decreasing as these lands are reclaimed for dry land activities needs to be explored. On this latter point, it is also possible that if reclaimed mangrove swamps, for example, are diverted into rice agriculture, methane emissions might increase in given areas.

What can be said about future atmospheric methane concentrations? Considerable weight must be given to evidence of recent rates of increase. Ehhalt *et al.* (1983), as noted earlier, concluded that a 0.5% per year increase occurred between 1965 and 1975 and a more rapid increase (1–2% per year) materialized after 1978. These trends suggest that the atmospheric concentration $f(t) = f_0 \exp a(t)t$, i.e., that the data can be fit with an exponential function of time, t, in which the growth exponential, a, is itself growing with time. If so, the *percentage* growth per year would increase with time. Given the length and quality of existing data bases, it is probably more defensible to fit the data with the form $f(t) = f_o \exp(a_o t)$ where a_o is a constant such that $0.005 < a_o < 0.015$, i.e., a constant percentage growth rate per year between 0.5% and 1.5% per year. This latter growth rate would lead f(2020 A.D.), the methane concentration in the year 2020, to lie between 1.98 and 2.86 ppm (starting from a global average of 1.65 ppm in 1983). At a constant rate of increase of 1.5% per year, global methane concentrations would double to 3.30 ppm by 2030 A.D.

More rapid future rates of increase of methane are a distinct possibility, because it may not be assumed that the atmospheric sink strength (concentration of gaseous OH molecules) will remain constant. Instead, because tropospheric OH acts to remove carbon monoxide (CO), methane and many hydrocarbons, it is easily conceivable that OH might decrease from its present levels; indeed, it might have done so already, providing part of the explanation for the rise in atmospheric methane. Two early studies on this subject published in 1977 (Chameides *et al.* 1977; Sze 1977) have been updated, by Thompson and Cicerone (1986).

Consequences of Increased Methane

In the first section of this chapter, it was stated that gaseous methane plays a number of roles in global atmospheric chemistry and climate. Because of the evidence summarized in this chapter that methane concentrations are growing and will continue to do so, the importance of methane in these roles is also increasing. In terms of human consequence, the potential climatic warming due to methane increases is perhaps the most important single effect. A doubling of atmospheric methane would lead to globally averaged warming of 0.2 to 0.3°K (Dickinson and Cicerone, 1986). Given that this doubling can occur by the year 2020 or 2030 A.D., the methane effect cannot be ignored compared to effects anticipated from carbon dioxide increases. From a physical point of view, our knowledge of perturbed global climate is quite weak concerning regional effects. Thus, although it is clear from existing simple models and from energy-balance considerations that global warmings are in store, other factors being constant, the ways in which important regions might change are not. Fortunately, efforts have begun to assess how civilization might adapt to global and regional climate changes (see Kellogg in this volume).

It must also be emphasized that methane is only one of several trace gases (nitrous oxide, tropospheric ozone, and chlorofluorocarbons—see Rowland in this volume) whose combined climate effects could be a pronounced global warming of the same magnitude as a possible doubling of CO_2 levels.

A second likely effect of methane increases would be an increase in tropospheric ozone concentrations. Once again, this would lead to a climatic warming because tropospheric ozone is a potent greenhouse gas due to pressure broadening of its 9.6 μ absorption band. Available theory and models indicate that increases in tropospheric hydrocarbons and nitrogen oxides cause increases in ozone concentrations but quantitative predictions are risky at best. A key uncertainty is due to our lack of knowledge of nitrogen oxide (NO_x) concentrations and their global distributions. The more NO_x, the higher the (O_3) response to methane or other hydrocarbon increases.

Potential climatic and chemical responses to methane changes can arise also from stratospheric changes (Blake and Rowland, 1988). First, upper stratospheric water vapor should increase in proportion to methane. Roughly speaking, each unit of stratospheric CH_4 gives two of H_2O (Singer 1971). While H_2O in the stratosphere does not influence surface climate directly, there is potential influence through ozone changes. Increased H_2O accelerates the destruction of ozone due to chlorine compounds while it slows ozone destruction by NO_x. This coupled system is understood in principle, and an unstable feedback is possible (CH_4 increases, H_2O increases, chlorine increases independently, O_3 decreases, the tropopause warms and H_2O increases more...). Further, as methane increases, the reaction $CH_4 + Cl \rightarrow HCl + CH_3$ will slow the effect of chlorofluorocarbon on ozone.

Another likely effect that can also involve destabilizing feedbacks begins in the lower atmosphere and centers on tropospheric OH, a key player in tropospheric chemistry. As CH_4 increases, OH will decrease near the earth's surface, but increase at higher altitudes (this prediction, based on current models, is sensitive to NO_x levels). If our activities cause a simultaneous increase of carbon monoxide, OH levels will be depressed everywhere. Because attack by OH is the principal removal mechanism for many atmospheric pollutants, e.g., chlorinated hydrocarbons, CO, organobromine compounds, most hydrocarbons including aromatics, and in some situations, SO_2 and NO_x, the residence times of many polluting gases will rise. Increased flows of chlorine to the stratosphere through CH_3Cl and $C_2H_3Cl_3$ would result and gas-phase oxidant levels for, e.g., H_2O_2 would be affected.

On a more cosmic note, the escape flow of hydrogen atoms from the earth into outer space should increase in proportion to increased stratospheric methane, water vapor, and hydrogen. While of no direct consequence to humans, this increased flow to outer space, resulting as it will from microbial production of methane, provides a powerful illustration of the degree of coupling between Earth biology, chemistry, and physics.

It is clear that assessments of the global environmental and climatic effects of a continued increase of atmospheric methane will require chemists, biologists, meteorologists and others to work together and to adopt more unified views of the earth system. Attempts by society to adapt to, or mitigate these effects, will require similar unity of purpose and understanding from social and physical scientists, politicians and society at large. Given the vital importance of the activities, e.g., domestication of cattle, rice agriculture, and forest clearing for cropland, that are causing certain of our global environmental problems, it is clear that methods of adapting civilization to environmental change are needed.

Coda

Helmut E. Landsberg

Assessment of the chemistry of trace elements in the stratosphere must take into account the transport and presence of water vapor in the stratosphere. Contrary to model assumptions this does not occur through diffusion but more often through direct transport by cumulonimbus clouds breaking through the tropopause. Events of this type have been frequently observed in the warm season in the United States where cloud tops have been estimated to reach occasionally 21 km. With the high number of thunderstorms on earth this must be a major mode of water vapor entering the lower stratosphere and a potential source of OH radical.

References

Blake, D.R., E.W. Mayer, S.C. Tyler, Y. Makide, D.C. Montague and F.S. Roland. 1982. "Global increase in atmospheric methane concentrations between 1978 and 1980." *Geophys. Res. Lett.* 9: 477–480.

Blake, D.R., and F.S. Rowland. 1988. "Continuing worldwide increase in tropospheric methane, 1978 to 1987." *Science* 239: 1129–1131.

Chameides, W.L., S.C. Liu and R.J. Cicerone. 1977. "Possible variations in atmospheric methane." *Geophys. Res.* 82: 1795–1978.

Cicerone, R.J., J.D. Shetter and C.C. Delwiche. 1983. "Seasonal variation of methane flux from a California rice paddy." *J. Geophys. Res.* 88: 11,022–11,024.

Cicerone, R.J. and R.S. Oremland. 1988. "Biogeochemical Aspects of Atmospheric Methane." *Global Biogeochemcial Cycles* 2: in press.

Craig, H. and C.C. Chou. 1982. "Methane: the record in polar ice cores." *Geophys. Res. Lett.* 9: 1221–1224.

Dickinson, R.E., and R.J. Cicerone. 1986. "Future global warming from atmospheric trace gases." *Nature* 319: 109–115.

Ehhalt, D.H. and L.E. Heidt. 1973. "Vertical profiles of CH_4 in the troposphere and stratosphere." *J. Geophys. Res.* 78: 5271.

Ehhalt, D.H. 1974. "The atmospheric cycle of methane." *Tellus* 26: 58–70.

Ehhalt, D.H., R.J. Zander and R.A. Lamontagne. 1983. "On the temporal increase of tropospheric CH_4" *J. Geophys. Res.* 88: 8442–8446.

Fraser, P.J., M.A.K. Khalil, R.A. Rasmussen and L.P. Steele. 1984. "Tropospheric methane in the midlatitudes of the southern hemisphere." *J. Atmos. Chem I*: 105–135.

Gold, T.J. 1979. "Terrestrial sources of carbon and earthquake outgassing." *J. Petrol, Geol. I*: 3–19.

Holzapfel-Pschorn, A., and W. Seiler. 1986. "Methane emission during a cultivation period from an Italian rice paddy." *J. Geophys. Res.* 91: 11, 803–811, 814.

Khalil, M.A.K. and R.A. Rasmussen. 1983. "Sources, sinks and seasonal cycles of atmospheric methane." *J. Geophys. Res.* 88: 5131–5144.

Koyama, T. 1963. "Gaseous metabolism in lake sediments and paddy soils and the production of atmospheric methane and hydrogen." *J. Geophys. Res.* 68: 3971–3983.

Levin, I., K.O. Münnich, and W. Weiss. 1980. "The effect of anthropogenic CO_2 and ^{14}C sources on the distribution of ^{14}C in the atmosphere." *Radiocarbon* 22: 379–391.

Lowe, D.C., C.A.M. Brenninkmeijer, M.R. Manning, R. Sparks, and G. Wallace. 1988. "Radiocarbon determination of atmospheric methane at Baring Head, New Zealand." *Nature* 332: 522–525.

Oremland, R.S. 1988. "Biogeochemicstry of Methanogenic Bacteria," in *Biology of Anaerobic Microroganisms*, J.B. Zehnder, ed. John Wiley & Sons, New York, pp. 641–705.

Pearman, G.I., D. Etheridge, F. de Silva, and P.J. Fraser. 1986. "Evidence of changing concentrations of atmospheric CO_2, N_2O and CH_4 from air bubbles in Antarctic ice," *Nature* 320: 248–250.

Prinn, R., D. Cunnold, R. Rasmussen, P. Simmonds, F. Alyea, A. Crawford, P. Fraser, and R. Rosen. 1987. "Atmospheric trends in methylchloroform and the global average for the hydroxyl radical." *Science* 238: 945–950.

Rinsland, C.P., J.S. Levine, and T. Miles. 1985. "Concentration of methane in the tropsphere deduced from 1951 infrared solar spectra." *Nature* 318: 245–249.

Scholz, T.G., D.H. Ehhalt, L.E. Heidt and E.A. Martell. 1970. "Water vapor, molecular hydrogen, methane and tritium concentrations near the stratopause." *J. Geophys. Res.* 75: 3049–3054.

Seiler, W. and P.J. Crutzen. 1980. "Estimates of gross and net fluxes of carbon between the biosphere and the atmosphere from biomass burning." *Climatic Change* 2: 207–247.

Sheppard, J.C., H. Westberg, J. Hopper, K. Ganesan and P. Zimmerman. 1982. "Inventory of global methane sources and their production rates." *J. Geophys. Res.* 87: 1305–1312.

Singer, S.F. 1971. "Stratospheric water vapor increase due to human activities." *Nature* 233: 543–545.

Stauffer, B., E. Lochbronner, H. Oeschger, and J. Schwander. 1988. "Methane concentration in the glacial atmosphere was only half that of the preindustrial Holocene." *Nature* 332: 812–814.

Sze, N.D. 1977. "Anthropogenic CO emissions: implications for the atmospheric CO-OH-CH_4 cycle." *Science* 195: 673–675.

Thompson, A.M., and R.J. Cicerone. 1986. "Possible perturbations to atmospheric CO, CH_4, and OH." *J. Geophys. Res.* 91: 10, 853–10, 864.

Whalen, S.C. and W.S. Reeburgh. 1988. "A Methane Flux Time Series for Tundra Environments," *Global Biogeochemical Cycles 2:* in press.

World Meteorological Organization. 1986. *Atmospheric Ozone 1985*. Geneva, pp. 476–496.

Zimmerman, P.R., J.P. Greenberg, S.O. Wandiga and P.J. Crutzen. 1982. "Termites: A potentially large source of atmospheric methane, carbon dioxide and molecular hydrogen." *Science* 218: 563–565.

SEVEN

CHLOROFLUORO-CARBONS, STRATOSPHERIC OZONE, and the ANTARCTIC 'OZONE HOLE'

F. Sherwood Rowland

The chlorofluorocarbon-stratospheric ozone question first arose in 1974, soon after the publication in *Nature* (London) of a short paper by M.J. Molina & Rowland (1974). Scientific consideration of the problem (or 'hypothesis', or 'controversy') quickly spread to acquire political and regulatory ramifications (IMOS, 1975), accompanied by extensive discussion also in magazines and books (*see* for example, Brodeur, 1975, 1986; Dotto & Schiff, 1978). The scientific understanding of this problem has focused primarily on the expectations that such chlorofluorocarbon (CFC) molecules as CCl_3F and CCl_2F_2 will (M.J. Molina & Rowland, 1974; Rowland & Molina, 1975):

(a) survive for many decades in the Earth's atmosphere;

(b) decompose only after reaching the mid-stratosphere at altitudes where they are exposed to short-wavelength (190–230 nm) radiation which does not penetrate to lower altitudes because of absorption by ozone (0_3), or by molecular oxygen 0_2;

(c) release atomic Cl which then initiates long, ozone-depleting ClO_x chains involving Cl and ClO as alternating reactants.

The ClO_x chains have been known since 1974 to be most effective, globally, in the depletion of ozone through reactions in the altitude range of from 35 to 45 km, for which eventual future ozone losses with continual emission of CFCs have been predicted in the 35% to 60% range. The only postulated circumstances under which such massive ClO_x-induced ozone losses would *not* occur were:

(1) The suggestion, considered briefly in 1974, that the CFC molecules would not be able to reach the stratosphere at all because they were: (a) heavier than air (true, but irrelevant in convective mixing); or (b) would be efficiently destroyed by chemical reactions such as those with positive or negative ions before reaching the altitudes of heavy ultraviolet (UV or u.v.) exposure;

(2) The more persistent insistence that the CFC molecules react in the troposphere on the time-scale of 15-20 years, greatly reducing the fraction surviving to reach the mid-stratosphere (Jesson & Glasgow, 1977; Jesson, 1982). This postulate of 'tropospheric sinks' can be formulated either as specific removal processes (e.g. decomposition on hot Sahara sand; freeze-out in Antarctic snow, etc.), or as not-yet-identified, single or multiple, destruction processes whose net effect is much more rapid removal of the CFCs than that calculated for stratospheric destruction alone.

The detection of CFCs in the mid-stratosphere (NAS, 1976) demonstrated in 1975 that the first postulate was not correct. Tests of each new specific removal process in the troposphere as it was proposed, demonstrated that none of these was important for molecules such as CCl_3F or CCl_2F_2, while measurements in the atmosphere itself have shown that the lifetimes of these molecules correspond with increasing precision to those predicted from stratospheric dissociation alone (Cunnold *et al.*,

1983*a*, 1983*b*, 1986). For those reasons, the expectation that increasing amounts of organochlorine compounds in the atmosphere would inevitably result, over a few decades, in decreasing average amounts of ozone in the *upper* stratosphere (above 30 km) has been a firmly-fixed aspect of these scientific discussions.

Over the past 14 years, estimates have varied widely of the effects to be anticipated upon ozone concentrations in the *lower* stratosphere from increasing amounts of atmospheric chlorine, and concern has been expressed during the 1980s about the effects on tropospheric ozone from increasing concentrations of other trace-gases such as CO_2, CH_4, and N_2O (NAS, 1976, 1979, 1982, 1984; WMO-NASA, 1986). The combination of decreased ozone in the upper stratosphere and increased ozone below 25 km, can result in one offsetting the other in the calculations of total ozone change. Such offsets can occur especially in the widely-used 1-D models (global atmospheric models with only one variable dimension—altitude—assumed to represent approximate global averages). In many instances, complicated reports of the variation in ozone effects at different altitudes were reduced to the simple statement of the calculated long-term change in total ozone in a 1-D model.

The situation has changed in the mid-1980s with the more widespread development of excellent 2-D models which provide average estimated changes for given latitudes instead of a global average (*see*, for example, Isaksen & Stordal, 1986). Yet the mathematical complexity—and absence of truly global experimental data—involve so much difficulty that general extension to complete 3-D models with full chemistry will probably not occur until the 1990s.

Regulations to control the emission to the atmosphere of CFC propellants for aerosol sprays were adopted in the 1976–1978 time-period in the United States, Canada, and Scandinavia (Dotto & Schiff, 1978), but not elsewhere. Very little further regulation of atmospheric emissions of CFCs was considered prior to the development of proposals under the auspices of the United Nations Environment Programme (UNEP) from 1985 onwards. By 1985, the US National Academy of Sciences reports (NAS, 1982, 1984) were indicating that experimental measurements in the atmosphere tended to show loss of ozone in the upper stratosphere in the altitude regions for which effects from the ClO_x-chain had always been predicted.

The initial reports also came in mid-1985 of extremely large ozone depletions which were first observed over Antarctica in 1977, then in 1979, and then even more in the 1980s (Farman *et al.*, 1985; Chubachi & Kajiwara, 1986; Stolarski *et al.*, 1986). The seminal data of Farman *et al.* (1985) are shown in Fig.7-1. Suggestions were immediately made (Farman *et al.*, 1985; Rowland, 1986) of the possibility of a strong causal link between the rapidly enlarging ozone depletions over Antarctica and the concentrations of organochlorine compounds in the atmosphere.

Subsequent developments in the remote geographical region have undoubtedly played an important role in stimulating international discussion of possible regulations for restriction of future emissions of CFCs. An international Convention for the Protection of the Ozone Layer was adopted in 1985, prior to the disclosure of the data demonstrating the heavy loss of ozone in the Antarctic. The first Protocol specifying some of the details of ozone-layer protection was agreed to in Montreal in September 1987 (Tolba, 1987). While the Montreal Protocol was specifically and repeatedly stated not to be dependent upon the observations over Antarctica, it is also certainly true that none of the delegates in Montreal was unaware of the latter's existence, and of the widespread suspicion—confirmed strongly two weeks later (Watson & Albritton, 1987)—that CFCs were playing an important role in this massive depletion of ozone in the springtime Antarctic.

The following article is divided into three sections, the first of which outlines the scientific background of the subject and the current status of ongoing investigations connected with stratospheric ozone and the atmospheric behavior of the CFCs. The second section describes the major technological uses of the CFCs and some of the available alternatives to them. The final section describes the regulatory activity involving present or proposed future restrictions on CFC emissions. No discussion is presented here about the further consequences of global ozone depletion.

The general areas of major concern focus on the twin roles of ozone as: (a) a stratospheric shield against the penetration

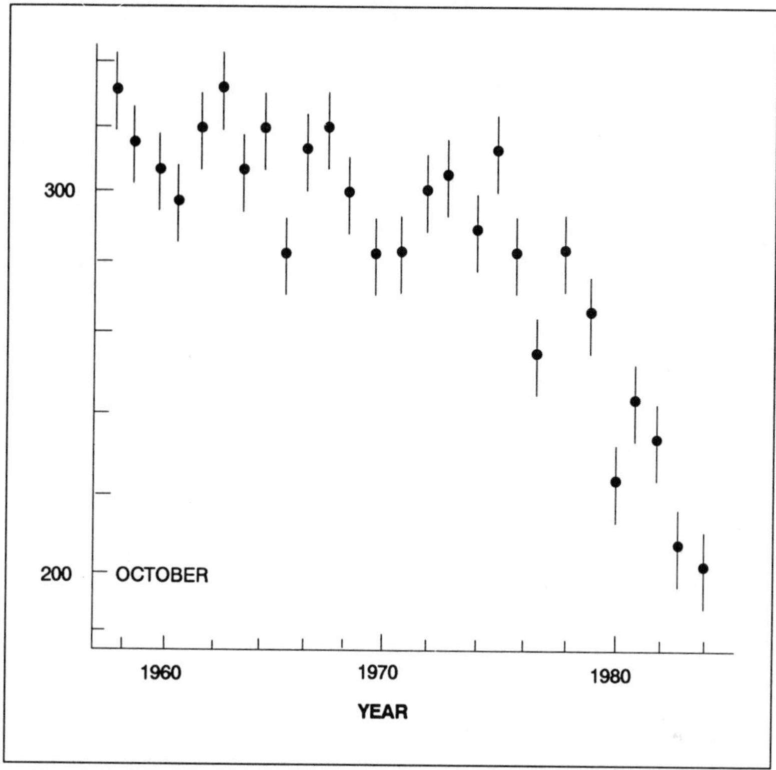

Fig. 7-1: Monthly means of total ozone at Halley Bay, Antarctica, for October of the years 1957 through 1984. (Data of Farman *et al.*, 1985.)

to the Earth's surface of biologically damaging ultraviolet radiation in the 280–320 nm wavelength band (designated as UV-B), and (b) the conversion of this energy absorbed by ozone into a stratospheric heat-source. A decrease in total ozone in the atmosphere permits increased penetration of UV-B in the ratio of -1% in O_3 +2% in UV-B. The major direct effects of increased UV-B on humans are increased incidences of skin cancer and eye cataracts, and perhaps some suppression of the immune system (NAS, 1976, 1979, 1982, 1984).

Effects of such enhanced UV-B penetration have also been demonstrated in many biological entities, including various species of animals, plankton, and agricultural crops. Moreover, alteration of the distribution of ozone within the stratosphere will change the heating pattern there. Because the absorption of UV radiation by ozone effectively determines the structure of the stratosphere, change in this heating pattern could result in climatic shifts, although the detailed nature of any such effects is not readily calculated. The first appearance of a possible climatic effect from ozone depletion has been the persistence of the Antarctic polar vortex in 1987, following the largest ozone losses yet found over that region in the springtime.

Scientific Status and Scientific Issues

Stratospheric Ozone, 'Greenhouse Effect', and the Radiation Balance of Earth

Ozone (O_3) comprises only about 3 parts in 10 millions of Earth's atmosphere, but nevertheless plays several critical roles in the radiation balance of the Earth. The atmosphere consists of a mixture of gases under constant bombardment by solar radiation of which the intensity peaks in the visible wavelengths (400–700 nm), while survival of each component gas is shaped by its photochemical properties. The most abundant atmospheric gases (N_2, 78%; O_2, 21%; Ar, 0.9%) are all transparent to visible radiation, and to most of the more energetic but less abundant ultraviolet (UV) radiation. However, UV radiation with wavelengths shorter than 242 nm is absorbed by O_2 and causes photodissociation into two O atoms (eq. 1). The usual fate of such O atoms is combination with other O_2 molecules to form ozone (eq. 2).

$$O_2 + \text{u.v.} \rightarrow O + O \quad (1)$$

$$O + O_2 + M \rightarrow O_3 + M \quad (2)$$

The extra energy of the nascent O_3 molecule can be transferred by collision to some other molecule, M, where it appears as kinetic energy and raises slightly the surrounding atmospheric temperatures. This process has been going on ever since O_2 became a permanent part of the atmosphere more than a thousand million years ago, and has made ozone an important atmospheric 'trace' component during the development of all the currently important biological species. Indeed, because it seems inconceivable that life as we know it could have evolved without an ozone shield, any serious modification of stratospheric ozone is fraught with biological dangers.

About 90% of atmospheric ozone is found in the stratosphere, with peak concentrations around 25 km in altitude above the Equator and 15 km near the poles. Ozone is especially effective in absorbing UV radiation with wave-lengths between 210 nm and 293 nm, releasing atomic oxygen by equation (3). Almost all of these O atoms react by (2), with the net effect of

$$(3) \qquad O_3 + \text{u.v.} \rightarrow O + O_2$$

UV absorption by O_3 involves the conversion of its energy into a heat-source in the upper atmosphere. This added heat-energy is the cause of the temperature inversion between 15 and 50 km altitude, which defines the stratosphere as having a temperature gradient from about 215 K near 15 km to about 275 K at 50 km. The temperature is at a minimum about 15 km, defined as the tropopause, and increases steadily down through the troposphere to the average surface reading of 288 K.

The chemical equations for the formation and removal of ozone in the atmosphere are often described in terms of total 'odd oxygen,' i.e. the sum of the concentrations of O atoms and O_3, in contrast to the 'even' O_2. Neither the photodissociation of O_3 nor the recombination of O with O_2 causes a change in total odd oxygen; nor does (3) permanently remove ozone from the atmosphere when followed by (2). The only important atmospheric source of odd oxygen, and therefore ultimately of ozone, is the +2 in concentration accompanying the photodissociation O_2 in (1). The only odd oxygen removal process initially identified was the occasional reaction of atomic O with O_3 in (4), an odd oxygen change of -2.

$$(4) \qquad O + O_3 \rightarrow O_2 + O_2$$

During the decade from 1964–1974, several additional removal sequences were identified, including the free-radical chain reactions of the type shown in reactions (5) and (6), with -1 in odd oxygen for each step.

(5) $X + O_3 \rightarrow XO + O_2$

(6) $XO + O \rightarrow X + O_2$

(X = HO, NO, or Cl)

The defining characteristic of a free radical is that it has an odd number of electrons (e.g. HO, 9; Cl, 17; NO, 15), in contrast to the generalization from observation that almost all stable molecules have an even number of electrons. The last, odd electron, unable to find a partner electron in its original molecule, is usually the reactive locus for free-radical attack on another molecule, and such reactions are usually very rapid. Each of the radicals HO, NO, and Cl, has a natural source, and each also has the possibility of supplementary sources resulting from the activities of humanity. Chemical sequences such as those in (5) and (6) are often described as the HO_x (HO, HO_2), NO_x (NO, NO_2), and ClO_x (Cl, ClO) chains.

The main inputs of radiation into the Earth's atmosphere are the incoming solar radiation (with peak intensity in the visible wavelengths around 500–600 nm), and the outgoing radiation emitted by the Earth. The intensity of this terrestrial radiation maximizes in the infrared (IR) around 10,000–15,000 nm, approximately in inverse ratio to the surface temperatures of the Sun (6,000 K) and the Earth (288 K). Most of the visible radiation (400–700 nm) and UV radiation (with wavelengths between 293–400 nm) penetrates throughout the stratosphere and troposphere, all the way to the Earth's surface. However, each unit, or *quantum*, of radiation which is absorbed, carries enough energy to cause photodecomposition for the gaseous molecule which intercepted it. Accordingly, chemical compounds which absorb in the 293–600 nm range have very short atmospheric lifetimes, and do not accumulate there. On the other hand molecules which are transparent to 293–600 nm wavelengths can be decomposed by UV radiation with wavelengths shorter than 293 nm, but are shielded from those rays in the troposphere by the ozone in the stratosphere above. Their atmospheric lifetimes can stretch into a century or more.

The IR (infrared) wavelengths include only a minor component of the incoming solar flux, but carry all of the outgoing terrestrial radiation. The absorption of IR radiation by a molecule depends upon a close match in energy with the permissible energies of its internal vibrations—if the energies match, absorption is strong; if not, no absorption occurs. Because molecules with three or more atoms have many more kinds of internal vibration than do the diatomic molecules N_2 or O_2, the natural atmospheric absorption of IR radiation is dominated by the triatomic molecules O_3, CO_2, and H_2O. Atmospheric absorption can be supplemented by other multi-atom molecules present in trace amounts, such as CH_4, N_2O, CCl_3F, and CCl_2F_2.

The total amount of solar energy entering the Earth's atmosphere must be effectively balanced by outgoing energy from the Earth. The absorption of some IR radiation by molecules in the atmosphere is followed by re-emission in all directions, including downward towards the Earth's surface. In the latter case, the failure of this reflected IR radiation to escape into space, forces a higher level of terrestrial IR emission to seek the necessary balance between outgoing and incoming energy. This enhanced IR emission is obtained through a higher surface temperature—the 'greenhouse effect.' Even before there were any known changes to the atmosphere from the activities of humans, the existing gaseous components were retaining sufficient IR energy to raise the Earth's surface temperature from a calculated 225 K for an airless Earth to the actual 288 K average. (For further discussion of the greenhouse effect and its consequences, see chapters by Covey and Kellogg in this volume.)

Atmospheric Chemistry of Chlorofluorocarbons

The first chlorofluorocarbon (CFC) compounds were synthesized in the 1920s in a successful attempt to find a satisfactory inert refrigerant. Dichlorodifluoromethane, CCl_2F_2 (CFC-12, or Freon-12 [DuPont], Frigen-12 [Farbwerke Hoechst], Genetron-12 [Allied-Signal], Arcton-12 [ICI], etc.) has long been the coolant used in home refrigerators and in automobile air-conditioners. Inert volatile compounds such as this CFC-12, CCl_3F (CFC-11), and CCl_2FCClF_2 (CFC-113), have also found extensive use as aerosol propellants (CFC-11/CFC-12 mixtures), in the expansion of plastic foams (CFC-11), in the cleaning of electronic components (CFC-113), and in a multitude of lesser applications. Another refrigerant gas, $CHClF_2$ (CFC-22), is used in home air-conditioners. In almost all of these applications, the gaseous CFC is sooner or later released intact to the atmosphere. Methylchloroform (CH_3CCl_3) is widely used as a solvent in the degreasing of metals.

The use of CFC-12 and CFC-11 expanded very rapidly in the United States after World War II, increasing by about 10% per year for almost three decades. European production followed along a few years behind. The yearly global release of CFC-12 and CFC-11 reached 100,000 tons in 1960 and 1965, respectively, and has remained in the range of 350,000 to 450,000 tons for CCl_2F_2 and 250,000 to 300,000 tons of CCl_3F since 1975. When the lesser production of several other CFCs is included, the total release to the atmosphere is approximately one million tons of CFCs per year.

The dominant factor in the atmospheric chemistry of CFCs is their inertness towards the common chemical and physical removal processes which strongly affect most other molecules (M.J. Molina & Rowland, 1974; Rowland & Molina, 1975; NAS, 1976, 1979, 1982, 1984; WMO-NASA, 1986). The most common tropospheric sinks for gaseous molecules are: (a) photodecomposition by visible radiation or by ultraviolet radiation between 293 and 400 nm; (b) rainout, after dissolution in rain droplets; or (c) chemical reaction in the troposphere with oxidizing agents such as hydroxyl (HO) radical or O_3. However, the CFCs are transparent between about 230 and 700 nm, and do not photolyze in the troposphere; they are essentially insoluble in water and do not rainout; and they are inert towards HO and other oxidant 'species'. Molecules such as $CHClF_2$ and CH_3CCl_3, however, are different from most other CFCs in a very important way, namely that they *are* reactive towards HO in the troposphere, as shown in (7), and therefore do have an important tropospheric sink which destroys most of the molecules before they can be decomposed in the stratosphere.

$$(7) \qquad HO + CHClF_2 \rightarrow H_2O + CClF_2$$

The conclusion that CFCs had no important tropospheric sinks was originally come to by M.J. Molina & Rowland (1974), and was hotly disputed for almost a decade, with numerous suggestions of supposed important but overlooked sinks: freezing out on Antarctic snow; photodecomposition while absorbed on Sahara sand; ion-molecule reactions at the tropopause; catalytic decomposition on stratospheric aerosols; etc. None of these suggestions held up as quantitatively important (NAS, 1976, 1979, 1982, 1984; WMO-NASA, 1986). Convincing evidence for the existence of tropospheric sinks would be the disappearance of the CFCs more rapidly than expected from stratospheric loss alone (Jesson & Glasgow, 1977; Jesson, 1982). Conversely, the primary proof for the negligible importance of such sinks is the observed rapid accumulation and long atmospheric lifetimes of the CFCs (Cunnold *et al.*, 1983*a*, 1983*b*, 1986; WMO-NASA, 1986).

Gases mix throughout the troposphere and stratosphere in large air-parcels, carrying along without gravitational separation the heavier-than-air CFCs (molecular weights: CCl_3F, 137.5; air, 28.9). As air mixes to an altitude of 30 km or higher, most of the O_2 (98%) and O_3 (about 70–90%) then lies below, and all of the gaseous components are exposed to solar UV radiation in the 200–230 nm range (M.J. Molina & Rowland, 1974; Rowland & Molina, 1975; NAS, 1976). The CFCs readily absorb such UV radiation and photodecompose with the loss of Cl, as in (8) for CCl_2F_2. The atomic Cl released in the stratosphere can then participate

(8) $$CCl_2F_2 + \text{u.v.} \rightarrow Cl + CClF_2$$

in the ClO_x chain-reaction (M.J. Molina & Rowland, 1974; Stolarski & Cicerone, 1974; Rowland & Molina, 1975; NAS, 1976; WMO-NASA, 1986), forming ClO in (5) and returning to Cl in (6). At still higher altitudes, the CFCs are exposed to even more intense short-wavelength UV radiation and their concentrations fall to very low levels. Measurements of more than ten different halocarbons in stratospheric air samples have clearly demonstrated both their upward penetration to high altitudes, and their decomposition there by UV radiation (WMO-NASA, 1986).

At any given time, most of the molecules in the atmosphere, including the CFCs, are well below 30 km altitude, and therefore shielded by stratospheric ozone from intense short-wavelength UV exposure. Consequently, the average atmospheric survival-time of molecules whose only sink is stratospheric photodissociation can be very long. Calculations in 1974 of the atmospheric lifetimes for CCl_3F and CCl_2F_2, with no tropospheric sinks, gave ranges of from 40 to 80 years for the former, and from 80 to 150 years for the latter (M.J. Molina & Rowland, 1974; Rowland & Molina, 1975). Close comparisons of the accumulated emissions of each compound *versus* the amount still present in the atmosphere, have always shown that relatively little of either has actually been decomposed, corresponding to average lifetimes of many decades or more for each. Present evaluations for CCl_3F suggest a lifetime of 75 years (Cunnold *et al.*, 1983*a*, 1986), while that of CCl_2F_2 is about 110 years (Cunnold *et al.*, 1983*b*, 1986). A lifetime of 110 years corresponds to 36% survival in the year AD 2100 of the CCl_2F_2 molecules in the atmosphere in 1988, with 6% expected still to be floating in the atmosphere in AD 2300.

The atmospheric concentrations of several chlorinated molecules are now in the range of 0.1–0.6 parts per thousand millions (10^9) by volume (ppbv). The most abundant organochlorine compounds in the 1988 atmosphere are, in decreasing order of total chlorine content (WMO-NASA, 1986): CCl_2F_2, CCl_3F, CH_3Cl (methyl chloride), CCl_4 (carbon tetrachloride), CH_3CCl_3, and CCl_2FCClF_2. Only CH_3Cl from this group has any important natural source, and it was probably the only quantitatively important chlorinated organic molecule in the atmosphere at the beginning of the 20th century. The total atmospheric content of organochlorine compounds has mushroomed over the past few decades from about 0.8 ppbv in 1950 to 1.0 ppbv in 1970 and 3.5 ppbv in 1987, with a present rate of increase exceeding 1 ppbv per decade. Fig. 7-2 shows the growth in these concentrations in the northern hemisphere since 1950.

Halocarbons can be conveniently separated into categories of long-lived or short-lived atmospheric 'species', dependent upon whether tropospheric sinks exist for each. Several groups can be identified:

(a) *Saturated perhalo chlorocarbons*, transparent in the UV for wavelengths longer than 293 nm and without other tropospheric sinks (e.g. CCl_2F_2, CCl_3F, CCl_4);

(b) *Olefinic molecules* containing C=C double bonds, susceptible to rapid destruction by hydroxyl attack in the troposphere, usually with lifetimes of much less than one year (e.g. $CHCl{=}CCl_2$, $CCl_2{=}CCl_2$);

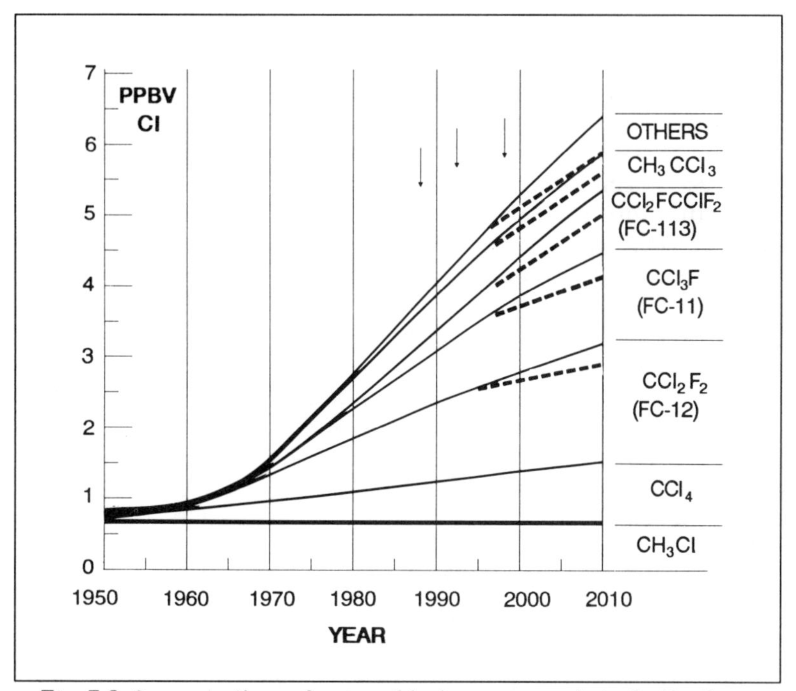

Fig. 7-2: Concentrations of organochlorine compounds in the Northern Hemisphere Atmosphere, 1950–2010. The three arrows (from left to right) point to: (a) the present date; (b) proposed UNEP date for 20% reduction in CFC emissions; (c) proposed UNEP date for 50% reduction in CFC emissions. Solid lines for years after 1988 indicate concentrations with constant emissions at 1986 levels. Dashed lines indicate concentrations with emission cutbacks at times indicated in (*b*) and (*c*).

(c) *Hydrochlorocarbons*, containing C-H bonds which can also be attacked by hydroxyl radicals, with quite variable lifetimes from a year or less to several decades, depending upon the reactivity of the individual C-H bonds (e.g. CH_3Cl, CH_3CCl_3, $CHClF_2$); and

(d) Molecules which have measurable absorption spectra at wavelengths longer than 293 nm, capable of absorbing and being photodissociated by UV radiation throughout the troposphere, and usually having lifetimes of a few years or less.

Molecules with no more than a single bromine atom bonded to any carbon atom do not have UV absorption beyond 293 nm, and are expected to have very long lifetimes. Molecules with two Br atoms or one Br and one Cl on a single carbon (e.g. CBr_2F_2, $CBrClF_2$) absorb photons with wavelengths beyond 293 nm, and can be photodissociated in the troposphere, having lifetimes of a decade or so (L.T. Molina *et al.*, 1982).

These observations provide a basis for dividing halocarbon molecules into three classes with respect to their potential for causing stratospheric ozone depletion. Classification is largely measured by their survival long enough to deliver Cl and Br to the stratosphere:

(a) *Very high ozone depletion potential:*—Almost all of the compounds in this class are saturated perhalo chlorocarbons, whose only important sinks are stratospheric, and which therefore release essentially all of their chlorine as Cl atoms in the upper reaches of the ozone layer. Among the technologically important molecules in this class are CCl_4, CCl_3F, CCl_2F_2, CCl_2FCClF_2, $CClF_2CClF_2$ (CFC-114), and $CClF_2CF_3$ (CFC-115). Others are brominated fluorocarbons with single bromine atoms attached to each carbon (e.g. $CBrF_3$ and $CBrF_2CBrF_2$), capable of stratospheric release of atomic Br, an even-more-effective ozone-depleting agent than Cl on a per-atom basis.

(b) *Very much lesser potential for ozone depletion:*—Halocarbons with tropospheric sinks, including those containing (i) a C=C double bond, e.g. $CHCl=CCl_2$ and $CCl_2=CCl_2$; (ii) a C-H bond, e.g. $CHClF_2$ (CFC-22), CH_3CCl_3, CH_3Cl, and CH_3Br; or (iii) a tropospheric photochemical sink, e.g. CBr_3F and CBr_2F_2. The release of atomic Cl or Br in the troposphere provides a negligible threat to stratospheric ozone because the rapid subsequent reactions of these halogen atoms lead to inorganic molecules which never reach the stratosphere because they are susceptible to tropospheric removal through such processes as rainout.

(c) *Negligible potential for ozone depletion:*—Perfluorocarbons (CF_4, C_2F_4, etc.), and hydrofluorocarbons, e.g. CH_2FCF_3 (FC-134a), are in this category. Atomic F has an ozone depletion potential of about one-thousandth the strength per atom of atomic Cl, because HF, when once formed, is a permanent *reservoir* molecule, remaining chemically unchanged until it diffuses into the troposphere and is removed by rain. However, the perfluorocarbons have atmospheric lifetimes estimated as thousands of years or longer, and frequently have very strong potential for enhancing the 'greenhouse' effect. The hydrofluorocarbons, on the other hand, have comparatively short atmospheric lifetimes because of their reactivity towards HO, the hydroxyl radical.

Estimated Atmospheric Lifetimes for Molecules with Tropospheric Sinks

The most important oxidizing agent in the troposphere is hydroxyl radical, which is capable of attacking almost all C-H bonds with the subsequent destruction of the entire molecule. The atmospheric lifetimes for the C-H-containing molecules can readily vary by a factor of 100, and can be satisfactorily estimated from the inverse ratio of the reaction rates with HO radical in comparison with a standard molecule, CH_3CCl_3. The CH_3CCl_3 molecule has been released to the atmosphere in large amounts (about 500 kilo-tons per year), primarily through its use as a degreasing solvent. Fairly accurate estimates of its accumulated release up through the late 1970s were available, and—in contrast to the observations with CCl_3F and CCl_2F_2—the atmospheric burden of CH_3CCl_3 was clearly only about half as large as the accumulated release, i.e. the other half of the CH_3CCl_3 molecules had already been destroyed.

A general consensus now exists that the atmospheric lifetime of CH_3CCl_3 is about 6 or 7 years (Makide & Rowland, 1982; Prinn *et al.*, 1987)—far shorter than those of CCl_3F and CCl_2F_2. The inverse ratios of the measured laboratory hydroxyl reaction rates *versus* CH_3CCl_3, lead to atmospheric lifetime estimates for $CHClF_2$ (CFC-22) of 17 years, for CH_2FCF_3 (FC-134a) of 11 years, and for $CHCl_2CF_3$ (CFC-123) of 2 years (Makide & Rowland, 1982). Each of these last three compounds is under active consideration as a technological substitute for a perhalo CFC.

Temporary Reservoir Molecules in the Stratosphere

While atomic Cl reacts very efficiently with ozone molecules, the stratospheric ClO_x chain can be temporarily stopped by diversion of Cl into HCl through reaction with molecules such as CH_4 or HO_2 (Rowland & Molina, 1975; NAS, 1976, 1979, 1982, 1984; WMO-NASA, 1986). In the mid-stratosphere, the time-scale for the reaction of Cl with O_3 is approximately one second, and for ClO and O about one minute, while the competing reactions to form HCl intervene on the average only after about 1,000 cycles of reactions (5) and (6). A molecule such as HCl can be described as a *temporary reservoir*, because its Cl atom is not attacking ozone while tied up in this chemical form, although it can easily be released again by HO attack to give H_2O and Cl.

A more complete description of stratospheric chlorine chemistry includes the formation of two additional important temporary reservoirs (WMO-NASA, 1986): chlorine nitrate ($ClONO_2$), from the reaction of ClO with NO_2; and hypochlorous acid (HOCl), from the reaction of ClO with HO_2. Most of these chlorinated molecules formed after the release of Cl from its original carbon compound, have actually been measured in the stratosphere. The chlorine chemistry of the normal stratosphere is believed to be reasonably well understood, even though the geographical spread of successful measurements is not global, and simultaneous measurements of many interacting chemical components are rather scarce.

Atmospheric Chemistry of Other Trace Gases

Careful measurements since 1957 have established the reality of an upward trend of the yearly average CO_2 concentration from 315 parts per million by volume (ppmv) in 1957 to about 347 ppmv in 1988 (Keeling *et al.*, 1984; WMO-NASA, 1986; Kellogg, this volume). This steady increase in CO_2 concentration provides additional polyatomic molecules capable of absorbing emitted terrestrial IR radiation, and therefore requires a gradual increase in the average temperature of the Earth to maintain a constant energy-emission from the top of the atmosphere. (Covey, this volume.)

The rapid growth in the concentrations of CCl_2F_2 and CCl_3F was recognized in 1975 as providing two additional 'greenhouse gases,' both of them capable of absorbing additional terrestrial IR radiation (Ramanathan, 1975). At the first impression, it may seem surprising that the 'greenhouse' effects from yearly CFC increases averaging about 0.017 ppbv for CCl_2F_2 and 0.011 ppbv for CCl_3F are not completely overshadowed by yearly CO_2 increases of 1,500 ppbv. The key is that the existing 347 ppmv of CO_2 molecules already absorb almost all of the terrestrially emitted IR radiation which matches the absorption frequencies of CO_2, so that such wavelengths are largely absent from the outgoing radiation, and incremental CO_2 has only a minor additional effect *per molecule*.

Through the vagaries of molecular structure, CCl_3F and CCl_2F_2 happen to absorb strongly in IR regions that are not blanked out in the atmosphere by CO_2, H_2O, or O_3. The incremental IR absorption *per molecule* by these CFCs is about 15,000 times as great in the atmosphere as would be that of an additional molecule of CO_2, making the CFCs important 'greenhouse' gases. Carbon dioxide, with its yearly increment of 1,500 ppbv, remains the major contributor to increased 'greenhouse' absorption, accounting for about one-half of the increase in total absorption during the 1980s and 1990s (Ramanathan *et al.*, 1985).

Regular measurements of the concentrations of CH_4 in remote locations have only been carried out since the late 1970s and show that CH_4 is now increasing in concentration at a yearly rate of about 16 to 17 ppvb or approximately 1% per year of the present concentration (WMO-NASA, 1986; Blake & Rowland, 1988; Cicerone, this volume). Retrospective measurements with air trapped in ice cores from Greenland and Antarctica show levels of CH_4 200 years ago of 0.7–0.8 ppmv—more than a factor of two below the present world-wide average. Our own measurements have shown an increase in methane of 1.52 ppmv in early 1978 to 1.69 ppmv in late 1987—an increase of about 11% in a decade, as shown in Fig. 7-3 (Blake & Rowland, 1988). A lesser yearly percentage rate of increase in N_2O (0.6 ppbv increase *versus* about 0.3 ppmv concentration = 0.2%) has also been observed (WMO-NASA, 1986). When the augmented IR absorption capabilities of these additional atmospheric trace species are considered together, the combined effect over the coming decades is approximately equal to that found for CO_2 by itself, and so the total estimate for the trapping of IR radiation by the greenhouse effect has been practically doubled over the previously anticipated CO_2 alone (Ramanathan *et al.*, 1985). Typical calculations suggest an increase in average global temperature of 3.0+/-1.5°C by the middle of the 21st century. The most recent evaluations of world-wide temperature records show a global increase of 0.5°C over the past century (Jones *et al.*, 1986).

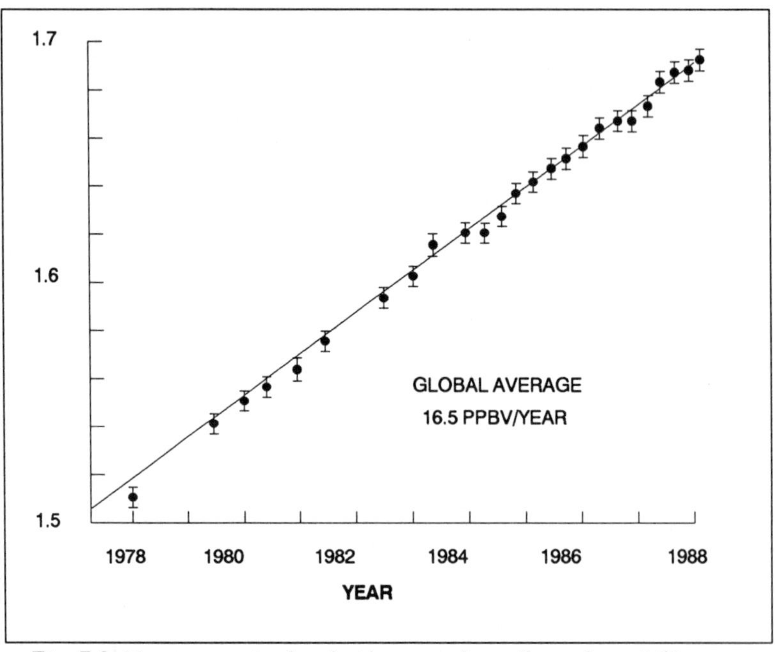

Fig. 7-3: Measurements showing increase in methane from 1.52 ppmv in early 1978 to 1.69 ppmv in late 1987—an increase of about 11% in a decade. (Data of Blake & Rowland, 1988.)

The Antarctic Ozone 'Hole'

The most striking change yet observed in atmospheric ozone concentrations has been the (southern) springtime ozone depletion found over the Antarctic since 1977. The ozone loss was first recorded with the ground-based instrument at the British Antarctic Survey station at Halley Bay on the Antarctic coast (Farman *et al.*, 1985), and showed a drop in the October average of the total ozone concentrations from about 320 Dobson Units (D.U.: 1 Dobson Unit ≅ 1 part in 10^9 in the atmosphere) in the 1957–1964 period to less than 200 D.U. in 1984 (Fig. 7-1). The ozone losses have been confirmed both by measurements from the US Nimbus-7 satellite (Stolarski *et al.*, 1986), and by balloon-borne instruments sent aloft by the Japanese (Chubachi, 1984; Chubachi & Kajiwara, 1986) and by the United States (Hofmann *et al.*, 1987).

The massive ozone losses begin with the first sunlight of the Antarctic spring in late August, and almost half of the total ozone disappears during the month of September. The satellite observations show that the area of major ozone loss is roughly the size of the United States, and frequently exhibits sharp gradients (Stolarski *et al.*, 1986) which have led to its description as an ozone 'hole.' Further ozone loss has basically stopped by mid-October, and strong air circulation from the south-temperate zone in mid-November causes the dissipation of the 'hole' by dilution with air that has not been so depleted of ozone. The Antarctic polar vortex, with its frigid stratosphere and associated deficiency in ozone, persisted much longer into November-December 1987 than had ever previously been recorded. The great reduction in the usual heat-source from ozone absorption of UV radiation, is presumably the cause of this delayed breakup of the vortex. The monthly average amount of ozone over Antarctica is now somewhat less during all of the sunlit months than had been observed in the corresponding months of the 1960s (Farman *et al.*, 1985).

Three classes of theoretical explanations have been put forth for the appearance of this massive Antarctic ozone depletion:

(a) Chemical reactions initiated by natural geographical processes, and, in particular, as a consequence of the 1980 peaking in the 11-years' solar sunspot cycle, postulated to have produced additional quantities of ozone-depleting NO_x (Callis & Natarajan, 1986*a*, 1986*b*);

(b) A dynamic, meteorological phenomenon of natural origin which did not actually destroy ozone, but instead shifted it to less-polar latitudes, or prevented the arrival of ozone-rich air (Mahlman & Fels, 1986; Stolarski & Schoeberl, 1986; Tung *et al.*, 1986).

(c) Chemical reactions initiated by the activities of mankind, based on unusual chlorine chemistry under the specialized physical conditions in the Antarctic. The 'standard' ClO_x-chain removal of ozone by reactions (5) and (6) cannot be very effective in the low-Sun conditions of the polar springtime, because the concentrations of atomic O are too low to make (6) an important process there. One or more alternate chain-reactions are needed, with special efficiency under Antarctic conditions (Crutzen & Arnold, 1986; McElroy *et al.*, 1986; S. Solomon *et al.*, 1986; L.T. Molina & Molina, 1987).

Two clarifying statements need to be made immediately: (1) the actual destruction of ozone in the stratosphere is *always a chemical process* involving some specific molecular reaction; (2) the geographical location in which ozone destruction takes place is a sensitive function of the meteorology, and ozone-rich or ozone-poor air can be readily transported by prevailing winds into a new place in which it can be observed and measured. The only meteorological or dynamic explanations which truly belong in category (b) involve theories postulating that *no* ozone is, actually, chemically converted back to diatomic oxygen, but rather that the observed changes in regional ozone are caused by transport in from somewhere of an unaltered air-mass with an already lower concentration of ozone. The division between (a)/(b) *versus* (c) is the distinction between naturally occurring events and those influenced by mankind, and has political and regulatory overtones. This division can be defined through the question: 'Would these very low values for springtime Antarctic ozone have occurred if the trace-gas content of the atmosphere, especially of chlorinated compounds, had not been substantially increased over the past decades?'

The natural sunspot cycle explanation (Callis & Natarajan, 1986*a*, 1986*b*) failed badly on two critical points: (a) three maxima in sunspots have taken place in the past 30 years (1958, 1969, 1980), but deep springtime Antarctic ozone loss occurred only after 1980; and (b) the postulated influx of NO_x runs counter to the observations over Antarctica that the NO_x concentrations accompanying the ozone-depleted air are extremely low, rather than unusually high as predicted by Callis & Natarajan. The observation of very low NO_2 actually was made and reported (McKenzie & Johnston, 1984) well before the solar-cycle theory was advanced, and has been abundantly confirmed since (Keys & Johnston, 1986; Farmer *et al.*, 1987; Mount *et al.*, 1987). Because of these major failures, this theory is no longer under active consideration.

The dynamic explanations have relied upon observations such as the near-constancy during 1980–1982 of total ozone south of 45°S during the August-October period (Stolarski & Schoeberl, 1986). However, previous measurements of total ozone south of 45°S over the years 1957–1975 have regularly shown a substantial increase from August to mid-October

(Londen & Angell, 1982). The amount of ozone north of 45°N latitude also regularly shows an increase during the northern springtime—six months out of phase with the southern observations.

The south polar data, taken alone, are consistent with either: (a) a change in the meteorology such that the 'standard' (i.e. 1957 to 1975) north-to-south flow of ozone at the time of the southern spring equinox has gradually been sufficiently reduced to level off the previously observed mid-October maximum in ozone south of 45°S; or (b) a new chemical removal process has been added which would approximately offset the 'standard' increase during 1980–1982 to produce a roughly constant amount of total ozone over that period (Elliott & Rowland, 1988).

Another dynamical theory postulated vertical motion upwards within the region of the ozone hole, carrying ozone-poor tropospheric air (Tung *et al.*, 1986). In these cases, as well as for the chemical theories, auxiliary questions of causality need also to be considered: 'Why did the process responsible for the springtime Antarctic ozone loss *not* occur between 1957 and 1975? Why did it appear first in the late 1970s, and then intensify in the 1980s?' The answer to these questions in the chemical theories is the rapid growth in total chlorine concentrations, as demonstrated in Fig. 7-2.

The deficiencies of the purely dynamical theories become much clearer when the chemical observations accompanying the Antarctic ozone hole are also included among the phenomena to be explained. The observation that CH_4 and N_2O levels over Antarctica were not different for strongly ozone-depleted air within the vortex and for air outside the ozone hole, eliminated the dynamical postulate of differential vertical upward transport for those two regions (Farmer *et al.*, 1987).

Some highly unusual chemical phenomena accompanied the air-masses that were depleted in ozone: extremely high concentrations of the radical chain-carrier ClO below about 25 km, and relatively normal amounts of ClO higher in the stratosphere (Zafra *et al.*, 1987; P. Solomon *et al.*, 1987); the first observations anywhere in the atmosphere of another chlorine radical, OClO, being present throughout September but gone by mid-October (S. Solomon *et al.*, 1987); very low concentrations of NO_x and of HNO_3 (Keys & Johnston, 1986; Farmer *et al.*, 1987; Mount *et al.*, 1987); dehydration of the air-parcels in the lower stratosphere (Watson & Albritton, 1987); concentrations of HCl much lower than in surrounding air-masses that were not showing large ozone losses (Farmer *et al.*, 1987); vertically very patchy losses of ozone, with layers of 90% ozone loss interspersed with other layers having only 20% to 30% loss (Hofmann *et al.*, 1987; Watson & Albritton, 1987); and concentrations of ClO within the Antarctic vortex up to 500 times greater than those found to the same altitudes outside the vortex (Watson & Albritton, 1987).

All of these evidences of extremely unusual free-radical chemistry accompanying the Antarctic ozone hole, compel us to the conclusion that large chemical losses of ozone are taking place through chlorine-driven chemistry, although important quantitative aspects remain to be settled.

The present indications are that the chlorine-based chemical depletion of ozone occurs over Antarctica under the following circumstances:

(a) the Antarctic polar vortex circulates tightly around the pole during the winter, with air-parcels remaining always in the frigid darkness. The stratospheric air becomes so cold that polar stratospheric clouds (PSCs) form (McCormick *et al.*, 1982), probably containing substantial amounts of water ice associated with nitric acid (Crutzen & Arnold, 1986), and perhaps sulfuric acid, in the cloud particles; (b) the air in the Antarctic stratosphere has travelled a long distance from its 'entrance' into the stratosphere from the troposphere through the tropical tropopause, and is substantially depleted in stable compounds such as N_2O and CH_4; (c) the nitrogen oxides are converted first to N_2O_5 in the darkness, as happens overnight in other latitudes, and then to nitric acid on the surfaces of the PSCs, with much of the HNO_3 remaining stuck on the particles (Crutzen & Arnold, 1986; Hamill *et al.*, 1986); and (d) the temporary reservoirs for chlorine, HCl, and $ClONO_2$, also strike the PSCs and stick on the surface, chemically reacting to convert the chlorine to Cl_2 and HOCl, with the nitrogen ending up as nitric acid (M.J. Molina *et al.*, 1987; Tolbert *et al.*, 1987).

The potential importance of the reactions of H_2O and HCl with $ClONO_2$ in (9) and (10) was raised in 1984

(9) $$H_2 + ClONO_2 \xrightarrow{PSC} HOCl + HONO_2$$

(10) $$HCl + ClONO_2 \xrightarrow{PSC} Cl_2 + HONO_2$$

through experiments which showed that both of these reactions occurred extremely rapidly under laboratory conditions, even when the surfaces of the reaction vessels were specially coated to be inert towards reactive gases (Rowland & Sato, 1984; Rowland *et al.*, 1986). These reactions, and the N_2O_5 analog of (9) in reaction (11), do not take place through gas-phase collisions,

(11) $$H_2O + N_2O_5 \xrightarrow{PSC} HONO_2 + HONO_2$$

but rather are catalytically facilitated by surfaces in the laboratory, or on particles in the stratosphere. Both reactions (9) and (10) were separately shown to have a profound effect on future estimates of stratospheric ozone loss, changing long-term evaluations of a predicted 4% loss into 25% to 32% decreases (Wuebbles *et al.*, 1984).

With this chemical and physical priming of the stratosphere, the first rays of the late August sunlight cause the release from both Cl_2 and HOCl of Cl atoms which immediately attack O_3 in (12) to form ClO (S. Solomon *et al.*, 1986). The HO radical from HOCl can also attack O_3 to form HO_2 in (13), and the further reaction of ClO with HO_2 in (14) reforms HOCl and O_2, completing a chemical chain converting $2\ O_3 \rightarrow 3\ O_2$. The photolysis of HOCl in (15) then starts the chains all over again.

(12) $$Cl + O_3 \rightarrow ClO + O_2$$

(13) $$HO + O_3 \rightarrow HO_2 + O_2$$

(14) $$ClO + HO_2 \rightarrow HOCl + O_2$$

(15) $$HOCl + hv \rightarrow Cl + HO_2$$

Alternatively, the ClO radical can react with a second ClO radical (L.T. Molina & Molina, 1987) to form a dimer, $(ClO)_2$, which can decompose either photochemically or thermally to O_2 and two Cl atoms that can then continue the chain. Still another cycle involving ClO postulates its combination with

BrO formed by Br attack on O_3. Subsequent reactions of the ClO and BrO radicals with each other release O_2 and both Cl and Br in atomic form (McElroy *et al.*, 1986).

All three of these cycles are capable of removing ozone through reactions summing to $2\ O_3 \rightarrow 3\ O_2$, avoiding any dependence upon the relatively low available concentrations of atomic O. These ClO-ClO and ClO-BrO reactions are also known from laboratory studies to possess branching pathways which form OClO (e.g. ClOClO + sunlight → Cl + OClO). All of these chains are able to continue rapidly throughout September because the usual chain-breaking steps of ClO reaction with NO_2, or Cl reaction with CH_4, are at a very low ebb, with the NO_x still tied up as HNO_3 and the methane in low concentration in this well-travelled air. The vertical patchiness in ozone loss may have its origin in the layered structure of the polar stratospheric clouds (Hoffman *et al.*, 1987).

A major question relative to the potential importance in the stratosphere of reactions such as (9) and (10), was whether or not molecules such as $ClONO_2$ and HCl could stick to cloud particles with an efficiency high enough to be significant. This problem has been settled by recent demonstrations (M.J. Molina *et al.*, 1987; Tolbert *et al.*, 1987) that both HCl and $ClONO_2$ stick to icy surfaces with very high efficiencies (≥ 0.01, in contrast to 'sticking' coefficients in the range of 10^{-6} for other chemical reactants on different surfaces). In this manner, stratopheric chlorine can be shifted from the reservoir molecule $ClONO_2$ to HOCl by (9), or from HCl and $ClONO_2$ to Cl_2 by (10). Furthermore, nitrogenous species such as NO and NO_2 are successively oxidized to form NO_3, then combined to form N_2O_5, and finally converted to $HONO_2$ through reaction (11). Much or all of the nitrogenous material which is usually found as gaseous NO_x is effectively embedded into cloud particles as $HONO_2$.

Finally, the special Antarctic conditions outlined above in (a), (b), and (c), have presumably taken place for decades or hundreds of years without extraordinary ozone losses. The difference in the 1980s occurs in the last step (d) because of the rapid increase in the organochlorine concentration in the atmosphere over the past decade, from 1.0 ppbv in 1970 to 3.5 ppbv in 1987, furnishing extra chlorine for the attack on ozone (Farman *et al.*, 1985; Rowland, 1986; cf. Dütsch, 1987).

Global Observations of Stratospheric Ozone

Measurements of ozone have been made with ground-based Dobson UV spectrophotometers in a few locations since the 1920s, and from many more since the International Geophysical Year in 1957 (including Halley Bay, Antarctica). Another ozone-measuring UV instrument provided data from the Nimbus-4 satellite in 1970–1972, and two separate UV instruments have been sending back ozone data since late 1978 from the Nimbus-7 satellite (Bowman & Krueger, 1985; Bowman, 1988). The Dobson instruments were installed primarily to look for ozone changes on the time-scale of a few days or weeks, not for years or decades, and much of the accumulated archival data still needs to be reprocessed to include all of the appropriate long-term corrections from periodic recalibrations. In addition, the quality of these stations is very non-uniform, and some of the recorded data will undoubtedly not be able to be satisfactorily corrected. Several statistical studies have been made of the archival data, without recalibration corrections or elimination of the less-well-run stations (WMO-NASA, 1986).

Our own examination of the Dobson ozone data taken with extreme care in Arosa, Switzerland, from 1931 to 1987 (Birrer, 1975; Dütsch, 1984, 1985, pers. comm. 1988) indicates a decrease in average ozone concentration of about 3% since 1970 (Harris & Rowland, 1988). These average ozone losses are not distributed uniformly throughout the year, but instead show as much as 6% drop-off in the winter months, and less than 1% decrease in late summer (Fig. 7-4). Comparably large decreases in winter and small ones in summer have also been found over the U.S. Dobson stations in Caribou (Maine) and Bismarck (North Dakota), both of which are at about the same latitude as Arosa. The earlier statistical studies of these data had made no allowance for the possibility of seasonal variations in ozone depletion (WMO- NASA, 1986). These ozone losses are as yet only established as the result of the statistical analysis of the accumulated ozone data, and provide no indications of the mechanistic explanation for the decrease. Because PSCs are also found over the Arctic, albeit in lesser amounts than over the Antarctic (McCormick *et al.*, 1982), some of the cloud-based chemical reactions which are active in the south polar region, may very well have partial counterparts also under northern polar conditions.

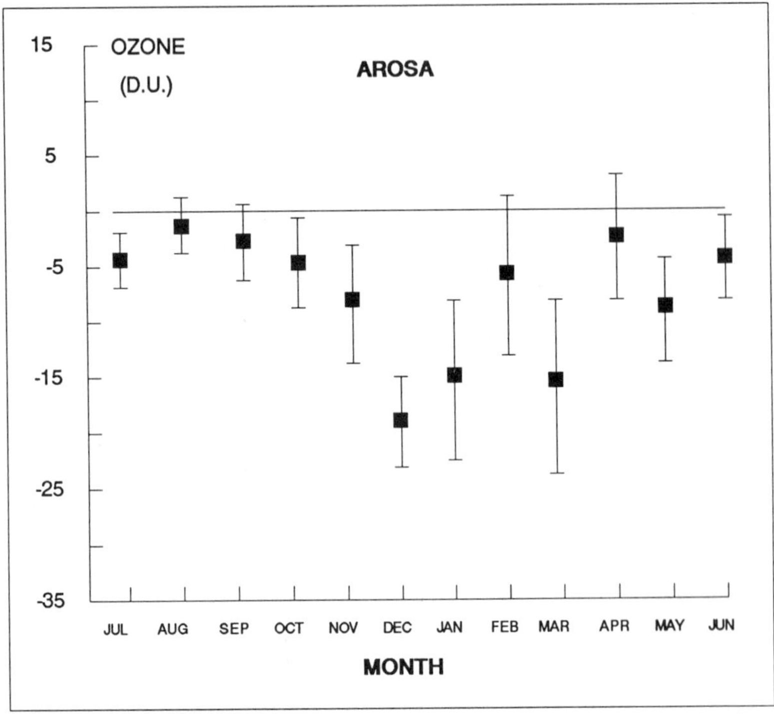

Fig. 7-4: Average monthly ozone measurements at Arosa, Switzerland. The average value for the period 1931–1969 has been subtracted from the average value for the period 1970–1986. Monthly values below the line indicate less ozone present during 1970–1986 than during the earlier 40-years' period. Data taken by the E.T.H., Zürich (courtesy of Hans U. Dütsch). D.U.= Dobson Units.

Both the ground-based Dobsons and the satellite instruments can provide some information about the vertical distribution of ozone. In each case, analysis of the data indicates a loss between 1970–1980 of several percentage points of ozone in the region around 40 km (WMO-NASA, 1986), an altitude for which atmospheric models have always predicted the maximum effect of the ClO_x-chain removal of ozone.

The total ozone data collected by the satellite instruments present a formidable calibration problem because the exposure of the instrument to the direct sunlight of outer space has caused appreciable change in some of its integral components.

One very useful calibration procedure has been the comparison of the ozone measurements from the satellite with those from the ground whenever Nimbus-7 passes over a Dobson station (Fleig *et al.*, 1986). The measured total ozone concentrations from the SBUV instrument on the Nimbus-7 satellite have been reported to show an average global ozone loss of several percentage points between 1978 and 1984, and further drop to 1986, prior to any correction needed for instrument degradation as judged during the Dobson station overpasses.

The ozone data from the TOMS instrument on Nimbus-7 are much more voluminous (*c.* 200,000 measurements per day of ozone over the whole sunlit globe for more than 9 years), and are in agreement with the results from the SBUV instrument in showing an appreciable loss of ozone between 1978 and 1986 (Bowman, 1988). However, the two instruments actually share the same 'diffuser plate'—the part of each instrument most exposed to solar degradation—so that some of their agreement does not represent independent verification. With the 11-years' solar cycle going from its maximum in 1980 to a minimum in 1986, the relative importance in the satellite data-set of any natural change in ozone from solar influences (e.g. if more UV radiation is emitted during sunspot maximum) is difficult to separate from the possibility of changes from the steadily increasing organochlorine content of the atmosphere. In contrast, the ground-based Dobson data cover the past 30 years in many locations—long enough to permit factoring out of any solar cycle effect that may be mixed into the long-term trends.

Technological Status and Issues: Chlorofluorocarbon Primary Uses

Refrigeration

Refrigeration and air-conditioning are virtual necessities of modern life, and the world-wide demand for both is continually growing, especially with advancing technology in developing countries. The only feasible response to the problem of ozone depletion by CFCs is, therefore, to search for a satisfactory substitute refrigerant. The major chlorofluorocarbon refrigerant gases in current production are CCl_2F_2 (CFC-12), $CHClF_2$ (CFC-22), and CCl_3F (CFC-11), together with some mixtures such as CFC-502 (CFC-22 mixed with $CClF_2CF_3$ [CFC-115]). While the eventual environmental solution may well lie with hydrofluorocarbons (FC-134a) which pose no ozone-depletion problem, the near-term choices include CFC-22, because the problems of large-scale industrial production have already been solved for it, and because its ozone depletion potential is certainly much less than that of CFC-12 for which it might well be a substitute.

Essentially all refrigeration devices are carefully engineered to match very closely the characteristics of the chosen refrigerant, and will work only poorly, if at all, with other refrigerant gases. In some large-scale uses ammonia is a competitor of CFC-12, but in small-scale uses—e.g. 0.5 to 3 pounds (227 grams to 1.36 kg) in automobile air-conditioning or in home refrigerators—the inert, non-toxic, non-odorous nature of CFC-12 has made it the almost universal choice as a refrigerant. On the other hand, CFC-22 is the overwhelming choice as refrigerant in home air-conditioners, and was an early choice for automobile air-conditioning before CFC-12 became the standard. Although synthetic chemistry has produced innumerable new classes of volatile compounds in the fifty years since the establishment of the CFC family of compounds, all of the substitute refrigerant gases under serious discussion to this point, are members of the fluorocarbon family, with one or more hydrogen atoms furnishing a point of attack for tropospheric decomposition.

The choice for replacing CFC-12 in its refrigerant uses thus falls between (a) redesign of the equipment to accommodate a substitute refrigerant; or (b) finding a refrigerant which is similar enough to CFC-12 to be simply 'dropped into' existing equipment. The boiling point of FC-134a (-26°C) is only slightly higher than for CFC-12 (-29°C), and its properties are close enough that tests are now being conducted to determine whether FC-134a is a 'drop in' substitute for CFC-12. The large-scale manufacture of FC-134a has not yet been attained, and comments from industry indicate some likelihood of production problems because of catalyst poisoning, etc. Another possible 'drop in' substitute might come from appropriate choice of the proportions of two refrigerants in a mixture—variations in composition of mixtures analogous to CFC-502 (although one of its components, CFC-115, has appreciable potential for ozone depletion).

Aerosol Propellants

A mixture of CFC-11 and CFC-12 was the world-wide favorite propellant for aerosol sprays in the mid-1970s. Several countries (United States, Canada, Norway, Sweden) then enacted severe controls on such uses in the late 1970s, but all of the other countries continued to choose CFCs as the propellant gas for a wide spectrum of applications. The experience of the US, Canada, and Scandinavia, has been that a delivery system based on CFCs can be replaced without much difficulty by either alternate propellants—usually hydrocarbons such as isobutane and propane—or by other delivery systems, e.g. 'roll-ons' and 'pump valves.' Economic studies following the US changeover in the 1970s showed financial losses in some industries (CFC manufacturers; can-making companies), and gains in others (hydrocarbon producers; glass manufacturers), with the gains apparently slightly larger than the losses. Some aerosol propellant uses, such as certain medical applications, are still permitted, and some uses in the US have been mysteriously classified as non-propellant (e.g. gases used for dusting photographic equipment); but most aerosol propellant uses of CFCs have disappeared from the US, Canada, and Scandinavia. The aerosol industry itself in the United States has, without CFCs, returned to approximately the sales volume (almost 3 thousand million cans per year) that was typical of the mid-1970s.

The refrigeration/air-conditioning and aerosol propellant usages illustrate the two general categories of substitution in the economic markets. In the former case, the properties required of the refrigerant gas are sufficiently restrictive that the market is almost a captive one for the chlorofluorocarbon industry—substitutes are possible, but they are almost always other molecules from the same general category, i.e. hydrochlorocarbons or hydrofluorocarbons.

Foam-blowing of Plastics

Plastic foams fall into two categories, rigid and open-cell. In the former case, the gas serves the functions of both expanding the plastic into a cellular structure, and then of being trapped within the component cells to reduce the heat-transfer capability of the product. These insulation materials have very many separate applications, and the possible substitutions for each use need to be evaluated individually. Modern refrigerators usually contain more CFC-11 trapped in the foamed insulation than CFC-12 in the refrigeration coils. 'Fast foods' such as hamburgers have often been packaged in insulated containers made from CFC-rigid foams. Cardboard is a competitor in this usage, and non-CFC insulation materials are available within the construction industry. But whereas the CFC-foamed plastics are usually better insulators per unit volume, they cost more than cardboard for a given amount of insulation protection.

Foamed plastics can be made by the expansion of other gases, such as CFC-22, in place of the perhalo CFC-11 and CFC-12, while other agents are available for blowing open-cell foam, including methylene chloride (CH_2Cl_2) and pentane. With each, however, important questions such as toxicity and flammability must be evaluated both for their potential effects on the convenience and cost of manufacture, and for potential restrictions on their ultimate usage.

Solvents

The most significant CFC solvent is CFC-113, because a large part of the semiconductor and related electronics industries have designed various critical components to be cleaned with this solvent. The most basic problem of cleaning electronics is the need to remove minute traces of extraneous materials from sensitive metallic components, while *not* attacking other parts such as plastic mountings, and CFC-113 has been chosen because of its specific solvent characteristics. In many situations, however, the hydrogen-containing chlorofluorocarbons are somewhat better solvents than CFC-113, and this can be a serious disadvantage when differential solution characteristics are sought.

Other cleaning agents are available, and are often used, but some of the best organic cleansers are much more expensive than CFC-113. If, as is sometimes proposed, a tax were to be placed on the use of CFCs for various industrial purposes, an added cost of US $5 per pound (454 gm) would immediately drive many of the uses into the alternatives and substitutes market. However, the tax might have to be as much as $50 per pound to cause the electronics industry to shift away from CFC-113 towards substitutes!

Another alternative that is often suggested is the recycling of the CFCs rather than emission to the atmosphere after a single use. Such recapture is much more feasible at the manufacturing than the use level, as in the trapping within an operating plant of the CFC gases that have been used in making open-cell foam. On the other hand, recapture of the individually small (but cumulatively rather large) amounts in discarded refrigerators or other consumer products appears not to be a very effective option, because of the huge logistic effort required.

Regulatory and Policy Status and Issues

Aerosol Propellant Regulations

Eventual national regulation of the CFCs as aerosol propellants was forecast by a 14-agencies US task-force in June 1975 (IMOS, 1975); forecast again by a US National Academy of Sciences committee in September 1976 (NAS, 1976); and jointly announced by the US Environmental Protection Agency and the Food and Drug Administration in October 1976, effective in 1978 (Dotto & Schiff, 1978). Changeover to pump valves, roll-ons, and substitute propellants, was already in progress in 1975–1976, and the imminent demise of the CFC propellant market spurred the economic competition towards capture of the substitute market. The switch away from CFC aerosol propellants occurred so swiftly and smoothly in the United States that it was almost unnoticed outside of the industries which were directly affected.

The atmospheric protection attained by the regulatory ban on CFCs as aerosol propellants began to erode almost immediately, as markets for other uses such as foam-blowing were actively sought by the industries, and those uses continued to expand rapidly. The now-widespread use of CFC-113 for cleaning electronic components has been developed almost entirely since the bans on CFC-11 and CFC-12 as aerosol propellants, even though it has been obvious all along that all three CFC molecules are roughly equivalent in their ability to deplete stratospheric ozone. In the original US announcements, the control of CFC usage as aerosol propellants was described as Phase One, with Phase Two—regulations covering all other uses—scheduled for 1978. The Phase Two schedule soon began to slip, and for all intents and purposes disappeared early in 1981 with the anti-regulation philosophy imposed upon the Environmental Protection Agency by the new presidential administration.

While Canada, Sweden, and Norway, also moved quickly towards legislation or regulation of the aerosol propellant uses, Japan and the major CFC-manufacturing countries of western Europe (Great Britain, West Germany, France, and Italy) strongly resisted any restrictions until early 1987. In the 1980s, the international position of the 'Canadian group' (essentially those with aerosol propellant restrictions—the United States, Canada, and Scandinavia) called for extension of the aerosol bans to the European countries as the desirable next step—in effect saying, 'we've had some pain, now it's your turn.'

The European Economic Community (EEC) group countered with the proposal that a 'cap' be placed on total production of CFCs for any use, with allocation among the various uses to be left aside as a national matter. A key point in the EEC argument was their correct contention that a ban on use of aerosol propellants as atmospheric protection was ineffective if other industries such as foam-blowing and the cleaning of electronic parts were allowed to pick up the slack in CFC production that had been left by the aerosol propellant ban. On the other hand, EEC position was very much a 'no pain' proposal, because the anticipated production 'cap' was scheduled to be placed at the existing CFC *production capacity,* which exceeded the actual EEC production in the early 1980s by 60%–70%. In essence this proposal called for essentially no change in current practice, with continued expansion of CFC production and markets for ten years or more until actual production caught up with already-existing production capacity.

Vienna Conference & Convention on Protection of the Ozone Layer

The Conference of Plenipotentiaries on the Protection of the Ozone Layer was convened by the Executive Director of the United Nations Environment Programme (UNEP) in Vienna, Austria, on March 18–22, 1985, in accordance with a decision adopted by the Governing Council of UNEP on May 28, 1984.

Thirty-six countries, including all of the major CFC-producing countries, accepted the invitation to participate in the Vienna Conference, and an additional seven countries sent observers. Article 2 of the 'Vienna Convention' adopted at this Conference established 'an obligation to take appropriate measures to protect human health and the environment against adverse effects resulting or likely to result from human activities which modify or are likely to modify the ozone layer.' The question of *which* appropriate measures were to be taken *when* was left for further work on a *protocol* to 'control equitably global production, emissions and use of CFCs.'

The subject-matter for that protocol was designated in a resolution on a Protocol Concerning Chlorofluorocarbons, and includes 'fully-halogenated chlorofluorocarbons (CFCs) and other chlorine-containing substances.' Although the Convention defined and discussed the possible effects on ozone from bromine released from brominated compounds, the protocol resolution did not mention them, thereby leaving open an enticing opportunity for international lawyers. Obviously, $CBrF_3$ is not a CFC because it contains no chlorine; but the question of whether $CBrClF_2$, a bromochlorofluorocarbon, was or was not a chlorofluorocarbon, was extensively discussed at the Geneva meeting 'for the Preparation of the Protocol on Chlorofluorocarbons' during December 1–5, 1986.

The basic US negotiating position during 1986 and early 1987 was strongly in favor of further restrictions on the release of CFCs to the atmosphere, with proposals for: (a) a 'cap' at 1986 CFC production levels, and (b) a later follow-on by a phase-out of as much as 95% of the CFC emissions on a time-scale of roughly one decade. The December 1986 international meeting on the Protocol in Geneva ended without agreement, and reports indicated that EEC group was having internal problems in maintaining a united position, with the Federal Republic of Germany taking the lead in asking for some actual reduction in yearly releases. During March 1987, a change in position by the EEC group was publicized, in which they had agreed to a 'cap' on CFC production at 1986 levels, to be followed by a phase-out of 20% of this within a few years. The renewed Geneva negotiations in May 1987 did agree to these two points, although with some delays in the timing.

Montreal Conference on the Chlorofluorocarbon Protocol

The Montreal Conference in September 1987 agreed to a Protocol for the control of future emissions of the chlorofluorocarbons (Tolba, 1987). The agreement represented a compromise between the two positions outlined above, and has the following essential characteristics:

(a) limitation to the 1986 rates of CFC emission, beginning in 1990;

(b) reduction in emissions by 20% below the 1986 level, taking effect during 1994;

(c) reduction in emissions by 50% below the 1986 level, taking effect during 1999; and

(d) permission for less-developed countries to develop capabilities for CFC production for an extra decade, with limitations affecting them delayed also by a decade beyond the time-scale for the developed countries.

The Protocol includes a specific list of CFC compounds which are covered: CFC-11, CFC-12, CFC-113, CFC-114, CFC-115. The list does *not* include the perhalo molecule CCl_4, nor some other perhalo CFCs (e.g. CFC-112) which are currently in minor commercial production. The calculated growth of chlorine concentrations in the atmosphere is shown in Fig. 7-2, with projections to the year AD 2010 on two different assumptions. The first of these is that the release of CFCs remains constant at 1986 levels for the next 25 years—baseline for observation of the effect of the cutbacks in emission when they are introduced; the second assumption is that the 20% and 50% cutback provisions come into effect in 1994 and 1999, respectively, as planned under the Montreal protocol. Both of these assumptions are conservative in the sense that they tend to *under*estimate the amounts of organochlorine compounds which might be present in the atmosphere in the future. There are several ways in which these tend to be underestimates:

(1) The baseline case assumes no *increase* in CFC emissions over the next 25 years, and this will be wrong if the Protocol is not ratified and put into effect, because the pressures and trends through the early 1980s have all been in the direction of increasing CFC releases rather than decreasing them.

(2) Both cases assume that there will be no increase in emissions during the period from 1987 to 1990, although such

increases are permitted under the Protocol, and are in fact occurring.

(3) The cutback assumptions assume that no major developed country will decide to ignore the Protocol, and that no developing country will add significantly to the world-wide CFC production—in other words, that the cutback in 1994 will actually correspond to a 20% decrease in CFC emissions then, and the 1999 cutback will represent a real 50% decrease in emissions. The *least* conservative calculation of the effect of the Protocol has been that the nominal 50% cutback in 1999 will actually be a 15% *increase* in emissions, primarily the result of 'successful' development of CFC industries in countries with negligible current production.

The graph in Fig. 7-2 shows that, with either constant emissions or the Protocol-mandated cutbacks of 20% and 50%, the concentrations of CFCs and of total chlorine in the atmosphere will continue to rise steadily throughout the next 25 years. The total organochlorine concentration in the atmosphere remained at the 0.6–0.7 ppbv level throughout the first half of the 20th century, reached 1.0 ppbv in the late 1960s, and was about 1.8 ppbv in the mid-1970s, when the chlorofluorocarbon-ozone problem became a matter of both scientific and public knowledge. In the intervening 14 years since the first scientific discussions, the total chlorine concentration has approximately doubled to 3.5 ppbv, with a rate of increase slightly larger than 2.0 ppbv per decade. Under the provisions of the Montreal Protocol, this rate of increase will be essentially maintained for at least another decade, and perhaps for longer (depending upon the growth of CFC industries in countries now in the initial stages of their development). The scientific function of the Montreal Protocol is essentially to maintain into the 21st century the rate of growth in chlorine concentrations which has characterized the past 15 years—to prevent an even more rapid increase than is already occurring.

One further physical aspect of the chlorofluorocarbon problem which has not been discussed yet in this article is the *timing delay between release in the troposphere and observed effects in the stratosphere*. The magnitude of such delays has been discussed from the beginning of consideration of the problem (M.J. Molina & Rowland, 1974; Rowland & Molina, 1975), and is illustrated in Fig. 7-5 (Rowland & Molina, 1976). This one-dimensional calculation carried out 12 years ago shows the predicted changes in concentration at 40 km altitude of the total chlorine (Cl_x) released from CCl_3F. This altitude is at the center of the band of most effective ClO_x removal of stratospheric ozone.

The calculations in Fig. 7-5 indicate the behavior to be expected for a molecule with an average atmospheric lifetime of 75 years (Rowland & Molina, 1976), and are in excellent agreement with current estimates of the actual lifetime (Cunnold *et al.*, 1986). In this model calculation, the release of CCl_3F grew exponentially until 1976, then levelled off for 15 years, and finally introduced an abrupt cut-off in further release in 1991. This scenario is rather similar to the actual emissions to date for the 12 years since its publication.

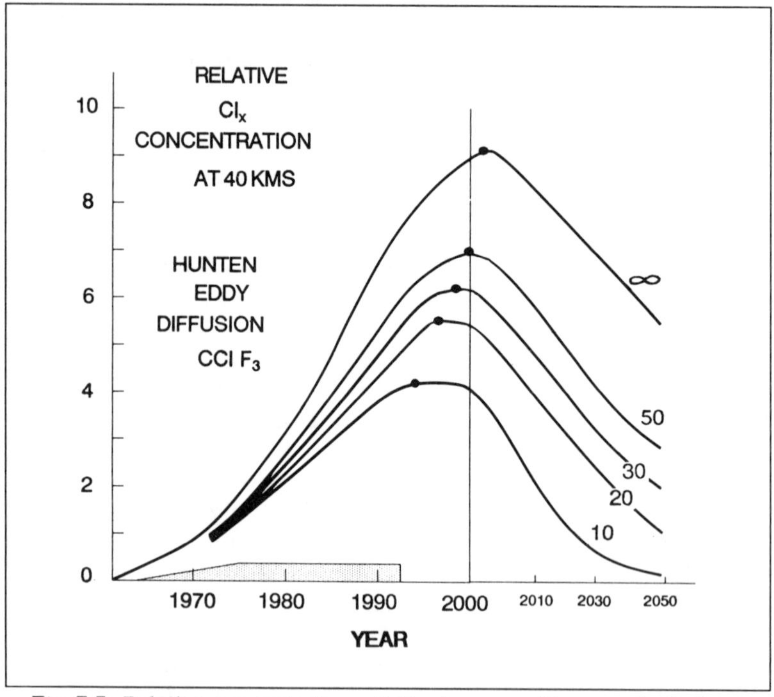

Fig 7-5: Relative concentration of Cl_x decomposition products from CCl_3F at 40 km altitude with various hypothetical atmospheric lifetimes for tropospheric sinks. Atmospheric release as shown by cross-hatched area. Maxima in calculated concentrations are indicated by black dots.

The data of Fig. 7-5 (uppermost curve) show that with complete termination of further release in 1991, the concentration of Cl_x at 40 km would reach a maximum about 15 years later, and then would gradually taper off over most of the 21st century. The growth to the Cl_x maximum at 40 km for the longer-lived CCl_2F_2 exhibits a similar increase for about 15 years, and then falls off much more slowly because of its lifetime of more than a century. In summary, then, the effects in the upper stratosphere from molecules such as CCl_3F and CCl_2F_2 reach a delayed maximum 10 to 15 years after termination of release, and persist for a century or more. The ozone depletions already observed to date over Antarctica and elsewhere are basically the consequence of releases into the mid-1970s. The full extent of ozone decreases from the total release of CFCs through 1988 will not be felt until about the year 2000.

The original calculations of Fig. 7-5 also exhibit the potential importance of tropospheric sinks, were they to exist (which they do not), for CCl_3F. The uppermost curve displays the anticipated changes in Cl_x at 40 km with a stratospheric lifetime of 75 years, and an infinite lifetime in the troposphere. The successively lower curves in Fig. 7-5 demonstrate the effects to be expected when a 75 year stratospheric lifetime is combined with various tropospheric lifetimes down to as short a time as 10 years. The same emission pattern shown at the bottom of the figure was used for all calculations. The concentrations at 40 km through 1988 show only a relatively small difference to date (maximum about a factor of two); the large reductions from the tropospheric sinks come, in the future, two or three decades ahead—small effects in 1988, big effects by AD 2050.

The tropospheric-sink curves are not actually applicable for CCl_3F, because the prediction of negligible tropospheric sinks (Molina & Rowland, 1974; Rowland & Molina, 1975) has been borne out by experience in the atmosphere (Cunnold *et al.*, 1986). However, the shapes of the curves are illustrative of the behavior to be expected for molecules such as $CHClF_2$, which has an estimated atmospheric lifetime of 17 years (Makide & Rowland, 1982). Comparison of $CHClF_2$ with CCl_3F shows an immediate reduction in Cl_x at 40 km, by a factor of 3, just from the number of Cl atoms in the molecule. During the steady emission period, another reduction in Cl_x can be seen by comparison of the curves for 20 years and infinite tropospheric lifetimes, but this difference is relatively small (about a factor of 2) in 1988. The really large saving in ozone depletion by Cl_x comes in the mid-21st century, by which time a 20-years' tropospheric lifetime has greatly reduced its residual stratospheric concentration.

The adoption of chlorinated substitutes, such as $CHClF_2$, for CCl_3F and CCl_2F_2, has an interesting ethical advantage: the damage to stratospheric ozone and hence to life on Earth would largely be confined to the lifetimes of those people under whose political control such release took place.

Summary

The momentous subject of chlorofluorocarbons (CFCs) and their effect on the biosphere's stratospheric ozone shield is treated rather generally but in sufficient depth where necessary in three main sections dealing with (i) scientific background and current status of ongoing investigation, (ii) the major technological uses of CFCs and available or foreseeable alternatives to them, and (iii) the policy status and regulatory activity involving present or proposed future restrictions in CFC emissions.

It being unlikely that life, at least as we know it, would have developed on Earth without an ozone layer in the stratosphere to 'filter off' harmful ultraviolet rays from solar radiation, the prospect of continuing manufacture in developing countries of its destroyers is highly alarming, especially as these destructive CFCs may take more than a decade from emission to reach the levels around 40 km altitude at which they do the most harm.

Acknowledgement

Research on these atmospheric problems has been supported by NASA Contacts NAGW-452 and NAGW-914.

References

Birrer, W. 1975. *Homogenisierung und Diskussion der Total-Ozon-Messreihe in Arosa 1926–71*. LAPETH-11, Doctoral thesis from the Laboratory of Atmospheric Physics, Zürich, Switzerland. [Not available for checking.]

Blake, D.R. & Rowland, F.S. 1988. Continuing worldwide increase in tropospheric methane, 1978 to 1987. *Science* 239: 1129–31.

Bowman, K.P. 1988. Global trends in total ozone. *Science* 239: 48–50.

Bowman, K.P. & Krueger, A.J. 1985. A global climatology of total ozone from the Nimbus-7 Total Ozone Mapping Spectrometer. *J.Geophys. Res.* 90: 7967–76.

Brodeur, P. 1975. Annals of Chemistry (Aerosol Sprays): Inert. *New Yorker,* April 7, 1975, pp. 47–58.

Brodeur, P. 1986. Annals of Chemistry (The Ozone Layer): In the Face of Doubt. *New Yorker,* June 9, 1986, pp. 70–87.

Callis, L.B. & Natarajan, M. 1986*a*. Ozone and nitrogen dioxide changes in the stratosphere during 1979–1984. *Nature* (London), 323: 772–7.

Callis, L.B. & Natarajan, M. 1986*b*. The Antarctic ozone minimum: relationship to odd nitrogen, odd chlorine, the final warming, and the 11-year solar cycle. *J. Geophys. Res.* 91: 10771–96.

Chubachi, S. 1984. Preliminary result of ozone observation at Syowa station from February 1982 to January 1983. *Mem. Natl. Inst. Polar Res.*, Spec. Issue, 34: 13–9.

Chubachi, S. & Kajiwara, R. 1986. Total ozone variations at Syowa, Antarctica. *Geophys. Res. Lett.* 13: 1197–8.

Crutzen, P. & Arnold, F. 1986. Nitric acid cloud formation in the cold Antarctic stratosphere: a major cause for the springtime 'ozone hole.' *Nature* (London) 324: 651–5.

Cunnold, D.M., Prinn, R.G. Rasmussen, R.A., Simmonds, P.G., Alyea, F.N., Cardelino, C.A., Crawford, A.J., Fraser, P.J. & Rosen, R.D. 1983*a*. The atmospheric lifetime experiment, 3: Lifetime methodology and application to 3 years of $CFCl_3$ data. *J. Geophys. Res.* 88: 8379–400.

Cunnold, D.M., Prinn, R.G., Rasmussen, R.A., Simmonds, P.G., Alyea, F.N., Cardelino, C.A. & Crawford, A.J. 1983*b*. The atmospheric lifetime experiment, 4: Results for CF_{12} based on 3 years of data. *J. Geophys. Res.* 88: 8401–14.

Cunnold, D.M., Prinn, R.G., Rusmussen, R.A., Simmonds, P.G., Alyea, F.N., Cardelino, C.A., Crawford, A.J., Fraser, P.J. & Rosen, R.D. 1986. Atmospheric lifetime and annual release estimates for $CFCl_3$ and CF_2Cl_2 from 5 years of ALE data. *J. Geophys. Res.* 91: 10797–817.

De Zafra, R.L. — see Zafra, R.L. De.

Dotto, L. & Schiff, H. 1978. *The Ozone War*. Doubleday & Co., Garden City, New York, NY, USA: 342 pp.

Dütsch, H.U. 1984. An update of the Arosa ozone series to the present using a statistical instrument calibration. *Quart. J.R. Met. Soc.* 110: 1079–96.

Dütsch, H.U. 1985. Total ozone in the light of ozone soundings, the impact of El Chichon. pp. 263–68 in *Atmospheric Ozone* (Eds. C.S. Zerefos & E. Ghazi). D. Reidel Co., Dordrecht, The Netherlands: [not available for checking].

Dütsch, H.U. 1987. The Antarctic 'ozone hole' and its possible global consequences. *Environmental Conservation* 14(2): 95–7, figs.

Elliott, S. & Rowland, F.S. 1988. Comment on "Further interpretation of satellite measurements of Antarctic total ozone". *Geophys. Res. Lett.*, 15: 196–7.

Farman, J.C., Gardiner, B.G. & Shanklin, J.D. 1985. Large losses of total ozone reveal seasonal ClO_x/NO_x interaction. *Nature* (London), 315: 207–10.

Farmer, C.B., Toon, G.C., Schaper, P.W., Blavier, J.-F. & Lowes, L.L. 1987. Stratospheric trace gases in the spring 1986 Antarctic atmosphere. *Nature* (London). 329: 126–31.

Fleig, A.J., Bhartia, P.K. & Silberstein, D.S. 1986. An assessment of the long-term drift in SBUV total ozone data, based on comparison with the Dobson network. *Geophys. Res. Lett.* 13: 1359–62.

Hamill, P., Toon, O.B. & Rurco, R.P. 1986. Characteristics of polar stratospheric clouds during the formation of the Antarctic ozone hole. *Geophys. Res. Lett.* 13: 1343–6.

Harris, N. & Rowland, F.S. 1988. *Unpublished Data Analysis*. Described briefly in testimony to US Senate by F.S. Rowland, October 22, 1987.

Hofmann, D.J., Harder, J.W., Rolf, S.R. & Rosen, J.M. 1987. Balloon borne observations of the development and vertical structure of the Antarctic ozone hole in 1986. *Nature* (London) 326: 59–62.

IMOS. 1975. *Fluorocarbons and the Environment: Report of Federal Task Force on Inadvertent Modification of the Stratosphere*. Council on Environmental Quality. Federal Council for Science and Technology, Washington, DC, USA: 150 pp.

Isaksen, I.S.A. & Stordal, F. 1986. Ozone perturbations by enhanced levels of CFCs, N_2O and CH_4: A two-dimensional diabatic circulation study including uncertainty estimates. *J. Geophys. Res.* 91: 5249–63.

Jesson, J.P. 1982. Halocarbons. pp. 29–63. in *Stratospheric Ozone and Man*, Vol. II (Eds F.A. Bower & R.B. Ward). CRC Press, Boca Raton, Florida, USA: [xiii+] 263 pp.

Jesson, J.P. & Glasgow, L.C. 1977. The fluorocarbon-ozone theory, II: Tropospheric lifetimes—an estimate of the tropospheric lifetime of CCl_3F. *Atmos. Environ.* 11: 499–510.

Jones, P.D., Wigley, T.M.L. & Wright, P.B. 1984. Global temperature variations between 1861 and 1984. *Nature* (London) 322: 430–4.

Keeling, C.D., Carter, A.F. & Mook, W.G. 1984. Seasonal, latitudinal and secular variations in the abundance and isotope ratios of atmospheric CO_2. *J. Geophys. Res.* 89: 4615–28.

Keys, J.S. & Johnston, P.V. 1986. Stratospheric NO_2 and O_3 in Antarctica: Dynamic and chemically controlled variations. *Geophys. Res. Lett.* 13: 1260–3.

London, J. & Angell, J.K. 1982. The observed distribution of ozone and its variations. pp. 7–42 in *Stratospheric Ozone and Man*, Vol.1 (Eds F.A. Bower & R.B. Ward). CRC Press, Boca Raton, Florida, USA: [xiii+] 217 pp.

McCormick, M.P., Steele, H.M., Hamill, P., Chu, W.P. & Swissler, T.J. 1982. Polar stratospheric cloud sightings by SAM II. *J. Atmos. Sci.* 3: 1387–97.

McElroy, M.B., Salawitch, R.J., Wofsy, S.C. & Logan, J.A. 1986. Reductions of Antarctic ozone due to synergistic interactions of chlorine and bromine. *Nature* (London) 321: 759–62.

McKenzie, R.L. & Johnston, P.V. 1984. Springtime stratospheric NO_2 in Antarctica. *Geophys. Res. Lett.* 11: 73–5.

Mahlman, J.D.& Fels, S.B. 1986. Antarctic ozone decreases: A dynamical cause? *Geophys. Res. Lett.* 13: 1316–9.

Makide, Y. & Rowland, F.S. 1982. Tropospheric concentrations of methylchloroform, CH_3CCl_3, in January 1978, and estimates of the atmospheric residence-times for hydrohalocarbons. *Proc. Nat. Acad. Sci. US* 78: 1366–70.

Molina, L.T. & Molina, M.J. 1987. Production of Cl_2O_2 from the self-reaction of the ClO radical. *J. Phys. Chem.* 91: 433–6.

Molina, L.T., Molina, M.J. & Rowland, F.S. 1982. Ultraviolet absorption cross-sections for several brominated methanes and ethanes of atmospheric interest. *J. Phys. Chem.* 86: 2672–6.

Molina, M.J. & Rowland, F.S. 1974. Stratospheric sink for chlorofluoromethanes: chlorine atom catalysed destruction of ozone. *Nature* (London) 249: 810–2.

Molina, M.J., Tso, T.-L., Molina, L.T. & Wang, F.C.-Y. 1987. Antarctic stratospheric chemistry of chlorine nitrate, hydrogen chloride and ice: release of active chlorine. *Science* 238: 1253–7.

Mount, G.H., Sanders, R.W., Schmeltekopf, A.L. & Solomon, S. 1987. Visible spectroscopy at McMurdo station, Antarctica, 1: Overview and daily variations of NO_2 and O_3, Austral spring, 1986. *J. Geophys. Res.*, 92: 8320–8.

NAS. 1976. *Halocarbons: Environmental Effects of Chlorofluoromethane Release*. Committee on Impacts of Stratospheric Change, ix + 125 pp.; *Halocarbons: Effects on Stratospheric Ozone*, Panel on Atmospheric Chemistry, xv + 352 pp. National Academy of Sciences, Washington, DC, USA.

NAS. 1979. *Stratospheric Ozone Depletion by Halocarbons: Chemistry and Transport*. Panel on Stratospheric Chemistry and Transport, xi + 238 pp.; *Protection Against Depletion of Stratospheric Ozone by Chlorofluorocarbons*, Committee on Impacts of Stratospheric Change and Committee on Alternatives for the Reduction of Chlorofluoro-carbon Emissions, xvii + 392 pp. National Academy of Sciences, Washington, DC, USA.

NAS. 1982. *Causes and Effects of Stratospheric Ozone Reduction: An Update*. Committee on Chemistry and Physics of Ozone Depletion and Committee on Biological Effects of Increased Solar Ultraviolet Radiation, xi + 339 pp. National Academy of Sciences, Washington, DC, USA.

NAS. 1984. *Causes and Effects of Changes in Stratospheric Ozone*, xi + 254 pp., National Academy of Sciences, Washington, DC, USA.

Prinn, R., Cunnold, D., Rasmussen, R., Simmonds, P., Alyea, F., Crawford, A., Fraser, P. & Rosen, R. 1987. Atmospheric trends in methylchloroform and the global average for the hydroxyl radical. *Science* 238: 945–50.

Ramanathan, V. 1975. Greenhouse effect due to chlorofluorocarbons: Climatic implications. *Science* 190: 50–2.

Ramanathan, V., Cicerone, R.J., Singh, H.B. & Kiehl, J.T. 1985. Trace gas trends and their potential role in climate change. *J. Geophys. Res.* 90: 5547–66.

Rowland, F.S. 1986. Chlorofluorocarbons and the Antarctic 'ozone hole.' *Environmental Conservation* 13: 193–4, fig.

Rowland, R.S. & Molina, M.J. 1975. Chlorofluoromethanes in the environment. *Rev. Geophys. Space Phys.* 13: 1–35.

Rowland, F.S. & Molina, M.J. 1976. Estimated future atmospheric concentrations of CCl_3F (Fluorocarbon-11) for various hypothetical tropospheric removal rates. *J. Phys. Chem.* 80: 2049–56.

Rowland, F.S. & Sato, H. 1984. *[Presentation at the International Meeting on Current Issues in Our Understanding of the Stratosphere and the Future of the Ozone Layer.]* Feldafing, West Germany, June 11–16, 1984.

Rowland, F.S., Sato, H., Khwaja, J. & Elliott, S.M. 1986. The hydrolysis of chlorine nitrate, and its possible atmospheric significance. *J. Phys. Chem.* 90: 1985–8.

Solomon, P., Connor, B., Zafra, R.L. de, Parrish, A., Barrett, J. & Jaramillo, M. 1987. High Concentrations of chlorine monoxide at low altitudes in the Antarctic spring stratosphere: Secular variation. *Nature* (London) 328: 411–3.

Solomon, S., Garcia, R.R., Rowland, F.S. & Wuebbles, D.J. 1986. On the depletion of Antarctic ozone. *Nature* (London) 321: 755–8.

Solomon, S., Mount, G., Sanders, R.W. & Schmeltekopf, A. 1987. Visible spectroscopy at McMurdo station, Antarctica, 2: Observations of OClO. *J. Geophys. Res.* 92: 8329–38.

Stolarski, R.S. & Cicerone, R.J. 1974. Stratospheric chlorine: A possible sink for ozone. *Can. J. Chem.* 52: 1610–5.

Stolarski, R.S. & Schoeberl, M.R. 1986. Further interpretation of satellite measurements of antarctic total ozone. *Geophys. Res. Lett.* 13: 1210–2.

Stolarski, R.S., Krueger A.J., Schoeberl, MR., McPeters, R.D., Newman, P.A. & Alpert, J.C. 1986. Nimbus-7 satellite measurements of the springtime Antarctic ozone decrease. *Nature* (London) 322: 808–11.

Tolba, M.K. 1987. Guest comment: The ozone agreement—and beyond. *Environmental Conservation* 14(4): 287–90.

Tolbert, M.A., Rossi, M.J., Malhotra, R. & Golden, D.M. 1987. Reaction of chlorine nitrate with hydrogen chloride and water at Antarctic stratospheric temperatures. *Science* 238: 1258–60.

Tung, K.K., Ko, M.K.W., Rogriguez, J.M. & Sze, N.D. 1986. Are Antarctic ozone variations a manifestation of dynamics or chemistry? *Nature* (London) 333: 811–4.

Watson, R.T. & Albritton, D. 1987. Press conference held on September 30, 1987 at Greenbelt, Maryland, to discuss the results from the 1987 ground-based expedition to McMurdo, Antarctica, and the 1987 aircraft expedition flying over Antarctica from Punta Arenas, Chile. No formal papers have been published by the scientific investigators themselves.

WMO-NASA. 1986. *Atmospheric Ozone 1985; A Statement of Our Understanding of the Processes Controlling its Present Distribution and Change.* World Meteorological Organization, Geneva, Switzerland: Report No. 16, 3 volumes, 1150 pp. all with prelim. and suppl. (reference list) pages and unnumbered illustrations.

Wuebbles, D.J., Connell, P. & Rowland, F.S. 1984. *[Presentation at the International Meeting on Current Issues in Our Understanding of the Stratosphere and the Future of the Ozone Layer.]* Feldafing, West Germany, June 11–16, 1984.

Zafra, R.L. de, Jaramillo, R.L., Parrish, A., Solomon, P., Connor, B & Barrett, J. 1987. High concentrations of chlorine monoxide at low altitudes in the Antarctic spring stratosphere: diurnal variation. *Nature* (London) 328: 408–11.

EIGHT

STRATOSPHERIC OZONE: SCIENCE & POLICY

S. Fred Singer

The discovery in 1985 of a "hole" in the atmospheric ozone layer near the South Pole has focused worldwide interest on what is happening to ozone, a minor yet vitally important constituent of the earth's atmosphere. It has also raised concern about possible depletion of ozone on a global scale and resulting health effects, particularly an increase in the skin cancer rate.

An issue dating back to the 1970 controversy about the effects of supersonic transport aircraft has resurfaced: To what extent are human activities producing ozone changes? In particular, the emission into the atmosphere of chlorofluorocarbons (CFCs) has raised fears of "destroying" the ozone layer and has led to demands that the production of these chemicals be curtailed or even abolished (See Rowland, 1989).

The case against CFCs is based on a plausible but still incomplete theory, whose predictions are in a state of flux; on observations of the Antarctic ozone hole (AOH), whose future is uncertain; and on an assertion, as yet unverified, that ozone has been declining on a global scale.

Theory

As understanding of the complicated ozone photochemistry has improved—it involves over 150 simultaneous reactions—estimates of the effects of the CFCs have fluctuated. In recent years, the calculated effects have diminished; for example, the National Academy of Sciences in 1979 calculated an 18 percent ozone depletion due to CFCs, a 9 percent effect in 1982, and only a 3 percent effect in their 1984 report. But more recent calculations suggest between 5 and 7 percent ozone depletion for a CFC scenario of continued production. We have also learned that other polluting gases released by human activities, such as NO_x, methane, and carbon dioxide, all tend to diminish the CFC effects—an interesting and entirely fortuitous circumstance.

A major criticism of the calculations is that they do not consider the input of chlorine and bromine from various natural sources—that would dilute the effects of CFCs: volcanoes (Symonds, Rose & Reed, 1988), and oceanic biota (Manley & Dastoor, 1987; Singh, Salas & Stiles, 1983) and salt particles (Finlayson-Pitts, Ezell & Pitts, 1989).

AOH

The Antarctic ozone hole was not predicted by current theory. It was discovered by British scientists operating an observing station on the Antarctic continent. After they reported their findings in 1985, NASA scientists searching their records of satellite data confirmed the effect. Indeed, the "hole" has been around since the mid-1970s, and getting larger every year, reaching a depletion of about 50 percent (Rowland, 1989) Concern has centered on the rapid increase and the fear that it may grow to engulf the whole globe.

(Of course, the "hole" isn't really a hole at all, but a temporary thinning in the stratospheric ozone layer. This phenomenon takes place for a few weeks, around October, in the region of the Antarctic. As far as we can tell, there have been no long-term changes in ozone elsewhere, although the evidence is not conclusive.)

In spite of recent discoveries related to the mechanism of the AOH, we do not yet have a sufficient scientific base to answer the important policy questions: Is the AOH a completely new phenomenon? Is it produced by human activities? What is its likely future behavior? Will it persist, grow, or weaken? And what can and should be done about it?

There is little doubt that chlorine chemistry is the *immediate* cause of the seasonal (October) ozone decrease at around 18km in the southern polar regions—rather than purely meteorological effects based on dynamics or direct solar influences related to the solar cycle (Rowland, 1989). It is also probable that the major source of the chlorine is man-made chlorofluorocarbons [CFC]—although no precise estimate exists of the chlorine contributed by various natural sources (see above).

Yet how does one explain the sudden onset and rapid growth of the AOH phenomenon? Starting from essentially zero in the mid-1970s, the thinning reached, within a few years, about 50 percent of the vertical ozone column—and essentially saturation in the lower stratosphere. This rapid change presents an important clue. The CFC content of the atmosphere has not risen quite so rapidly, nor should one expect any trigger effect related to the chlorine concentration. The research results suggest that, in addition to the chlorine, ozone destruction requires the presence of ice particles, "polar stratospheric clouds," that can form in the coldest part of the earth's atmosphere, the lower Antarctic stratosphere (See Rowland, 1989). But it is highly unlikely that the water vapor content could have increased so suddenly within a few years' time—although increased emissions of methane should lead to increased injection of water vapor into the stratosphere (Singer, 1971). Recent measurements have confirmed and extended this hypothesis (Blake & Rowland, 1988).

This line of reasoning leads me to propose that the *trigger* for the AOH has been a gradual cooling of the stratosphere, which took the temperature below the freezing point; this cooling could have taken place as part of a general climate fluctuation of the earth (Singer, 1988). And indeed, there has been an unusual surface temperature increase since about 1975; under some theoretical models of climate change, such a surface warming should be accompanied by a cooling of the upper atmosphere. (Ramanathan, 1988).

If this hypothesis is borne out by appropriate measurements, then the AOH should disappear—or at least become less pronounced—if the stratosphere warms again, perhaps in conjunction with a cooling of the earth's surface. Conversely, a further cooling of the stratosphere could induce an Arctic ozone hole and a larger Antarctic hole.

The policy implication is that the AOH would not be much affected by further slow increases of atmospheric CFC, nor could it be removed if the CFC concentration were to decrease. In other words, the AOH phenomenon should be reasonably insensitive to stratospheric chlorine concentration, somewhat more sensitive to stratospheric water vapor concentration, and extremely dependent on the exact value of the temperature minimum in the lower Antarctic stratosphere. (Singer, 1988.)

Global Ozone Decrease?

In March 1988, the Ozone Trends Panel of NASA, after a massive re-analysis of data from ground stations and satellites, announced the existence of a declining trend in northern hemisphere ozone over the period of 1970 to 1986. A press release was issued, but as of this date (June 1989) the underlying analysis has not been released for independent examination. A news story in *Science* (Kerr, 1988) quotes an average decline of -0.2 percent per year, which is greater than predicted from the current CFC-ozone theory. Since the decline is "worse than expected," the surprising, and not quite logical, conclusion was reached that CFCs must now be phased out completely. (A more logical conclusion might have been that the analysis, or the theory, or possibly both, are incorrect.)

While the NASA panel's report is not yet available, a parallel report from the Center for Applied Mathematics of Allied-Signal, Inc. was distributed at a UNEP Ozone Science Meeting at the Hague in October 1988. The Allied study (Bishop, Hill & Marcucci, 1988) carries out a sophisticated regression analysis of the same data as the NASA study. After correcting for many natural variations, including the 11-year solar cycle, they derive a decline of -1.9 percent over the period 1970 to 1986—which is only 1-1/2 solar cycles. But their sensitivity analyses show that the result depends on the time interval under consideration, suggesting therefore that the solar cycle correction was not adequate and that the decline is at least partly an artifact of the analysis. (Singer, 1989).

(I have also suggested that the decline rate will diminish, or even disappear, if the analysis is extended to include data from the years 1987 to 1990—as we reach a solar cycle maximum. A more definitive answer about the reality of global ozone depletion should therefore be forthcoming.)

The Skin Cancer Issue

The possible connection of skin cancer with stratospheric ozone first gained public attention during the SST controversy in 1970. Certain forms of common skin tumors, basal cell and squamous cell carcinomas, have a much greater incidence at lower latitudes. They are more than twice as common in south Texas than in Minnesota—presumably because of the greater UV exposure in Texas due to the steeper sun angle there. With ozone weakened, more UV would reach the earth's surface everywhere, causing more tumors. It was this emotional skin cancer issue, more than anything else, that persuaded Congress to cancel the SST program. It is ironic that on the basis of current theory, SSTs flying in the lower stratosphere are believed to enhance ozone there rather than destroy it. (Kinnison & Wuebbles, 1988).

Further relevant information adds important perspectives to the skin cancer discussion.

1. For example, the increase of UV-B radiation that is feared to result from the thinning of ozone does in fact occur simply as a result of moving closer to the equator. Thus ozone decreases of the order of 5 percent would correspond to a move of less than 100 miles.

2. Also omitted is the observation that UV-B has *decreased* at all 8 locations where measured during the last 11 years. (Scotto *et al.*, 1988).

3. Melanoma is said to have increased at an "alarming" rate from 1974 to 1983. Actually, its incidence has increased some 700 percent since 1935, when proper records were first taken. Why?

4. Non-melanoma skin tumors [nearly 100 percent curable] do show an increase towards lower latitudes—the only one of the effects to show a latitude dependence. But it is not sure that all of the increase is due to UV-B intensity, as generally assumed when predicting future increases due to stratospheric ozone depletion. Much of the increase probably comes from greater exposure and other lifestyle factors at the warmer locations.

Summary

The science of stratospheric ozone is at an interesting crossroads. The CFC-theory is not yet good enough to explain the observations; and the observations are not yet good enough to confirm the theory. The policy question is whether drastic worldwide controls should be instituted immediately or whether one should wait for a better scientific understanding.

References

Blake, D.R., and F.S. Rowland. 1988. Worldwide increase in tropospheric methane. *Science* 239: 1129.

Bishop, L., W.J. Hill, and M.A. Marcucci. "An Analysis of the NASA Ozone Trends Panel Dobson Total Ozone Data over the Northern Hemisphere." Ctr. for Appl. Math., Allied-Signal, Inc., Aug. 3, 1988.

Finlayson-Pitts B.J., M.J. Ezell, and J.N. Pitts, Jr. 1989. "Formation of chemically active chlorine compounds by reactions of atmospheric NaCl particles with gaseous NO and ClONO." *Nature* 337: 241.

Kerr, R.A., "Research News." *Science*, March 25, 1988.

Kinnison, D.E. and D.J. Wuebbles, "A Study of the Sensitivity of Stratospheric Ozone to Hypersonic Aircraft Emissions." Livermore National Lab. UCRL-98314 (preprint), Sept. 1988.

Manley, S.L., M.N. Dastoor. 1987. Methyl Halide (CH_3X) Production from the Giant Kelp, *Macrocystis*, and Estimates of Global CH_3X Production by Kelp. *Limnol Ocean.* 32(3):709-715.

Ramanathan, V. 1988. The greenhouse theory of climate change: A test by an inadvertent global experiment. *Science* 240: 293.

Rowland, F.S. 1989. This volume.

Scotto, J. *et al.* 1988. "Biologically effective UV radiation: Surface Measurements in the US, 1974 to 1985." *Science* 239: 762.

Singer, S.F. 1971. Stratospheric water vapor increase due to human activities. *Nature* 223: 543.

Singer, S.F. 1988. "Does the Antarctic Ozone Hole have a future?" *Eos* 69, No. 47 (Nov. 22)

Singer, S.F. "What is Causing Global Ozone Depletion." Submitted to *Nature*, May 1989.

Singh, H.B., L.J. Salas, R.E. Stiles. 1983. Methyl Halides in and over the Eastern Pacific (40N-32S). *J. Geophys. Res.* 88:3684-3690.

Symonds, R.B., W.I. Rose, and M.H. Reed. 1988. "Contributions of Cl- and F-bearing gases in the atmosphere by volcanoes." *Nature* 334:415.

NINE

COMMENTS on CHLOROFLUOROCARBONS and STRATOSPHERIC OZONE

Andrew A. Lacis

We have analyzed the radiative impact on the equilibrium surface temperature of the earth due to changes in the vertical distribution of ozone with the help of a 1-D radiative/convective equilibrium model similar to that previously used by Lacis *et al.* (1981). The model uses 33 layers to cover the atmosphere from ground to 90 km. We assume a 50% cloud cover comprised of low, middle, and high clouds, and include global average amounts of H_2O, CO_2, O_3, aerosols, and other minor trace gases. We start with an equilibrium reference model. We then add small increments of ozone at specified heights in the atmosphere and compute a new equilibrium surface temperature with a time marching procedure. For clarity of comparison, we allow *no* feedbacks to operate. Thus, the computed surface temperature differences, ΔT_0 correspond directly to the radiative forcing by the incremental ozone at the specified height.

In the real world, strong positive feedbacks exist which magnify the initial radiative perturbation by a factor f. Thus the equilibrium surface temperature change realized in the presence of feedbacks is

$$(1) \qquad \Delta T_S = f\Delta T_0$$

where ΔT_0 is the equilibrium temperature change that would be attained without feedbacks. The principal feedback processes due to water vapor, clouds and snow/ice yield a feedback factor f = 3.3, based on feedback analysis with the GISS 3-D climate model (Hansen *et al.* 1983) which has a surface temperature sensitivity of 4°C for doubled CO_2.

It has been known for some time that the surface temperature is less sensitive to stratospheric ozone than it is to tropospheric ozone (Ramanathan *et al.* 1976). Ramanathan *et al.* attributed the sensitivity difference to pressure broadening of the 9.6 μm ozone band which would make tropospheric ozone a more efficient absorber of thermal radiation. To examine more closely the nature of the surface temperature sensitivity to vertical ozone changes, we made separate computations for the radiative effects in the UV, visible and IR spectral regions.

To help understand the surface temperature sensitivity results in Figure 9-1, we note that pressure broadening does indeed account for a reduction in absorption efficiency by about a factor of four between the ground and the stratosphere. However, the strength of the greenhouse effect is proportional to the difference in temperature of the radiation that the ozone absorbs (ground temperature) and the radiation that the ozone emits (local atmospheric temperature). Thus ozone near the ground may be an efficient absorber, but since it re-radiates at a temperature that is nearly the same as the ground temperature, its greenhouse strength is very small. Accordingly, ozone added near the tropopause is most efficient for increasing the surface temperature. The greenhouse contribution of the 9.6 μm ozone band remains positive but diminishes in strength as the ozone increment is moved upward in the stratosphere partly because of reduced pressure broadening, but also because the stratospheric temperature increases with height, forcing the ozone to radiate at a higher temperature.

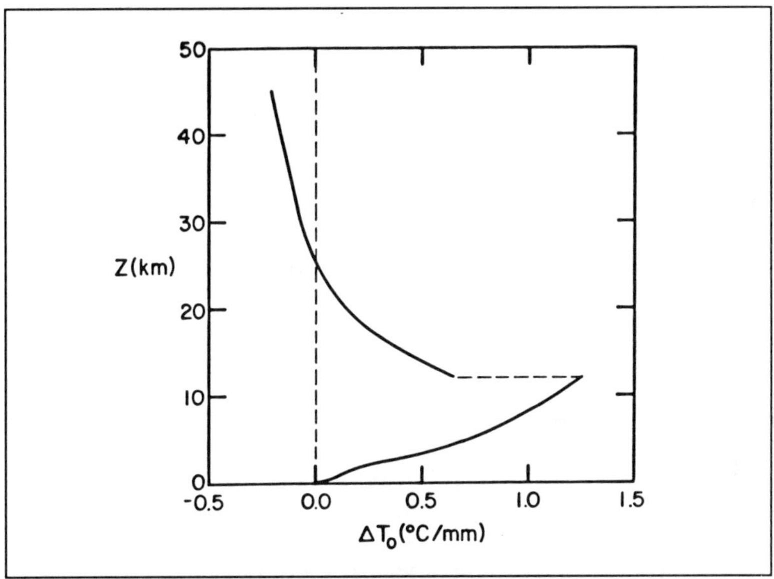

Figure 9-1: Sensitivity of surface temperature to changes in vertical ozone distribution. The solid line shows the change in equilibrium surface temperature, ΔT_0, that would be produced if 1.0 mm (STP) of ozone were added to the atmosphere at height Z. No feedbacks are included in ΔT_0.

The discontinuity at 12 km in the surface temperature sensitivity is an artifact of 1-D models. 1-D models operate with a fixed critical lapse rate which has the effect of binding together the tropospheric layers. Thus, to approach equilibrium, a layer in the stratosphere can warm individually, causing a small increase in surface temperature due to downward radiation, but a layer in the troposphere is forced to share its energy directly with the whole troposphere and ground. Non-radiative energy transport occurs in 3-D models as well, but there the tropopause is a function of latitude and time, so that in 3-D models this discontinuity would be smoothed out.

A negative contribution to surface temperature arises from the UV and visible spectral regions when ozone is added high in the stratosphere above ~25km. This is because ozone added in the stratosphere robs the troposphere and ground of some solar energy which would go directly into warming the ground.

The extra solar energy that the added ozone absorbs in the stratosphere goes to heat the local stratospheric layer. The stratospheric layer is then able to re-radiate some of the absorbed solar energy back to the ground as thermal radiation, but at reduced efficiency, because it must radiate to space at a higher temperature. When the additional ozone is placed in the troposphere, the solar radiation that is absorbed goes directly to heating the surface. Moreover, the ozone added in the troposphere retrieves some of the radiation normally absorbed in the stratosphere by reducing the reflected component from clouds and Rayleigh scattering. Unlike the 9.6 μm IR band, absorption in the UV and visible spectral regions is largely independent of pressure (Inn and Tanaka 1953).

Figure 9-2 shows the sensitivity of the surface temperature to changes in total ozone amount without changing the vertical profile. As before, no feedbacks are allowed to operate. To obtain equilibrium temperature changes with feedbacks, ΔT_0 should be multiplied by 3.3 as in equation (1). We note that, unlike CO_2, the surface temperature changes are fairly linear since the absorption in the visible and at 9.6 μm is relatively weak. The hook in the curve at about 0.1 shows where the UV absorption finally becomes unsaturated, allowing more solar radiation to be absorbed in the troposphere and also in allowing the temperature of the stratosphere to become cold. This acts to warm the surface temperature, producing the upswing in the ΔT_0 curve as ozone is reduced below 10% of its normal amount.

The projected changes in global equilibrium surface temperature due to anthropogenic activity are summarized in Figure 9-3. The lower curve is the surface temperature change due to stratospheric ozone reductions caused by increasing CFC's as computed by Wuebbles *et al.* (1983). The upper curve shows that the impact of NO_x, N_2O, and CO_2 causing an increase in tropospheric ozone is climatologically far more important than the effect of CFCs alone. Again, the computed change in surface temperature due to the vertical redistribution of ozone does not include feedback contributions.

The computed greenhouse warming for the reported increases of CH_4, N_2O, CCl_2F_2, and CCl_3F for the decade of the 1970s was found to be about 70% as large as that due to the

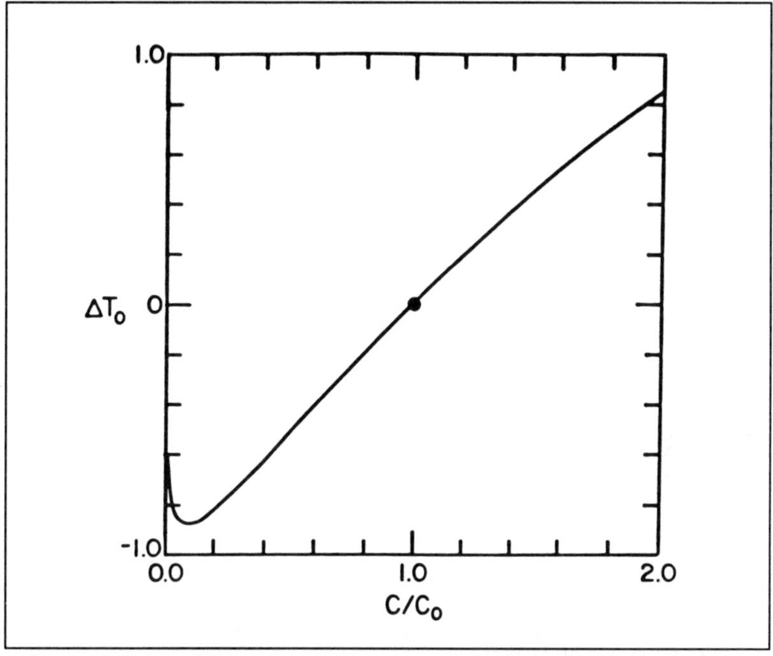

Figure 9-2: Sensitivity of surface temperature to changes in total ozone amount without change in vertical distribution. The dot at the center refers to a mid-latitude ozone distribution with a column amount of 3.43 mm (STP). ΔT_0 is the equilibrium surface temperature change for a fractional change of total ozone amount. No feedback effects are included in ΔT_0.

CO_2 increase during the same decade (Lacis *et al.* 1981). Referenced to a climate sensitivity of 4°C for doubled CO_2, the equilibrium surface temperature changes gained during the 1970s was 0.19°C due to CO_2 and 0.13°C due to the other trace gases combined.

Based on the 1-D photochemical model results of Wuebbles *et al.* (1983), the changes in vertical ozone distribution predict an equilibrium warming of about 0.08 °C during the 1970s and a 1.0 °C warming during the 1980s (see Figure 9-3). Although the computed changes in total column ozone amount are small, it is the predicted oxone increases in the upper troposphere and lower stratosphere that produce the bulk of the warming effect.

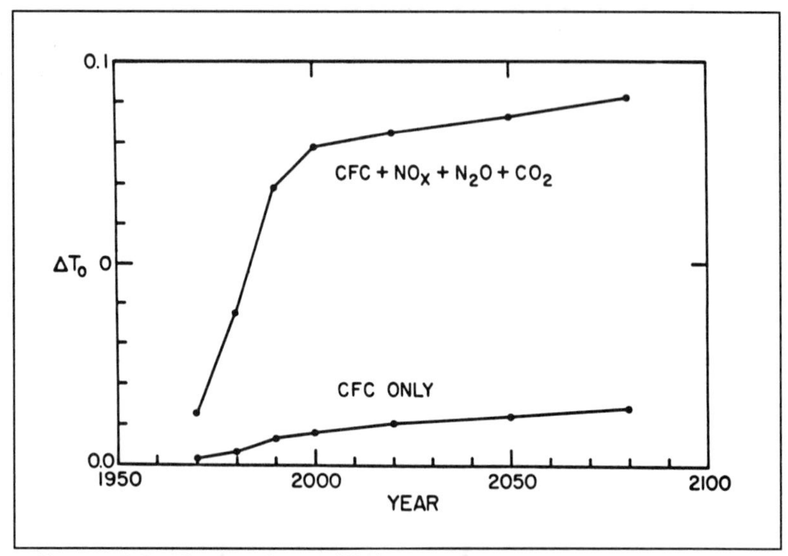

Figure 9-3: Changes in equilibrium surface temperature caused by CFC-induced changes in vertical ozone distribution. ΔT_0 is the cumulative change in equilibrium surface temperature (excluding feedback effects) due to changes in the vertical distribution of ozone computed by Wuebbles *et al.* (1983).

Coda

The large natural variabilityof atmospheric ozone, combined with calibration problems associated with the various measurement techniques, has made detection of long term trends difficult. Analysis of available Umkehr and ozonesonde data indicate that ozone decreased during the 1970s by about 3% in the 40-50 km region (Reinsel *et al.*, 1984) and by 5% in the 10-20 km region (Tiao *et al.*, 1986). Tiao *et al.* also found increases in surface and lower tropospheric ozone of 5-10% during the 1970s. Unfortunately, the ozone trend measurements are available only for the northern mid-latitudes. Nevertheless, on the strength of the ozone decrease in the 10-20 km region, the measured trends produce a regional surface cooling equal in magnitude to about half of the warming contributed by CO_2 increases for the same time period (Lacis *et al.*, 1989).

The disagreement between 1-D photochemical model results and the observed trends is due to the inability of 1-D models to capture the observed reduction in ozone in the lower stratosphere and upper troposphere. Prediction of 2-D photochemical models suggest that ozone changes in the lower stratosphere depend strongly on latitude with ozone decreases at middle and high latitudes, but with increases in the tropics (WMO, 1986). As partial corroboration of the 2-D model predictions, the ozone decrease in the lower stratosphere produces a local cooling of 0.3-0.4 °C in the 15-20 km region (Lacis *et al.*, 1989) that is consistent with the cooling trend observed in the lower stratosphere by Angell (1986) and the warming found by Labitzke *et al.* (1986) at 10°N at 24 km.

The role of ozone as an important contributor to the greenhouse forcing by atmospheric trace gases is clearly established. Its contribution is strongly latitude-dependent, with cooling at middle to high latitudes and a possible warming effect in the tropics. However, large uncertainties still exist in both the magnitude and latitude dependence of the climate forcing by ozone because of persistent measurement and modeling difficulties.

References

Angell, J.K. 1986. *Mon. Weather Rev.* : 114, 1922.

Hansen, F., G. Russell, D. Rind, P. Stone, A. Lacis, S. Lebedeff, R. Ruedy, and L. Travis. 1983. *Mon. Weather Rev.* 111: 609.

Inn, E.C.Y. and Y. Tanaka. 1953. *J. Opt. Soc. Amer.* 43: 870.

Labitzke, K., G. Brasseur, B. Naujokat, and A. De Rudder. 1986. *Geophys. Res. Lett.*: 13, 52.

Lacis, A., J. Hansen. P. Lee, T. Mitchell, and S. Lebedeff. 1981. *Geophys. Res. Letters* 8: 1035.

Lacis, A.A., D.J. Wuebbles, and J.A. Logan. 1989. *J. Geophys. Res.*, (in press).

Ramanathan, V., L.B. Callis, and R.E. Boughner. 1976. *J. Atmos. Sci.* 33: 1092.

Reinsel, G.C., G.C. Tiao, J.J. DeLuisi, C.L. Mateer, A.J. Miller, and J.E. Frederick. 1984. *J. Geophys. Res.*: 89, 4833.

Tiao, G.C., G.C. Reinsel, J.H. Pedrick, G.M. Allenby, C.L. Mateer, A.J. Miller, and J.J. DeLuisi. 1986. *J. Geophys. Res.* 91: 13121.

Wuebbles, D.J., F.M. Luther, and J.E. Penner. 1983. *J. Geophys. Res.* 88: 1444.

World Meteorological Organization (WMO). 1986. *Atmospheric Ozone 1985, Volume III, WMO Global Ozone Research and Monitoring Project - Report No. 16*, WMO, Geneva: 1095.

PART THREE:

HYDROSPHERE

TEN

ACID RAIN

Kenneth Mellanby

The expression "acid rain" has in recent years aroused the interest of the media, and we have articles in the press and programs on television purporting to explain the subject. Much of the public, in Europe and in America, believe that this is a new and a dangerous phenomenon. They believe that it is caused by industry, particularly the electrical industry, emitting large amounts of sulfur into the atmosphere, with the result that the rain is made acid and that this affects trees, fresh waters, and the fish population. They believe that the problem could easily be solved by the control of the pollution from electric power stations, and that this control is not exercised because of the greed or the lack of concern of the operators. Unfortunately, this is a gross oversimplification of the situation. There is indeed a problem, but its causes are complex, and we do not always know how best to try to solve it.

Many of those working in this field think we should stop talking about acid rain. We should really be considering the effects of several types of air pollution which may often act as gases and not through the effects of precipitation. This chapter will deal with all the ways in which the emissions from the burning of fossil fuels may affect the environment.

The production of heat and energy has always produced pollution in the form of smoke particulates and toxic gases of which sulfur dioxide has been the main culprit. Damage in urban areas has been known for hundreds of years. Plants were killed or damaged, mainly by the sulfur dioxide, which also rotted buildings and corroded metals. Human health was affected, probably by a combination of smoke and sulfur dioxide. It was the polluted air which did the damage, though the rain was also dirty—and more than a hundred years ago in 1872 R.A. Smith, the first British alkali inspector (that is, the first government official appointed to control air pollution) published a book on acid rain. In this instance he really meant acid rain, for he collected it in various localities and measured its acidity.

In Britain, and in most Western countries, this gross pollution has been controlled or at least reduced. Smoke is seldom emitted in large amounts, and the famous London "pea-soup" fogs are a thing of the past. The improvement is largely caused by the reduction in the use of raw coal. The use of natural gas, cleaned and de-sulfurized before distribution, has reduced sulfur dioxide output. Oil, some of which is low in sulfur, has helped. And electric energy, even when produced from coal, has been generated at a distance so that most of the urban consumers escape its worst environmental effects. Also nuclear power, our cleanest form of energy, is increasingly important in safeguarding the environment.

The decrease in emission is considerable, but a great deal of sulfur dioxide is still produced. In Britain this reached a maximum of some six million tons a year in 1970. The output has since been substantially reduced, to less than four million tons. This decrease is important, as will be seen when we look at the possible harmful effects of the gas.

Figure 10-1 illustrates diagrammatically the fate of sulfur dioxide emissions from the urban and industrial source on the left of the picture. Some of the SO_2, particularly from domestic houses and vehicles, remains in a fairly concentrated form near to its source. Here part of it is deposited (dry deposition) and can cause damage to plants and buildings. This accounts for the absence of lichens, which are very sensitive to SO_2, from towns and the vicinity of factories and power stations with low chimneys. With high chimneys the gas mixes into the atmosphere at a higher level, so local deposition is much reduced. Further from the source the pollutant is mixed more thoroughly into the atmosphere, so the concentration is lower. As a result the amount of dry deposition, and the damage it causes, decreases, eventually becoming negligible.

However, as the air is blown away from the pollution source, chemical changes take place, turning SO_2 to sulfate and sulfuric acid. These substances are readily removed by precipitation and fall to the ground as wet deposition or acid rain. Rain near to the source will be mainly contaminated with small amounts of the untransformed pollutants recently emitted.

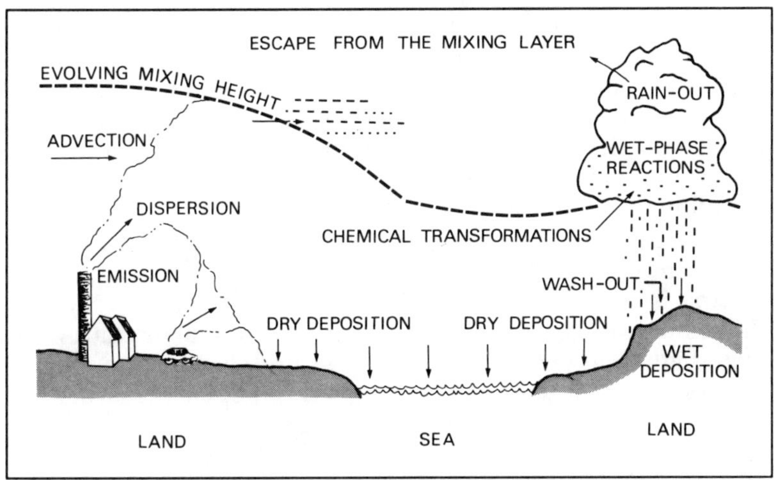

Figure 10-1: The dispersion and disposition of atmospheric sulfur dioxide.

The discharge of an increasing proportion of the polluted flue gases from industry at high levels from tall chimneys has done much to reduce urban pollution. This has produced a spectacular decline in the ground level concentration of sulfur dioxide in urban areas. This has allowed sensitive plants to be grown in places where 50 years ago they could not have survived. A less welcome effect has been that the fungal disease of roses, black spot, has returned to city parks. Roses themselves are rather resistant to pollution, and so they benefitted from this pollution of the air.

The effect of the tall chimneys is that the sulfur dioxide and other toxic gases are quickly mixed with the air and are diluted down to levels which are much less phytotoxic. We are finding indications that this dilution, though important, may not be sufficient to prevent all damage to crops and other plants in a region near to the source of emission, but this is strictly a local problem, restricted to a hundred or perhaps two hundred miles. So the direct effects of gaseous pollution usually occur within the country which produces this pollution; it may not give rise to international problems.

There is no doubt that the policy of discharge through high chimneys is beneficial to those living near to the source, but there are often complaints that this is the cause of damage in more distant localities. There is, however, little truth in this allegation. It is true that the sulfur which would have damaged the vegetation near the source is not absorbed there, as would have happened with lower chimneys. This is important to those who would have suffered, but the amount of sulfur dioxide that would have been removed is very small—even with low chimneys most is in the mass of air above the ground, and so does not come into contact with the vegetation. The process of mixing is such that within 50 or 100 miles there is very little difference within the air mass whether the initial discharge was from high or low chimneys, and if this finds its way into the rain the results are little different. High chimneys are a local boon which does little harm to distant regions.

At one time we thought that this policy of "dilute and disperse" had completely solved the problem. As already mentioned, we now know that the dilution may be sufficient to prevent acute phytotoxicity, but that chronic effects may occur, caused not only by sulfur dioxide, but by other gases including oxides of nitrogen and ozone. We are finding that these gases, either alone or in combinations, may be more damaging than we previously imagined. Nevertheless, this is still a local problem; the gases are quite quickly diluted to harmless levels. Unfortunately, this is not the end of the matter. We now know that these very dilute gases, harmless in that form, may be transformed slowly into other chemicals which may be more easily washed out of the air by the rain, and which may then damage fresh waters, and possibly crops, trees, and the soil many hundreds of miles from their source. This damage is usually attributed to acid rain. In fact, of course, almost all rain, clean or polluted, is acid. Very clean rain may have a pH of 5.6, being in equilibrium with the carbon dioxide in the air. Even where pollution is minimal, a pH of 5 or lower is not uncommon.

I have already mentioned that very acid rain, caused by emissions from fuel burning and factories, has been recognized in Britain for more than a hundred years. However, this is not what is worrying the Canadians and Scandinavians, who complain of the effects of pollution produced in the United States or in Britain and borne by the wind to their territories. I believe that it may be useful, and may help to reduce confusion, if we distinguish between what I will call *primary acid rain* and *secondary acid rain*.

Primary acid rain is caused by the washout from the atmosphere by falling rain of substances as they are emitted in urban and industrial areas. It may contain many different chemicals, for instance, hydrochloric acid. It may be very acid—levels of pH 3 have been recorded. However, there is little evidence that this primary acid rain does much harm to plants or to buildings. It is the gases which do this.

It should be noted that rain may, at different times, contain very different amounts of pollutants. A shower after a long dry period will be particularly contaminated. The first rain to fall will generally contain more pollution than that which falls at the end of a rainstorm. This means that plants may be washed clean by this later, purer water. Primary acid rain may affect soils and fresh water in some cases, as all the pollution is bulked in these situations. It should again be noted that primary acid rain is essentially a local phenomenon. As we move away from the source of the emission, the rain generally becomes less and less acid.

Secondary acid rain is something quite different. It is produced when the oxides of sulfur and nitrogen have been transformed in the air to sulfuric and nitric acid, and when these are removed, by rainout and washout. The transformation to acid, particularly of sulfur, is a slow process, and generally takes several days in which the pollution may have travelled many hundreds of miles. Were there no transformation to acid, there would be no acid rain problem. The gases are much too dilute to do direct harm to the environment.

A great deal of research is going on regarding this production of acid. We know that the reaction is affected by temperature, sunlight, and the presence of other substances such as oxidants and hydrocarbons in the atmosphere. If we could control this transformation, perhaps by controlling some of the other substances which take part, then there would be little point in removing the sulfur dioxide from the gases emitted by power stations.

Secondary acid rain may be important because it contains the acid from such large volumes of air. The rain brings down onto the ground the acid in a column of air reaching up to many thousands of feet. Also, in mountainous areas, rain generated over a wide area may come down heavily in a much smaller area.

Secondary acid rain is much too dilute to have any direct phytotoxic effects, even on very sensitive species of plants like lichens. In fact we find a rich flora of leafy lichens in most areas where acid rain damage has been reported. This includes the Black Forest in Germany, Southern Norway, and various localities in Scotland. This also shows that in these places, gaseous sulfur dioxide levels are very low and are not the cause of the damage.

I believe that my division of primary from secondary acid rain is useful, and helps to explain what happens, but it is probably unduly simple. In many areas we will find a mixture of primary and secondary rain, and one may grade into the other. The important point is to realize that what happens hundreds of miles downwind is quite different from what happens near the source. My conception also explains why some workers have said that acid deposition is proportional to output, and others find no such close correlation. Near a pollution source, where dry deposition and primary acid rain are involved, deposition is likely to parallel emission, while at a great distance, where the rate of transformation of the pollutants may be a major factor, it is not surprising that no close correlation can be detected.

So far I have given the impression that sulfur dioxide is the main cause of acid rain. Fifty years ago the main phytotoxic pollutant arising from the use of fossil fuels was, in fact, sulfur dioxide. Today, motor vehicles and high-temperature boilers produce a growing quantity of oxides of nitrogen, which contribute to the production of oxidants, including ozone, in cities like Los Angeles. Ultimately oxides of nitrogen are the precursors of nitric acid. In what I have called secondary acid rain, only about two-thirds of the acid is sulfuric, the remainder being nitric. There seems to be a tendency for the nitric fraction to increase. The concern of the media with sulfur, as though it was the only cause of acid rain, is misleading.

Dry deposition of sulfur dioxide and other substances, at high doses, damages plants as already described, and different species show very different susceptibilities. However, dry deposition also contributes to the transfer of acids to the soil. There is some dry deposition on bare soil itself, but more to vegetation, and the greater the surface of the plants the larger the amount deposited. Thus trees pick up considerable quantities. These deposits may be washed off by the rain—a process known as "throughfall." Even where concentrations of gas are so low as to have no recognizable effects on the vegetation, the throughfall may add as much of the pollutants as the rain itself.

Secondary acid rain certainly has no direct harmful effect on plants, as it is so dilute. However, it may contribute to acidity of fresh water, and may affect plants via the soil, where it is deposited year after year, and the effect may thus be cumulative.

In my opinion there is little doubt that acid precipitation, the acidity being caused by gaseous emissions from burning fossil fuels, has contributed to an increase in acidity in many lakes and rivers in Scandinavia, Scotland, and North America. These changes have been most serious where the rocks are granitic, with low calcium levels in the soils, and poor buffering capacity in the fresh water. There has, so far, been little evidence of damage in well buffered waters in areas with sedimentary rock rich in calcium.

However, the situation is far from simple, and much more research will be needed before the exact relationship between output of pollutants and damage to fish and aquatic life is understood.

The output of sulfur dioxide fell by nearly 40 percent in Britain between 1970 and 1987. Recently, a significant effect has been observed in lakes in Scotland which had showed evidence of acidification. This has been reversed; evidence of this is based both on chemical analyses and the reappearance of species of diatoms which are particularly susceptible to acids. On the other hand, no such improvement appears to have been found in Norwegian lakes, which supports the view that emissions from Britain are not the major contributors to their acidification.

In countries with very cold winters, snow poses a particular problem. All the rain, and the pollution it contains, may be stored as snow for six months. When it thaws the acidic materials come out in the first fraction of the melt, which may thus be very concentrated. If the ground is still frozen, this water will run over the surface straight into the streams and have a drastic, if temporary, effect. At other times the rain has to pass through the soil before it reaches the stream, and may be modified as it does so.

This whole question of the increasing acidity in streams and lakes is not as simple as is often imagined. There are lakes which are naturally very acidic. Acidification has often occurred in the past quite independently of human-induced air pollution. Deforestation, the clearing of forest, the burning of vegetation, and so on all have effects which complicate the issue. Acidification, if it is caused by pollution, may be related to the output of sulfur, but other substances such as nitrogen also play a part. When we examine a series of lakes, though the most acidic generally are devoid of fish and the least acidic support good populations, we find many anomalies. Fish survive where they might be expected to have disappeared, and lakes whose water, at least as far as pH is concerned, would appear suitable may be fishless. Often it appears that it is not only the acidity, but the levels of metals like aluminum, which may be toxic, which are the controlling factors. Nevertheless, it is generally agreed that if we could more effectively control the output of air pollution throughout the world, it would be generally beneficial to fresh waters and to their fisheries.

The situation may be ameliorated, and possibly cured, by the use of lime. In Sweden, lime has been added to many fresh waters for some years, with promising results. In lakes where there is not a rapid replacement of the water, and in slow running streams, lime has proved an effective, if temporary, measure, but it needs to be reapplied at least once every year.

At Loch Fleet in Scotland the catchment area rather than the water has been treated with powdered limestone. The water in the loch has become much less acid, and it appears likely that the result will continue for at least ten years. Incidentally, in the past many areas of upland Britain which were grazed by sheep were regularly treated with lime; the costs were reduced by a government subsidy on lime. This has now been discontinued, and little liming now takes place. This has undoubtedly contributed to the problem of growing acidity in lakes in previously-limed areas.

Acid rain is also thought to be endangering the forest in many countries. Here we have a complicated problem. In Britain fifty years ago it was impossible to grow coniferous trees in many parts of the country, because of the high levels of sulfur dioxide. Today conifers are growing well on the Pennine hills, where previously they were killed. Similar effects of sulfur dioxide may still be observed in Eastern Europe. Here it is dry deposition of the toxic gas which was the cause of the damage.

It is difficult to obtain a clear picture of what is happening to trees in other areas, where sulfur dioxide levels are low. In Britain the Forestry Commission is unable to associate damage with levels of pollution. Their extensive surveys show that in all areas of woodland, among the majority of healthy trees, there are individuals with symptoms of disease, of damage from excessive frost in cold winters, and of old age. After a dry summer, some trees, especially beech trees, lose their leaves prematurely in autumn. However, in most cases the beech trees appear to be in the best of health during the next growing season, particularly if rain is plentiful. There is little doubt that acid rain as such (that is, secondary acid rain as described in this paper) is seldom if ever directly phytotoxic, to trees or to agricultural crops. However, it is possible that poorly buffered soils may be affected, and so, indirectly, trees growing in these soils.

Readers of many newpapers have been given the impression that the Black Forest in Germany has been almost completely devastated by "acid rain." I visited this area in September 1987 with a party of horticulturalists who were inspecting European gardens. They were very surprised to be able to recognise no obvious damage whatever in the Black Forest, which appeared to be as beautiful as it had ever been. In fact it is possible to find quite a lot of damage, particularly at high altitudes, and some conifers appear to be losing their needles earlier than is thought to be normal. This damage is certainly not caused by sulfur dioxide, as lichens are very abundant, and these are good indicators of low sulfur levels. The most likely cause of the damage is ozone, the end product of a reaction in which the output of oxides of nitrogen by automobiles is a starting point. In this case the drastic efforts of the German government to reduce the output of sulfur dioxide in that country will clearly have little effect—however desirable this reduction may be for other reasons.

It is often suggested that too little is being done to solve problems relating to air pollution and acid rain, and some European critics suggest that Britain is particularly backward in such actions. It is suggested that better pollution control is always worthwhile, whatever the cost. However, the situation is not always quite so simple. We do not always know what are the most important causes of damage, when they occur. If we reduced sulfur dioxide by fifty per cent, something possible but costly, it might have little effect, and might discourage the introduction of other more effective controls. It may be more effective to put more effort into controlling the output of oxides of nitrogen, or of hydrocarbons, which also contribute to the present damage.

Finally, pollution control may have its own harmful effects on the environment. Most of the effective methods of reducing sulfur dioxide output use large amounts of lime or of limestone. To obtain this limestone, extensive quarrying is needed, and the most convenient quarries in Britain are in a National Park. It is perhaps ironical that those who are most vocal about the need to reduce sulfur dioxide output are also the strongest opponents to such "desecration" of areas considered to be an important part of our national heritage. Also, pollution control can produce other potential pollutants, in this case great volumes of toxic slurry which may be difficult to dispose of without further damage to the environment. With all these difficulties, it is sensible to try to identify the real nature of all pollution problems so that all our efforts to control them are effective.

This also brings us back to the whole energy question. Clearly the most effective way to minimize environmental damage is to be more economical and to reduce our use of all types of fossil fuel which produce so many different pollutants. We should make the greatest possible use of renewable and non-polluting sources of energy like water power, solar energy, and the wind—not forgetting the bitter opposition by conservationists to the hydroelectric schemes in Tasmania, that solar collectors to supply industry would have to cover vast areas of the countryside with a great effect on the environment, and that huge windmills would be noisy and intrusive. This is why I believe those who wish to preserve the environment must come down in the end to advocating the careful use of much more nuclear power, as being the cleanest and least intrusive form of energy at present available to mankind.

References

Beament, J., A.D. Bradshaw, P.B. Chester, M.W. Holdgate, and B.A. Thrush. 1984. "Ecological effects of deposited sulfur and nitrogen compounds." *Philosophical Transactions of the Royal Society of London* Series B.

Buckley-Golder, D.H. 1984. "Acidity in the environment." Energy Technology Support Unit, Harwell, England.

The Causes and Effects of Acidic Deposition (4 volumes). National Acid Precipitation Assessment Program (NAPAP). Washington, D.C., October 1987.

Drables, D. and A. Tollan (eds.) 1980. *Ecological Impact of Acid Precipitation*. SMSF, Oslo, Norway.

Mellanby, K. (ed.) 1988. *Air Pollution, Acid Rain and the Environment* Report 18. Watt Committee on Energy, Elsevier Applied Science Publishers.

Saure Niederschläge:Ursache und Wirkung. 1984. *VDI-Berichte* 500. Verein Deutscher Ingenieure, Dusseldorf, Germany.

ELEVEN

The COMPLEXITY of SURFACE WATER ACIDIFICATION by ACID RAIN

A.G. Everett

In order to gain some perspective on the problem of acid rain, it is important to realize the geologic antiquity of the phenomenon of acid rain, or acidic precipitation. The atmospheric scavenging process involves removal of chemical constituents of the atmosphere by both rainfall and snowfall. In addition, removal is accomplished by dry deposition, about which even less is known than is known about acid formation and precipitation involving the aqueous phase. The Barberton volcanics of South Africa indicate the presence of volcanic activity at the earth's surface in excess of 4 billion years ago, prior to the evolution of the earth's atmosphere to a composition close to that known today. Rubey (1955) has argued persuasively that the earliest primitive atmosphere was composed primarily of CO_2 and N_2 although others have proposed an initial atmosphere composed predominantly of CH_4 and NH_3 (Urey 1952). Under such primitive atmospheres, the processes of chemical weathering would have depended upon carbonation and hydration reactions in contrast to oxygen and acids that are the critical chemical components of weathering reactions today (Holland, 1984). Holland has indicated that petrologic considerations indicate that early volcanic gases following the Earth's formative period of accumulation most probably consisted of H_2, H_2O, CO, H_2S, and N_2. Recent work by Hattori *et al.* (1983) has estimated that the atmosphere became oxidizing on the order of 2.2 billion years ago. The significance of the oxidizing atmosphere is that sulfur compounds from volcanic emissions and nitrogen oxides from lightning would be oxidized to sulfate and nitrate ions, lowering the pH below that of the theoretical atmospheric carbon dioxide-water system having a pH of 5.65. About 400 million years ago, during the Devonian period, land plants became a major component of terrestrial surface cover. Decay of aquatic and terrestrial organic matter provides additional major natural sources of sulfur and nitrogen compounds to the atmosphere. Thus, the frequently cited anthropogenic origin of the acid rain phenomenon more properly should be characterized as an anthropogenic *addition* to precipitation that is inherently acidic as a result of natural sources of sulfur and nitrogen compounds.[1]

Acid Rain pH—A Variable in Time and Space

The natural sources vary in their intensity both spatially and seasonally, as do the anthropogenic sources. There is, therefore, no single value of hydrogen ion concentration that typifies rainfall. It is variable in both time and space as a function of a variety of natural and human-made emissions. Estimates of the most likely midpoint of the range of pH in the absence of anthropogenic contributions have been given by Miller and Everett (1981) as 4.5 ± 0.5 pH units and by Charlson and Rodhe (1982) as "well below 5" in some clean, remote areas. In a sampling network covering 1800 km^2 for 22 storm events during two years, Semonin (1976) has reported a range of up to 4.8 pH units variation in a single storm. Stensland and Semonin (1982) have looked at the variability of pH in precipitation in data from Illinois and found a change from 6.5 pH units in 1953 to approximately 4.3 in 1983. However, going further, they examined the meteorology of the 1950s, finding a drought occurring that elevated the acid rain pH to 5.4 by providing wind-blown soil components composed of calcium and magnesium salts, which buffered the rain upwards. Correcting on an equal ion concentration basis for calcium, they found a change from the 1950s to the present of only about 0.5 pH units. In addition to temporal variability, the spatial variability of the composition of airborne dust east of the Rocky Mountains provides buffering that results in a general decrease in pH from west to east except for sites very near the Atlantic Ocean (Smith *et al.*, 1970).

The frequent early citation of pH 5.65 as the baseline that control programs should be expected to achieve, and from which estimates of anthropogenic effects have been measured, has often obscured proper definition of the problem. The magnitude of human effects on atmospheric chemistry has been overestimated; and, despite the mammoth research effort on acidic precipitation, we have an inadequate data base on natural sources so that we can be able to tell by how much.

Estimates of changes in sulfur dioxide emissions in the eastern United States for the period of the 1970s range from an estimated increase by the National Academy of Sciences (1986) to a 20 percent decrease by the U.S. Environmental Protection Agency (Gschwandtner *et al.*, 1985). Although reliable data on emissions do not exist, based on the emissions controls required and the emissions inventories performed by the U.S.E.P.A., their estimates appear to be reasonable. However, despite decreases in emissions during the last 15 years, existing data will not permit one to determine 1) how much control will result in how much of a decrease in acidic deposition of sulfur oxides, or 2) where the reductions in acidic deposition will take place (see commentary by S.F. Singer, Chairman, Acid Rain Panel, National Advisory Committee for Oceans and Atmospheres, *Science*, 232:563 [1986]). Just as there is a great deal of uncertainty in the historical data base for atmospheric emissions, similar uncertainty in historically accurate surface water data plagues the interpretation of actual surface water responses to acidic precipitation.

Natural Sources Are Underestimated

Natural emissions have been significantly underestimated, leading to overestimation of the effect of anthropogenic sources. A significant example of the problem of underestimates of natural sources has been the historic estimates of volcanic emissions and exhalations to atmospheric sulfur loading. Cadle *et al.* (1979) noted that Stoiber and Bratton recognized that estimates of volcanic emissions were about a factor of 10 too low prior to 1979 as a result of previous analytical and monitoring techniques. An additional frequent source of underestimation of volcanic gaseous contributions has been the focus on emissions only during the spectacular eruptive events, ignoring non-eruptive fuming or degassing. The factor of 10 change would increase worldwide estimates of atmospheric volcanic contributions of sulfur gases from being about 15% (10 Tg/year of SO_2) of anthropogenic emissions to nearly 150%. Such analytic and methodological problems continue to complicate recent estimates and perhaps are the basis for Berresheim and Jaeschke (1983) estimating volcanic emissions prior to 1979 at 15.2 Tg SO_2 but stating that volcanic emissions subsequent to 1979 may be on the order of a factor of 10 greater than those from 1960 to 1979. Data on volcanic eruptions since 1979 do not support a tenfold increase in SO_2 into the atmosphere from volcanic (or any other) sources (Sedlacek *et al.* 1983; Simkin *et al.* 1981; Dr. L. Siebert, Smithsonian Institution, personal communication, 1983). These considerations suggest the reasonableness of Stoiber and Bratton's observation on the probable order of magnitude of error in earlier estimates of volcanic emissions.

Supporting the idea that the atmospheric contribution of volcanoes may be much larger than previously estimated is the finding by Sedlacek *et al.* (1983) that "at least 58% of the sulfate aerosol present in the lower stratosphere between 1971 and 1981 was of volcanic origin." Given that only a small portion of volcanic emissions, and probably little if any non-eruptive volcanic and fumarolic degassing reaches the stratosphere, the magnitude of tropospheric sulfate from volcanic sources clearly needs reassessment. Even less attention has been given to the atmospheric burden of acidic chlorine compounds that are of volcanic origin. Johnston (1980) pointed out that a single 1976 eruption in Alaska released 5 x 10^{11} grams of chlorine, a substantial portion of which went into the stratosphere.

Although there are indications that in the northeastern United States both sulfate emissions and wet deposition have been decreasing over the last decade, the decline in wet deposition does not appear to be linearly proportional to the decline in emissions. An evaluation of trends in acidity in precipitation in New York for the period 1965–1978 found that "...calculated trends in hydrogen ion concentration do not correlate with measured trends in sulfate and nitrate...(Peters *et al.*, 1982, p. 1)." The same researchers also concluded that the "...hydrogen ion concentration of surface water is a function of many geochemical and biochemical reactions, which makes it virtually impossible to relate trends to any simple cause (*ibid.*, p. 26)."

A larger-than-previously-assumed natural background may be one reason for that conundrum. Not only volcanic contributions are poorly known, however. The rapidly emerging literature on dimethyl sulfide emissions from the oceans indicates a source of geographically variable emissions comparable in volume to estimates of anthropogenic emissions. Such biological sources are variable temporally as well as spatially. Terrestrial analogues of this variable oceanic source also exist. The Environmental Protection Agency (EPA, 1982) has shown a limited number of New York and New England samples of soil and water emissions of organic sulfur compounds with concentrations nearly five times greater than the eastern United States average emissions from such sources.

For example, Nraigu *et al.* (1987) have shown that biogenic sulfur sources from boreal wetlands, bogs, and swamps, which comprise 30 percent of the surface area of Ontario, "...can account for up to 30 percent of the acidifying sulfur burden in the atmosphere in remote areas of Canada." The compounds included both dimethyl sulfide, (DMS) and hydrogen sulfide (H_2S). The concentration of H_2S was at least comparable to that of DMS. Winner *et al.* (1981) have previously reported that the oxidation of biogenic H_2S may contribute to formation of acid rain. Hitchcock (1976) inferred that up to 50 percent of the sulfate in some urban areas may be of biogenic origin; Hitchcock and Black (1984) demonstrated the presence of a substantial amount of biogenic sulfur from salt marshes in the Chesapeake Bay area. Accumulating evidence of the biogenic sulfur input to acid rain will require a rigorous quantitative evaluation in order that control programs for acid rain precursors can be designed to achieve the desired reductions in acidic deposition. A very comparable analogy is that of the role of biogenic hydrocarbons in the formation of ozone, the traditional neglect of which has led to overestimation of the effectiveness of ozone abatement strategies based solely on reducing hydrocarbon emissions of anthropogenic origin (Chameides *et al.*, 1988). A similar conclusion was reached nearly a decade earlier on the basis of then-ongoing research on biological contributions to air pollution (Budianski, 1980). Recent conclusions of the National Acid Precipitation Assessment Program (NAPAP) regarding the role of ozone in acidic deposition (NAPAP, 1988) will undoubtedly require a quite comprehensive understanding of natural sources of sulfur, nitrogen oxides, and volatile organic compounds in order to fashion effective controls for both ozone and acid precipitation.

Is Acid Rain the Culprit?

The Adirondacks Region as an Example

The acid rain problem has been much oversimplified by the popular press as well as by a number of scientists. Of course, it is not the first environmental issue about which there were "doomsday" predictions at an early date based on a slim and inadequate data base. For the most part, the researchers who were first to recognize the possible problem and its possible ecological consequences were apparently unaware of the substantial amount of literature on the chemical reactions of water passing through terrestrial environments. Many hydrogeochemical papers deal with processes and reactions that occur in surface water systems dominated by bicarbonate anions, but the bulk of published water research has dealt historically with waters having water chemistries acceptable for potable or irrigation use.

Freeze (1972*a*, *b*) pointed out that the importance of the subsurface response of watersheds had been largely insufficiently considered in most watershed studies, with both base flow, derived from deep percolation of infiltrated water, and subsurface storm flow (interflow), from infiltration and lateral flow in upper soil horizons, being significant. As early as 1971,

Kennedy had documented that fact for watershed geochemistry and it had been demonstrated to be significant in eastern forested watersheds by the Vermont studies of Dunne and Black (1970*a, b*). Recent studies are verifying the frequently extensive involvement of watersheds' ground water geochemistry as a significant element in understanding the extent of surface water acidification effects of acidic deposition (Booty *et al.*, 1988; Driscoll *et al.*, 1988; NAPAP, 1988, undated).

There is relatively little information on the hydrology of waters in glacial terrain (Born *et al.* 1979), yet the previously glaciated regions of the northern hemisphere have been shown to be most affected by acidic deposition (NAPAP, 1988). These terrains are characterized by boreal forests with acidic podzol soils, bogs, and wet conifer swamps (Larsen 1982; Foth and Schaefer 1980; and Duchaufour 1978). Surface water environments in these circumboreal regions commonly have higher concentrations of dissolved organic constituents, including dissolved organic acids, than do other terrains. The result is low alkalinity water poorly buffered by bicarbonate but frequently well buffered by other chemical systems, such as organic acids or aluminum complexes leached from soils in the watersheds. The waters are of a mixed inorganic-organic type whose chemical behavior still is not well understood, but research of the last eight years has increased the data base.

Traditional inorganic water analyses are not sufficient to characterize the chemical behavior of such waters. For example, during our Adirondacks water survey performed in the late autumn of 1981, Dr. James Kramer found that when samples titrated during Gran analyses were allowed to sit, hydrogen ion was slowly released for periods of up to 24 hours. These reactions, of apparently organic origin, have slow kinetics in comparison to typical inorganic acid-base reactions that usually characterize such titrations. Stevenson (1982) has concluded that the acidic properties of fulvic and humic acids are not explained entirely by simple ionization reactions but that they also involve formation of carbon ions and carbonium ions. Structural reorganizations of the organic anions may be involved as well. Thus, Gran titrations and positive excess ion balances are only qualitative, not quantitative, indications of the presence of organic acids in natural waters. Other qualitative indicators of significant organic constituents in water are:

yellow to yellow-brown water color, although we have found indications that even clear water streams can be acidified by organic acids, a finding that has recently been verified by Krug in Connecticut waters; substantially increased time of filtration, frequently with the deposition of a clear, waxy material on filter surfaces; high concentrations of aluminum and iron, out of equilibrium with the inorganic chemistry of the sample; depressed dissolved oxygen concentrations under high flow conditions; low positive to negative alkalinity; and very poor correspondence of pH and conductivity.

The waters of the Adirondack region contain a wide spectrum of waters ranging from quite organic to quite bicarbonate rich. Under the heavy rainfall conditions of our field sampling in the autumn of 1981, the bulk of the surface waters in streams, lakes and ponds were a mixture of organic and inorganic constituents. They are not adequately characterized solely by inorganic chemical analysis nor will their response to atmospheric strong inorganic acid addition be predictable solely from inorganic considerations. During dry weather, low flow conditions, such waters can be expected to be much more inorganic as a result of the dominance of base flow from the more mineral-rich horizons of the soil profile, glacial strata, and bedrock.

Not only is the Adirondack region receiving acid precipitation, but other factors are operating as well that influence surface water chemistry. Changes in land use practices since the early part of the century have led to a great increase in the area under forest cover, with a concomitant decrease in the number of acres burned annually in forest fires, as seen in Figure 11-1 (Duhaime *et al.* 1983). The result has been a substantial build-up of acidic organic litter through which precipitation must pass on its way to becoming stream flow. The decrease in forest fires has removed a source of carbonate buffering for the throughflowing waters. Older ash layers become more leached with time and thus are less effective in providing bicarbonate buffering. The long-term effect is that of increased amounts of organic material on the forest floors of the watersheds to provide organic acids upon leaching.

During those eight decades of increased forest cover and litter accumulation, there has been change in the mean precipitation regime. Quinn (1981) has shown a regional trend of

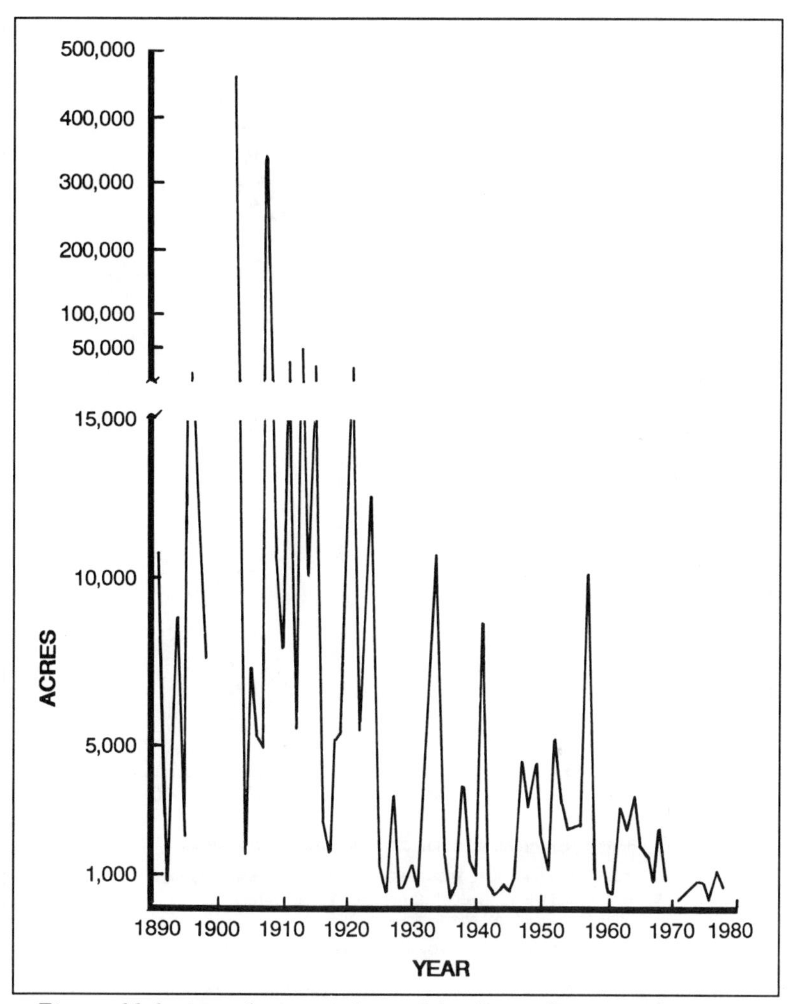

Figure 11-1: Annual acreage burned by forest fires in Adirondack counties, 1891-1978 (Source: Duhaime *et al.* 1983).

increased rainfall since the beginning of the century for the Great Lakes basin, which encompasses most of the area of the northeastern United States and northeastern Canada now being studied for acid rain effects. Quinn (personal communication, 1982) provided the regional data for Lake Ontario,

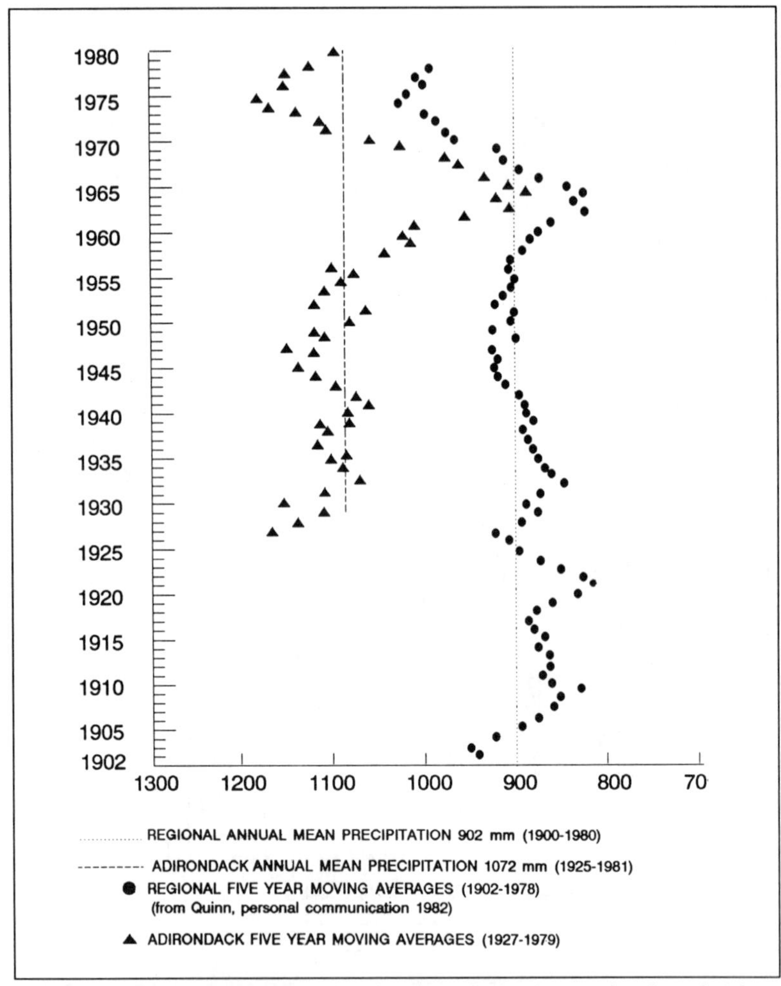

Figure 11-2: Comparison of Lake Ontario basin overland and Adirondack precipitation (in mm per year). Five year moving average (centered).

which includes the western two-thirds of the Adirondack region. Figure 11-2 contains a smoothed plot of that regional data in comparison with data from a set of five stations located in the central part of the Adirondack Park, centered in the area of the autumn 1981 surface water reconnaissance study. The close correspondence of the regional and park data is readily apparent, despite the orographic effect that increases mean Adirondack precipitation above that of the region. It is apparent that increased precipitation, hence increased flushing of naturally organic soils, wetlands, and forest floors, is present in the Adirondack region as well as in the Lake Ontario basin as a whole since about 1940. The difference in mean precipitation during the first four decades of the century in contrast to that of the last four is coupled with the substantial changes in land use and the amount of organic matter in the path of precipitation falling on and moving through inherently acidic watersheds to surface waters. It is to be expected that such concurrent changes would produce increased surface water acidification even in the absence of other contributing factors.

The soils of the Adirondack region are most commonly spodosols, acidic soils that produce a low pH in waters passing through them. The characteristics of spodosol soils and their effect on throughflowing waters have been reviewed by Krug and Frink (1983). Such soils are enormous reservoirs of hydrogen ion and are quite resistant to further increases in acidity by acid precipitation. The mean hydrogen ion concentration of our twelve soil zone and shallow ground water samples is 73 microequivalents (pH=4.06). Heimburger (1934) examined 264 soil samples from the Adirondack region and found that 79 percent had a pH of 5.0 or less. Our examination of soil: water measurements relative to raw organic carbon shows a very strong positive correlation of hydrogen ion to organic carbon. Thus, the water passing through organic-rich spodosol soils is strongly buffered to a low pH by organic acids, providing a regionally widespread mechanism for surface water acidification. The substantial buffering capacity of organic-rich soils has been demonstrated elsewhere as well (Litaor and Thurman, 1988). These acid soils are themselves only minimally susceptible to degradation by acid deposition because of their large reservoir of hydrogen ion (Frink and Voight 1977; Wiklander 1973–74; Krug and Frink 1983). Instead they are a large acidic buffering component in the hydrogeochemical system, and are commonly associated with surface waters having little, if any, bicarbonate buffering capacity.

In the surface water reconnaissance study, Everett *et al.* (1983) found indications that ground water oxidation of sulfide minerals in soils, glacial sediments, and bedrock is contributing to the sulfate loading of Adirondack surface waters in excess of that provided by acidic deposition. Pyrite and pyrrhotite are sulfides of common occurrence in the Adirondack region and gossans, frequently referred to as "rusty gneisses," are present in a number of localities associated with elevated sulfate levels in water discharging from rock fractures and soil horizons. Subsequent work in the Cascade Lake watershed under dry conditions also showed a significant association of sulfur with carbon in fine particulate matter, indicating a substantial aqueous particulate burden of organic sulfur derived from terrestrial sources (Everett, 1985). This finding is in agreement with the work of Mitchell *et al.* (1984) in two other Adirondack lakes in which carbon-bonded sulfur is the predominant sulfur-bearing constituent of lake sediments.

In the 40 lakes and ponds surveyed, sulfate and hydrogen ion are negatively correlated (that is, as sulfate increases, hydrogen ion tends to decrease). That is the opposite relationship to that of the region's rainfall and opposite to what would be expected if surface water acidification were solely the result of incident rainfall. The study, made under autumn heavy rainfall conditions, should have encountered maximum rainfall-sulfate relationship.

Mean sulfate concentrations in surface and subsurface waters discharging at the surface exceeded precipitation sulfate input, a finding in agreement with those of Mollitar and Raynal (1982) and David (1983). The statistical correlation of sulfate to alkalinity is positive for both the soil-zone and ground water discharge samples and for shallow, internally-drained ponds perched in glacial till. The latter were the waters that were thought to be most likely to be affected by acid precipitation at the outset of the study, thus this would reflect acidification and reduction of alkalinity by precipitation. The positive association of increased sulfate ion with increased alkalinity in lake waters, the obverse of what would be predicted if surface water acidification were the sole or principal result of atmospheric sulfate deposition, is also shown by statistical correlation of the aggregate chemical data reported for the 26-month long regional ILWAS sampling program (Driscoll and Newton, 1985).

In addition to the sulfide minerals, scapolite is widespread in Grenville gneisses and other rocks in the Adirondacks (Buddington, 1939). Scapolite is a sodic to calcic aluminosilicate mineral series that contains chloride, sulfate, and carbonate anions within its framework (Shaw, 1960). The three anions are released to solution upon weathering of the mineral, which weathers in a manner comparable to calcic plagioclase under Adirondack area conditions (Everett, 1985). The distribution of chloride and sulfate in groundwater discharges and associated surface waters sampled in the autumn 1981 survey indicates that scapolite is the source of most, if not all, of the chloride and some of the sulfate that is in excess of atmospheric deposition (Everett, 1985). Data indicate that both sulfide mineral oxidation and scapolite weathering contribute to the excess sulfate ion in surface waters, with some ground water acidified by sulfide mineral oxidation discharging to surface water before it has been neutralized by reaction with bedrock. Rapid biochemical transformation of sulfur in soil and vegetative reactions precludes accurate partitioning of the total sulfate pool into the portions contributed by mineral weathering and by the atmosphere.

Among the various terrestrial environments sampled were some of the Adirondacks' many wetlands. In all of the nine pairs of water samples taken from streams flowing through wetlands, the hydrogen ion concentration was higher in the downstream sample. The increases ranged from 0.2 to 0.8 pH units, showing that water passing through the very extensive wetland areas along streams, lakes, and ponds in the Adirondacks increases in hydrogen ion concentration. In addition to being areas of lush organic growth, these wetlands are also frequently areas of soil zone and ground water discharge.

Fisheries management practices in the Adirondacks have changed substantially during the past years with the result that many fewer fish are planted (Figure 11-3). One major reason given for the reduction in planting was that the planted trout did not successfully reproduce in numerous acidic Adirondack waters. This reduction in fish planting is coincidental with the reported reduction in available fish in Adirondack ponds and lakes during a period of heavy recreational fishing pressure, as reflected in part in the data on public campground use (Figure 11-4).

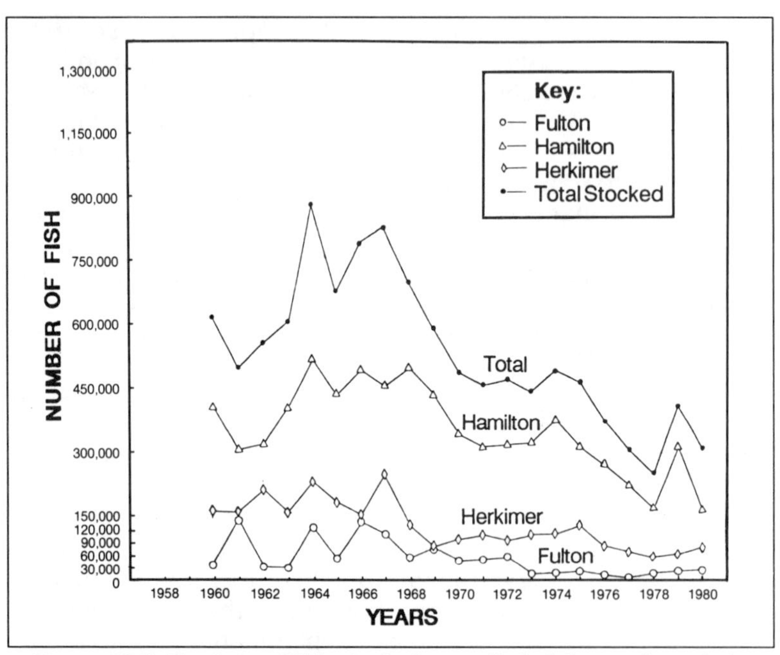

Figure 11-3: Numbers of fish stocked annually in lakes and ponds in three Adirondack counties (Source: Retzsch *et al.*, 1982.).

An additional complicating factor that may have had serious adverse consequences to fisheries, the food chain, and the development of chronic toxicity in various organisms, was extensive use of DDT to suppress the population of black flies and other biting insects in the Adirondacks. We found that DDT and its metabolites are present in concentrations ranging from tenths to several parts per million, concentrated primarily in the surface organic layer of soils. Organic-rich lake sediments also contain DDT metabolite residues in smaller concentrations. Soil pHs are lower than adjacent water pHs for the DDT-containing samples, thus it is quite possible that DDT metabolites are remobilized into the food chain as they are eroded from the soil profile into higher-pH surface waters. Based on the ratio of DDT metabolite isomer, O,P'-DDT to P,P'-DDT, we are of the opinion that the DDT is aged and has undergone extensive metabolic transformation, thus it is not of

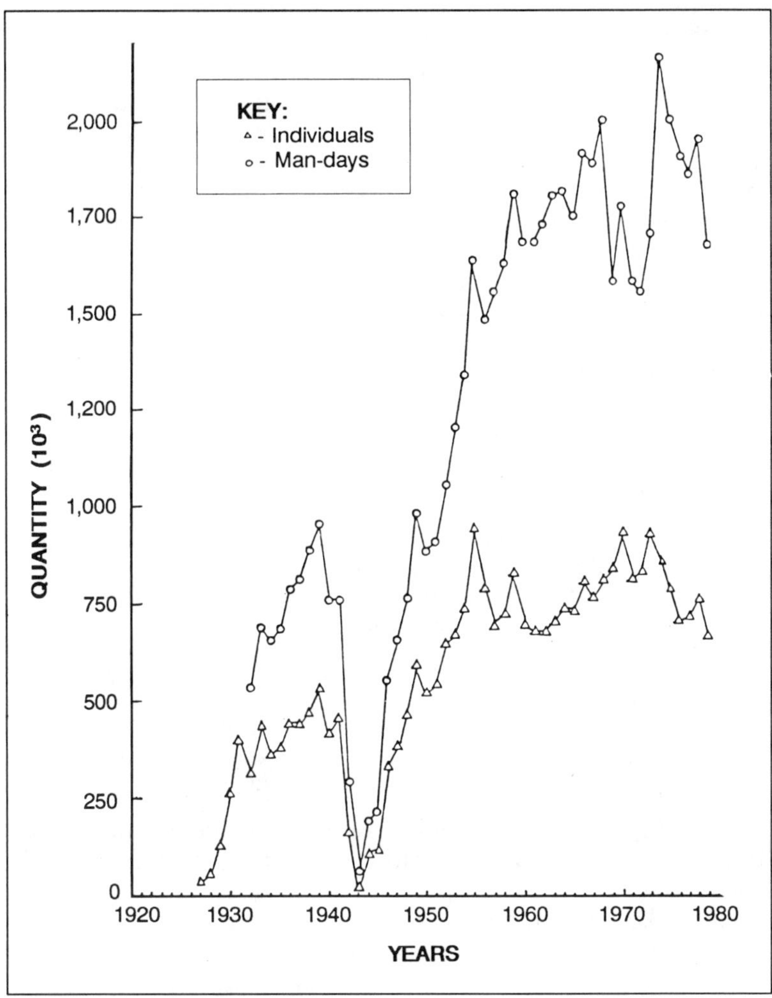

Figure 11-4: Adirondack Public Campground Occupancy, 1927–1979 (Source: Retzsch *et al.* 1982).

recent origin, but is the residuum of the extensive 1950s and 1960s period of heavy use. Such concentrations have been identified as sufficient to cause chronic toxic problems in a variety of organisms.

Conclusion

Using a related set of studies of the history of Adirondacks land use, surface water chemistry, and fisheries management practices (Duhaime *et al.* 1983; Everett *et al.* 1983; Retzsch *et al.* 1982; A.G. Everett, unpublished data), it is apparent that surface water acidification processes in the Adirondacks are much more complex than simply the deposition of man-made atmospheric acids.

The difficulty of relating atmospheric deposition of man-made pollutants to surface water chemistry exists not only in the northeastern United States but in other locations as well, such as in the western United States, where there are significant SO_2 emissions from smelters (Campbell and Turk, 1988). This complexity is being documented by the extensive research resulting from the National Acidic Precipitation Assessment Program (1987, 1988, undated).

> Surveys were conducted in potentially sensitive regions of the country to determine the chemical status of surface waters. Surveys of lakes larger than 4 ha in the East and 1 ha in the West sampled during fall turnover show there are essentially no lakes or reservoirs in the mountainous West, northeastern Minnesota, and the Southern Blue Ridge Province of the Southeast with pH less than 5.0 (at a pH of 5.0, most clear water lakes do not support sports fish) and very few with pH <5.5. Most other sampled subregions show less than 0.5 percent of the lake area and less than 1 percent of the number of lakes with pH less than 5.0 (NAPAP, undated).

Although Adirondack field studies have shown that the extent of surface water acidification appears to be greater in the spring than in the fall (35 percent as opposed to 11 percent), the spring lake outlet data do not necessarily reflect the chemistry of the entirety of a lake's water body because of stratification in the spring (NAPAP, 1988). The results of the extensive NAPAP research in acid sensitive areas of the United States clearly indicate the extensive involvement of watershed areas in the chemistry of through-flowing waters, with only the high elevation, boreal, watersheds showing significant acidification. To date, virtually none of the studies have taken the *volume* of throughput in watershed spodosol soils into account in the evaluation of either inorganic or organic sources of acidity. The baseline for most attempts at historic trend evaluation begins during drought periods during which throughput was much diminished and of higher alkalinity and pH as result of reaction with the watershed.

Not only is the problem of acidic deposition on watersheds complex, but a similar situation exists for the effects on forests (NAPAP, 1988), with air pollutants such as ozone showing possibly significant effects on forest vigor. Although consideration of forest responses is outside of the scope of this paper, the development of the complexity of that problem in parallel with that of surface water acidification indicates that natural systems responses to acidic deposition are still inadequately understood to provide a useable data base for an intelligent abatement program. It is clear, however, that aquatic and terrestrial ecosystems are not changing rapidly in a deleterious manner.

The complexly interrelated components of the terrestrial portion of the hydrogeochemical system continue to require study before all of the relationships, and their variability in time and space, can be worked out. Only then can the actual amount of surface water effects caused by the atmospheric acidic deposition be determined. Although the picture of a complex interrelated hydrogeochemical system has been sketched here only for the Adirondack region, similar patterns are emerging for parts of New England, Scandinavia, the New Jersey Pine Barrens, and other areas originally believed to be extensively adversely affected by acid precipitation.

It has taken time to develop an understanding of the interrelationships of hydrogeochemical systems, a complexity not discussed in many of the early reports that have influenced widespread political reaction. Now we find that preemptory political decisions are moving in directions divergent from the accumulating scientific information. Such expedient political actions, taken in the absence of understanding about what will produce actual ameliorative effects, are doomed to be wasteful and ineffective. Should such action also be accompanied by decreased research emphasis on still unknown relationships of acid deposition and surface water response, it is probable that the long-term situation will be worsened rather than improved.

Acknowledgements

I would like to thank Stephen D. Etter, James R. Kramer, Will M. Ollison, Walter D. Retsch, and S. Fred Singer for their contributions to this paper, noting however, that interpretations and opinions are those of the author.

Coda

Helmut E. Landsberg

My own work on acid rain began in the late 1940s in the vicinity of Washington, D.C., and later continued observations near Boston. The data showed that even then, precipitation in the northeast United States was acid. The average pH near Boston was 4.5 and it is still about the same. There is little doubt that in the major urban areas it has been acid for a long time without perceptible trends. The tall-stack emissions, which place pollutants at higher levels, permitted their drift to greater distances and rainout in rural areas undoubtedly contributed to the acid rain in areas such as the Adirondacks. My collaborators and I have, in recent years, made studies of the pH from mountains in low-level clouds. In the Catoctin Mountains of Maryland the values in a short series were around 5.0 and in some measurements in the Blue Ridge Mountains of Virginia during the summer of 1983 alkaline values above 6.0 prevailed. The latter were probably influenced by local limestone sources of dust.

Notes

1. pH is the negative base-10 logarithm of hydrogen ion activity, with a customary range values from 0 (representing one mole of hydrogen ion) to 14 (a concentration of hydrogen ion of 1×10^{-14}), with 7 representing neutrality (representing [H+] = [OH-]) (Hem, 1970). Values less than 7 are acidic whereas those greater than 7 are alkaline.

References

Berresheim, H., and W. Jaeschke. 1983. "The Contribution of volcanoes to the global atmospheric sulfur budget." *J. Geophys. Res.* 88: 3732–3740.

Booty, W.G., J.V. DePinto, and R.D. Scheffe. 1988. Drainage Basin Control of Acid Loadings to Two Adirondack Lakes. *Water Resources Research.* 24:1024–1036

Born, S.M., S.A. Smith and D.A. Stephenson. 1979. "Hydrogeology of glacial-terrain Lakes, with management and planning applications." *J. Hydrol.* 43: 7–43.

Buddington, A.F. 1939. *Adirondack Igneous Rocks and Their Metamorphism. Geological Society of America, Memoir 7, 354 p.*

Budianski, S. 1980. Biological Contributions to Air Pollution. *Environmental Science and Technology.* 14: 901–903.

Cadle, R.D., A.L. Lazrus, B.J. Huebert, L.E. Heidt, W.I. Rose, Jr., D.C. Woods, R.L. Chaun, R.E. Stoiber, D.B. Smith, and R.A. Zeilinski. 1979. "Atmospheric implications of studies of Central American volcanic eruption clouds." *J. Geophys. Res.* 84: 6961–6968.

Campbell, D.H., and J.T. Turk. 1988. Effects of Sulfur Dioxide Emissions on Stream Chemistry in the Western United States. *Water Resources Research.* 24: 871–878.

Chameides, W.L., R.W. Lindsay, J. Richardson, and C.S. Kiang. 1988. "The Role of Biogenic Hydrocarbons in Urban Photochemical Smog: Atlanta as a Case Study " *Science* 241:1473–1475.

Charlson, R.J., and H. Rodhe. 1982. "Factors controlling the acidity of natural rainwater" *Nature* 295: 683–685.

David, M.B. 1983. "Organic and Inorganic Sulfur Cycling in Forested and Aquatic Ecosystems in the Adirondack Region of New York State." Unpublished Ph. D. dissertation, State Univ. of New York, Coll. of Environ. Sci. and Forestry, Syracuse, N.Y., 268 p.

Driscoll, C.T., N.M. Johnson, G.E. Likens, and M.C. Feller. 1988. "Effects of Acidic Deposition on the Chemistry of Headwater Streams: A Comparison Between Hubbard Brook, New Hampshire, and Jamieson Creek, British Columbia." *Water Resources Research.* 24: 195–200.

Driscoll, C.T., and R.M. Newton. 1985. "Chemical Characteristics of Adirondack Lakes," *Environmental Sci. and Tech.* 19:1018–1023.

Duchaufour, P. 1978. *Ecological Atlas of the Soils of the World.* Masson Publishing USA, Inc., 178 pages.

Duhaime, P.F., A.G. Everett, and W.C. Retzsch. 1983. "Adirondack land use: A commentary on past and present impacts on terrestrial and aquatic ecosystems." Prepared for the American Petroleum Institute, Washington, D.C., 96 pages.

Dunne, T. and R.D. Black. 1970. "An Experimental Investigation of Runoff Production in Permeable Soils," *Water Resources Research.* 6:478–490.

Dunne, T., and R.D. Black. 1970. "Partial Area Contributions to Storm Runoff in a Small New England Watershed," *Water Resources Research.* 6:1296–1311.

E.P.A. 1982. *Critical Assessment Document: The Acidic Deposition Phenomenon and Its Effects*. Draft, May 10, 1982; United States Environmental Protection Agency, Washington, D.C.

Everett, A.G. 1985. "Sulfur Isotopes in the Cascade Lake Watershed, Adirondack Mountains, New York." Research report to the Department of Health and Environmental Sciences, American Petroleum Institute, Washington, D.C., 45 p.

Everett, A.G., W.C. Retzsch, J.R. Kramer, I.P. Montanez, and P.F. Duhaime. 1983. "Hydrogeochemical characteristics of Adirondack waters influenced by terrestrial environments." In *Proceedings* of the Second National Symposium on Acid Rain, Pittsburgh, PA, October 1982, 18 pages.

Foth, H.D., and J.W. Schafer. 1980. *Soil Geography and Land Use*. John Wiley & Sons, 484 pages.

Freeze, R.A. 1972*a*. "Role of Subsurface Flow in Generating Surface Runoff, 1. Base Flow Contributions to Channel Flow. *Water Resources Research* 8: 609–623.

Freeze, R.A. 1972*b*, "Role of Subsurface Flow in Generating Surface Runoff, 2. Upstream Source Area." *Water Resources Research* 7: 1272–1283.

Frink, C.R. and G.K. Voight. 1977. "Potential effects of acid precipitation of soils in the humid temperate zone." *Water, Air, and Soil Pollution* 7: 371–388.

Gschwandtner, G., K.C. Gschwandtner, and K. Eldridge. 1985. "Historic Emissions of Sulfur and Nitrogen Oxides in the United States from 1900 to 1980," v. 1, U.S. Environmental Protection Agency, Washington, D.C., EPA-600/7-85-009a.

Hattori, K., H.R. Krause, and F.A. Campbell. 1983. "The start of sulfur oxidation in continental environments: About 2.2 x 10^9 years ago." *Science* 221: 549–551.

Heimburger, C.C. 1934. *Forest Type Studies in the Adirondack Region*. Memoir 165, Cornell University, Agric. Exper. Sta.

Hem, J.D. 1970. "Study and Interpretation of the Chemical Characteristics of Natural Water." *U.S. Geological Survey, Water Supply Paper 1473*, 363 p.

Hitchcock, D.R. 1976. "Atmospheric Sulfates from Biological Sources." *J. Air Pollution Control Assn.* 26: 210–215.

Hitchcock, D.R., and M.S. Black. 1984. $^{34}S/^{32}S$ "Evidence of Biogenic Sulfur Oxides in a Salt Marsh Atmosphere." *Atmospheric Environment.* 17: 1–17.

Holland, H.D. 1984. *The Chemical Evolution of the Atmosphere and the Oceans.* Princeton University Press, 582 p.

Johnston, D.A. 1980. "Volcanic contribution of chlorine to the stratosphere — More significant to ozone than previously estimated?", *Science* 209: 491–493.

Kennedy, V.C. 1971 "Silica Variation in Steam Water with Time and Discharge." *Nonequilibrium Systems in Natural Water Chemistry*. Amer. Chem. Soc., Adv. in Chem. Series 106, p. 94–130.

Krug, E.C. and C.R. Frink. 1983. "Acid rain on acid soil: A new perspective." *Science* 221: 520–525.

Larsen, J.A. 1982. *Ecology of the Northern Lowland Bogs and Conifer Forests*. Academic Press, 307 pages.

Litaor, M.I., and E.M. Thurman. 1988. "Acid neutralizing processes in an alpine watershed, Front Range, Colorado, U.S.A. — 1: buffering capacity of dissolved organic carbon in soil solutions." *Applied Geochemistry* 3: 645–652.

Miller, M.L. and A.G. Everett. 1981. "History and trends of atmospheric nitrate deposition in the eastern U.S.A." Published in *Formation and Fate of Atmospheric Nitrates*, H.M. Bernes, ed., United States Environmental Protection Agency, EPA-600/9-81-025, pp. 162–178.

Mitchell, M.J., D.H. Landers, D.F. Brodowski, G.B. Lawrence, and M.B. David. 1984. "Organic and Inorganic Sulfur Constituents of the Sediments in Three New York Lakes: Effect of Site, Sediment Depth, and Season." *Water, Air and Soil Pollution* 21: 231–245.

Mollitar, A.V., and D.H. Raynal. 1982. "Acid Precipitition and Ionic Movements in Adirondack Forest Soils." *Soil Science Society of America Journal* 46: 137–141.

NAPAP. 1987. "Annual Report to the President and Congress." The National Acid Precipitation Assessment Program, Washington, D.C., 76 p.

NAPAP. 1988. "Annual Report to the President and Congess." The National Acid Precipitation Assessment Program, Washington, D.C., 94 p.

NAPAP. Undated. "Interim Assessment of Causes and Effects of Acidic Precipitation." The National Acid Precipitation Assessment Program. Washington, D.C., 4 vols. (Issued in Oct. 1987)

National Academy of Sciences. 1986. "Acid Deposition: Long Term Trends." National Academy Press.

Nraigu, J.O., D.A. Holdway, and R.D. Coker. 1987. "Biogenic Sulfur and Acidity of Rainfall in Remote Areas of Canada," *Science* 327: 1187–1192.

Peters, N.E., R.A. Schroeder, and D.E. Troutman. 1982. "Temporal Trends in Acidity of Precipitation and Surface Waters of New York."

U.S. Geological Survey, Water Supply Paper 2188, 35 p.

Quinn, F.N. 1981. "Secular changes in annual and seasonal Great Lakes precipitation, 1854–1979, and their implications for Great Lakes water resource studies." *Water Res. Res.* 17: 1619–1624.

Retzsch, W.C., A.G. Everett, P.F. Duhaime, and R. Nothwanger. 1982. "Alternative explanations for aquatic ecosystems effects attributed to acidic deposition." For Utilities Air Regulatory Group, Washington, D.C., 77 pages + appendix.

Rubey, W.W. 1955. "Development of the hydrosphere and atmosphere, with special reference to probable composition of the early atmosphere." In *Crust of the Earth*, A. Poldervaart, ed., Geol. Soc. Amer., Spec. Paper 62, p. 631–650.

Sedlacek, W.Q., E.J. Mroz, A.L. Laxrus, and B.W. Gandrud. 1983. "A decade of stratospheric sulfate measurements compared with observations of volcanic eruptions." *J. Geophys. Res.* 88: 3741–3776.

Semonin, R.G. 1976. "The variability of pH in convective storms." *Water, Air, and Soil Pollution* 6: 395–406.

Shaw, D.M. 1960. "The Geochemistry of Scapolite." *Journal of Petrology*, 1: 218–260, 261–285.

Simkin, T., L. Siebert, L. McClelland, D. Bridge, C. Newhall, and J.H. Latter. 1981. *Volcanoes of the World*. Smithsonian Institution, Washington, D.C., 232 pages.

Smith, R.M., P.C. Twiss, R.K. Krauss, and M.J. Brown. 1970. "Dust Deposition in Relation to Site, Season, and Climatic Variables." *Soil Science Society of America Proceedings* 34: 112–117.

Stensland, G.J., and R.G. Semonin. 1982. "Another interpretation of the pH trend in the United States." *Bull. Amer. Met. Soc.* 63: 1277–1284.

Stevenson, F.J. 1982. *Humus Chemistry: Genesis, Composition, Reactions.* John Wiley & Sons, 439 pages.

Urey, H.C. 1952. "On the early chemical history of the earth and the origins of life." *Proc. Nat. Acad. Sci.* 38: 351–363.

Wiklander, L. 1973-1974. The acidification of soil by acid precipitation." *Grundforbattring* 26: 155–164.

Winner, W.E., C.L. Smith, G.W. Koch, H.A. Mooney, J.D. Bewley, and H.R. Krouse, 1981. "Rates of emission of H_2S from plants and patterns of stable isotope fractionation." *Nature* 289: 672–673.

TWELVE

WORLDWIDE POLLUTION of the OCEANS

P. Kilho Park

You must not lose faith in humanity.
Humanity is the ocean.
If a drop of it becomes dirty,
The ocean does not become dirty.
Mahatma Gandhi (1869–1948)

You must not lose faith in humanity.
Humanity is the ocean.
If a part of it becomes dirty,
Humanity will clean it up in unison.
Momji (1931–)

When we stand by the ocean it is difficult to fathom that we people are collectively capable of altering the ocean to the detriment of humankind. The ocean appears so vast and our anthrosphere so benignly minuscule. Conversely, we have witnessed methyl mercury poisoning of fish and shellfish from industrial wastes discharged into Minamata Bay, Japan, in the 1950s; over 100 people died and many more were permanently harmed.

The fact is, that from time immemorial we have been disposing of anthropogenic waste materials into the ocean.

Agricultural pesticides enter the sea through rivers and streams. Lead emitted from automobiles into the atmosphere also precipitates over the ocean. Within the last century we have witnessed that some semiconfined bodies of water, such as Tokyo Bay, have been seriously degraded. Even in the open sea far from land, some pollutants, such as plastic litter, are distributed worldwide.

As global industrialization intensifies and population increases, the pressure to dispose of our wastes into the ocean, deliberately or inadvertently, will increase. This requires careful and critical evaluation of the ocean's capacity to accommodate the waste materials. Such evaluation is paramount to ensure that valuable resources of the sea are preserved and protected for future generations. The science of oceanography must be nurtured as we use the ocean more for our waste disposal; otherwise it will not be possible to attain a long-term solution.

Definition of Marine Pollution

The Joint Group of Experts on the Scientific Aspects of Marine Pollution (GESAMP), comprised of eight United Nations organizations, offers us the following definition:

> Marine pollution means the introduction by man, directly or indirectly, of substances or energy into the marine environment (including estuaries) resulting in such deleterious effects as harm to living resources, hazards to human health, hindrance to marine activities including fishing, impairment of quality for use of seawater and reduction of amenities.

The substances mentioned here are: toxic metals such as mercury and cadmium; synthetic organic compounds such as pesticides, including DDT (dichlorodiphenyl-trichloroethane); petroleum hydrocarbon compounds; radioactive wastes; solid wastes such as floatable plastic pellets; sewage and industrial wastes; and dredged material. The introduction of energy into the ocean may cause thermal pollution where, for instance, steam electric power plants using large amounts of cooling water elevate the seawater temperature locally; this may not be optimal for marine life inhabiting the area. Conversely, the introduction of energy may also cause thermal enhancement for culturing fish, shrimp, and eel. Similarly, sewage nutrients, such as nitrogen fertilizers, may help oceanic primary production. Therefore, it is important to view the substances and energy mentioned above as potential resources whenever possible.

The Ocean as a Waste Repository

Our civilization, past and present, has used the ocean as a cesspool. What we have practiced is the method of dilution and dispersion for waste management. The GESAMP scientists make an analogy of the smoke-filled room to describe the oceanic dilution and dispersion concept.

Assume that there are several people in a room and one starts to smoke. A plume of smoke drifts and the people nearby are affected by the smoker's antisocial act. However, as the turbulent currents of air gradually carry and mix the smoke to the farthest corners, the background level of smoke begins to build up. After a while the rate of increase of smoke concentration diminishes as the processes removing it begin to take effect; the removal processes include ventilation and deposition on curtain fabrics and in people's lungs. For the ocean the removal processes include water exchanges among various oceans and seas, sedimentation, and incorporation into marine organisms.

Although the ocean is complicated due to stratification of water layers, the rotation of the earth, chemistry, biology, geophysics, and geochemistry, the smoke-filled room analogy describes the major processes of dilution and dispersion and the processes of fluctuations and the steady-state in the near and far fields.

Since the onset of the nuclear age in 1944, we have also used the deep-sea floor for the disposal of low-level radioactive wastes. For instance, the United States placed 107,000 canisters of nuclear wastes on the sea floor of the Atlantic and Pacific Oceans in the period from 1946 to 1970. (The United States has not ocean-disposed of nuclear waste since.) The estimated total radioactivity at the time of disposal was 4.3×10^{15} disintegrations per second (about 100,000 curies).

The wastes were carefully packaged in metal drums with cement to isolate and contain the wastes from the deep-sea water and organisms. Many of these drums are still intact; in them, radioactive decay continues.

This method of waste management is called isolation and containment and has been the preferred method to cope with highly toxic and radioactive wastes. Since time and place of repository are known accurately, we can study the fate and effects of the wastes scientifically. In addition, when necessary, we can also retrieve the wastes to make them harmless or to redeposit them in safer places.

The Chemistry of the Ocean

The ocean is dynamic. Human-originated marine pollution is governed by the basic processes of the ocean. We must understand these processes to determine and forecast people's impact on the ocean.

Water moves. So does the sea floor. Scientists now know that the deep Atlantic Ocean water is about 200 years old, while that of the Pacific Ocean is 500 years old. Sea floor spreading renews the bottom of the ocean; the oldest marine sediments in the Atlantic are 200 million years old, a relatively young age compared with the ages of the earth and ocean of several billion years.

Water evaporates from the ocean into the atmosphere. Human-made carbon dioxide in the atmosphere exchanges with oceanic carbon dioxide; the seawater carbon dioxide reservoir is about 50 times greater than that of the atmosphere. At the air-water boundary heat exchanges also. Practically all the rain we receive comes from the ocean. The rain in turn changes into runoff and returns back to the ocean. In the runoff, eroded continental rocks, soil, and washed-out agricultural pesticides are carried into the sea.

The ocean may be considered a gigantic acid-base titration setup. Acids ejected from volcanoes, vents on the sea floor, and sulfur-rich coal burning are neutralized by alkaline substances from rocks. Today the pH of the ocean is about 8, a slightly alkaline solution which is the product of a gigantic acid-base titration with better than 99% completion of the natural geochemical neutralization. Nature has proven to be a superb chemist.

We owe the present stable oceanic pH range to the silicate and carbonate minerals in the oceanic environment. The buffer capacity of sea water, which resists pH change, is about two milliequivalents per liter (it takes about two milliliters of one normal hydrochloric acid to neutralize a liter of seawater). Lake water's buffer capacity is much less than that of seawater (about 0.8 milliequivalents per liter or less). Thus, the acid rain problem is severe in lakes but not so in the ocean. Recently Professor Iver W. Duedall of the Florida Institute of Technology calculated the maximal contribution of 12×10^{18} kg (12 billion billion kilograms) of coal burning in neutralizing the alkaline oceanic waters; he obtained a value of 30% reduction of the oceanic buffer capacity. His concurrent calculation on the elevation of seawater temperature by the total coal burning was 0.05°C. It appears the ocean can survive the onslaught of acid rain in terms of pH and energy.

For many terrestrial and freshwater organisms the present chemical makeup of the ocean is not hospitable. Pristine seawater already possesses toxic substances in solution and deposited in sediments. In our quest to understand the processes of marine pollution, we must accurately know the baseline concentration of the oceanic chemical elements and the processes leading to the establishment of the present makeup.

In general, seawater contains about 35 grams of salts in a kilogram of seawater (3.5% by weight). Its composition is shown in Table 12-1. Some of those numbers will be modified as we become more proficient in chemical analyses. Surprisingly the chemical composition is uniform among different oceans and regional seas. An analogy we can make here is the same as the smoke-filled room mentioned earlier; after many hours the smoke is widely and evenly distributed.

Seawater exerts over 20 atmospheres of osmotic pressure. Across biological membranes such as skin, osmoregulation is an important process to maintain life, without which the chemical makeup of body fluids cannot be maintained. Osmosis forces the water in body fluids to be transported toward seawater. Ancient civilizations discovered the use of salt to dry and preserve herring. The organisms which have adapted so as to osmoregulate can survive in the ocean.

Rivers add dissolved and turbid substances into the sea. Particle sedimentation and flocculation remove some of them; they accumulate on the sea floor. To illustrate the dynamics of the chemical substances, let us focus on copper as an example.

Table 12-1: Concentration of Chemical Elements in Seawater

	mg/l		µg/l		ng/l		ng/l
Chlorine	18,800	Zinc	4.9	Xenon	50	Lead	2
Sodium	10,770	Argon	4.3	Cobalt	50	Tantalum	2
Magnesium	1,290	Arsenic	3.7	Germanium	50	Yttrium	1.3
Sulfur	905	Uranium	3.2	Silver	40	Cerium	1.0
Calcium	412	Vanadium	2.5	Gallium	30	Dysprosium	0.9
Potassium	399	Aluminum	2.0	Zirconium	30	Erbium	0.8
Bromine	67	Iron	2.0	Bismuth	20	Ytterbium	0.8
Carbon	28	Nickel	1.7	Niobium	10	Gadolinium	0.7
Strontium	7.9	Titanium	1.0	Thallium	10	Praseodymium	0.6
Boron	4.5	Copper	0.5	Tin	10	Scandium	0.6
Silicon	2	Caesium	0.4	Thorium	10	Holmium	0.2
Fluorine	1.3	Chromium	0.3	Helium	7	Thulium	0.2
Lithium	0.18	Antimony	0.2	Hafnium	7	Lutetium	0.2
Nitrogen	0.15	Manganese	0.2	Beryllium	6	Indium	0.1
Rubidium	0.12	Selenium	0.2	Mercury	5	Terbium	0.1
Phosphorus	0.06	Krypton	0.2	Rhenium	4	Samarium	0.05
Iodine	0.06	Cadmium	0.1	Gold	4	Europium	0.01
Barium	0.02	Tungsten	0.1	Lanthanum	3		

The concentration of chemical elements in seawater. The units given are mg/l (milligram per liter), µg/l (microgram per liter), and ng/l (nanogram per liter). One gram (g) is 1000 mg, or 1,000,000 µg, or 1,000,000,000 ng (Courtesy of Dr. Peter G. Brewer).

Each year rivers add about 250 thousand tons of copper to the sea. Scientists discovered recently that the ionic form of copper is very toxic to marine life. Fortunately, vigorous photosynthesis occurs near river mouths. The ionic copper thus becomes bound to organic substances; these complex compounds do not possess the toxicity of ionic copper. Thus in reality, while our coastal fishing industry prospers, so is reduced the toxicity of river-borne copper. Mother Nature mysteriously cleans the sea for marine life.

The copper concentration of seawater is 0.5 micrograms per liter (parts per billion). Where does the seawater-borne copper go when rivers, submarine volcanoes, and vents continually eject copper into seawater? The majority of it will eventually settle onto the sea floor. The copper content of the Pacific deep-sea ferromanganese nodules can exceed one gram per kilogram (part per thousand) as seen in Figure 12-1. Mother

Nature again continues to purify seawater for marine life. Incidentally, the copper-rich manganese nodules are commercially exploitable, and when it happens we must study the reentrance of sediment-bound copper and other toxic substances into the marine ecosystem.

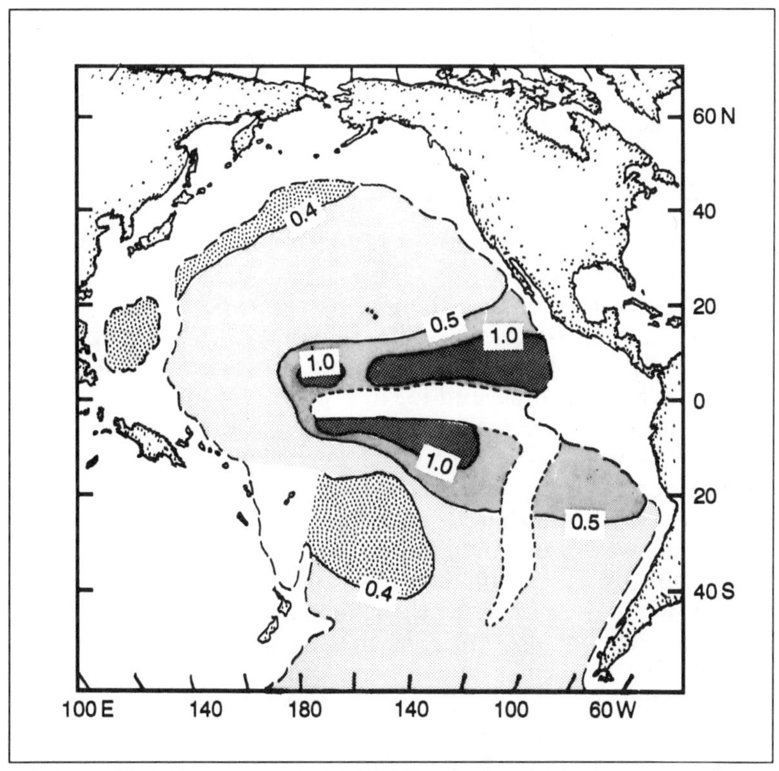

Figure 12-1: Copper content of Pacific deep-sea manganese nodules. The lines enclose locations of nodules containing more than stated copper concentrations. Lower-grade nodules are found throughout the copper-rich areas. The numbers on lines are in weight percent. Reprinted by permission of John Wiley & Sons, New York.

I personally want to stress here that the toxicity of an element is governed by its chemical forms, not solely by the elemental concentration. For copper it is ionic copper that we must deal with carefully. For cadmium, another toxic element, it is also its ionic form; when freshwater-borne cadmium mixes with seawater it forms chemical complexes with the chloride present in seawater resulting in reduction of its toxicity. So, cadmium is more toxic in freshwater than in seawater. For mercury, it is methyl mercury (an organic compound) that killed over 100 people and maimed over 700 others in Japan. Scientists are studying the chemical speciation and complex formation of toxic elements in the ocean. Some important breakthroughs are being made. For instance, we now know the mechanism of a detoxifying protein, metallothionein, that sequesters toxic metals, such as mercury and cadmium, in marine organisms.

The Biological Makeup of the Ocean

The purpose of our collective effort to minimize marine pollution is to assure the safety of people and the marine ecosystem, and to ensure the integrity of the marine environment for the continuation of life. Life in the sea is not correctly depicted by the placid display of aquariums all over the world. Struggles for existence and for the continuation of species take many different forms. The late John Dove Isaacs (1913–1980) said, "It is not in the terrestrial experience continuously to inhale the young, eggs, sperm, food and excreta of all our fellow creatures.... Although it may seem repulsive to us, it, nevertheless, is the way of marine life."

As the human population has burgeoned ahead, we have added two major complexities to marine life. We overfish and pollute. We may also enhance marine life by adding human excreta into the sea. The dynamics of marine biota is affected by all of these activities. Sharon A. MacLean of Oxford Laboratory of National Marine Fisheries Service, in Maryland, offers the following relationship to ponder:

$$V = F + P + E + M \quad (1)$$

Where
V = population variability
F = fishing pressure
P = predation
E = environment stress factors
(e.g., temperature, chemical pollutants)
M = natural mortality

We humans have collectively proven that we can overfish; the past exhaustive whaling is a notable example. Environmental factors are both natural and human-induced. The El Niño phenomenon along the Peruvian coast destroys both anchovy fishing and guano production by predator-birds. Finrot disease occurring among winter flounder off New York City is related to the magnitude and distribution of chemical pollutants in the marine sediments; it likely raises the fish's susceptibility to predation. It appears that people control the population variability of fishes in many ways, directly and indirectly.

Professor Robert M. Garrels of the University of South Florida once asked me, "Is the ocean a producing or consuming environment?" His profound question urges us to determine the inflow of organic substances from the land biosphere into the ocean as compared with their removal by sedimentary deposition and overfishing. The present marine biomass is estimated at 100 billion tons. The land biosphere produces over 10 times that each year. The annual yield of fishing from 1975 to 1978 was about 70 million tons. If we assume that there has been a delicate balance between the land and oceanic organic reservoirs, how will alteration of the land biosphere by people affect the oceanic biosphere? If the ocean were a consuming environment before, what would be the effect of less inflow of organic material due to land pollution and alteration? If the above relationship is significant, then land pollution will affect the welfare of the ocean in a fundamental way.

Dr. Leo S. Gomez of Sandia National Laboratories and his colleagues give us a drawing showing representative marine biota (Figure 12-2.). Of the marine biomass of 100 billion tons, 67 billion tons are bottom-dwelling organisms, of which 55 billion tons are distributed in the shoal waters of the continental shelves. Below 3000 meters, 0.56 billion tons of bottom-dwelling organisms cover the vast sea floor. Therefore, these great depths of the oceans are comparable to that of terrestrial deserts.

Man's Impingement

John A. Young, president of Hewlett-Packard, wrote an important editorial entitled "Quality: The competitive strategy" in *Science*, November 4, 1983. He wrote, "... [that] pursing quality is a cost-competitive strategy and that efforts to achieve

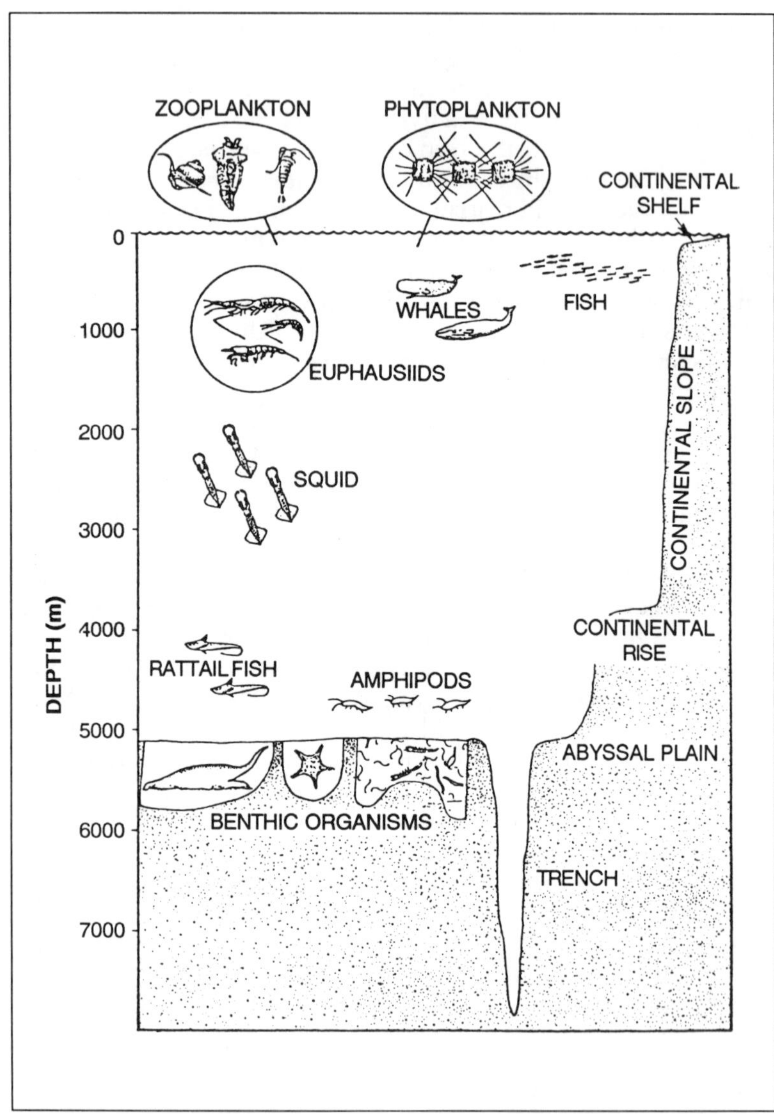

Figure 12-2: Representative oceanic biota (not drawn to scale). (From L.S. Gomez *et al.*, 1983, "Biological ramifications of the subseabed disposal of high-level nuclear wastes." In: *Wastes in the Ocean, Volume 3*, 411–427. Reprinted by permission of John Wiley & Sons, Inc.)

quality must begin in the design phase of a product. Some of our greatest improvements have been the results of designers working closely on processes with people in manufacturing, on parts specifications with our vendors, and on applications needs with our customers." His statement is directly applicable to our quest to cope with worldwide marine pollution. Waste management experts including oceanographers must work in the design phase of industrialization rather than be left at the back-end of it. Chess games are often decided by the end-games in which winners earlier strategically have placed their pawns well; these pawns may be the waste management specialists. In the case of the Minamata Bay mercury poisoning, the Chisso company by 1975 had paid indemnities totaling more than $80 million to the people affected; their inadequate waste management had cost dearly.

The majority of the present marine pollution problems are due to our shortsightedness and inexperience. In addition, the ocean, an international resource, does not have a comparable constituency to what we witness in coastal protection, urban air pollution, drinking water fluoridation, and noise abatement. The ocean is not protected as vigorously and urgently as is the land. Humankind has been fortunate to rely on the yet-to-be-quantified oceanic ability to cleanse and renew itself.

There are several modes for pollutants to enter the ocean: river discharge; atmospheric transport followed by precipitation into the sea; coastal discharge through outfalls; sea-going vessel generated pollution; and ocean dumping. Contrary to our earlier perception, DDT and its degradation products and PCBs (polychlorinated biphenyls) primarily enter the ocean from the continents via the atmosphere, in addition to river runoff.

Professor Edward D. Goldberg of the University of California at San Diego estimated in 1976 that we produce three billion tons of wastes annually; that is about 10 times the total human body weight on earth, requiring much care and energy to manage. In addition, we are in the midst of an epoch in human history of fossil-fuel exploitation (Figure 12-3). Fossil-fuel burning of over two million billion kilowatt-hours (65% from burning two billion tons of coal) during the next several hundred years will add further oxides of carbon, sulfur, and nitrogen into the atmosphere as pollutants. A substantial amount of these wastes eventually reaches the ocean. I also

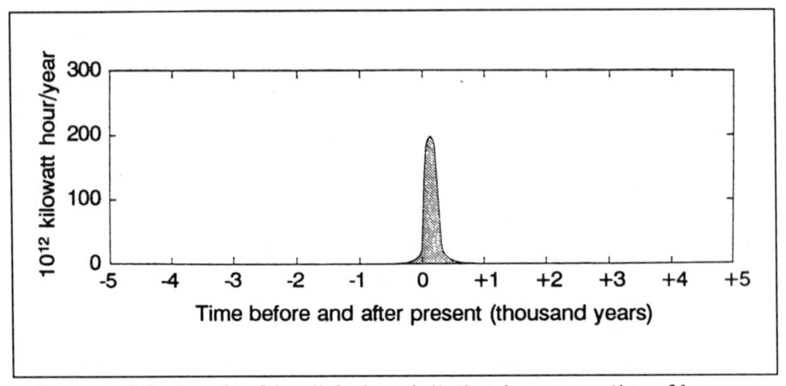

Figure 12-3: Epoch of fossil-fuel exploitation in perspective of human history from 5,000 years in the past to 5,000 years in the future. (From M.K. Hubbert, 1973, "Survey of world energy resources," *Canadian Mining and Metallurgical Bulletin, 66*, 37–53. Reprinted by permission of the Canadian Institute of Mining and Metallurgy.)

should mention the impact of the change in land use by people. Deforestation and construction of roads and cities augment the entrance of human-induced pollutants into the sea via the atmosphere and runoff.

The components of the earth system are all connected and interwoven. The amount of materials people move over the surface of the earth, three billion tons per year, is equivalent to almost 10% of nature's activity through the weathering cycle; it is the same magnitude of substances that glaciers currently transport into the ocean.

Metals

The tragic deaths of over 100 people by methyl mercury poisoning in Minamata have taught us many lessons. We have learned that we must pay attention to the toxicity of metals and their various ionic and complex substances, the persistence of these in the marine environment, and marine food web magnification (bioaccumulation).

Virtually all metals are found in the ocean (Table 12-1). Some metals such as mercury, silver, and copper are extremely toxic, while others are innocuous to marine organisms. Dr. Michael Waldichuck of the Canadian Department of Fisheries and Oceans gives us the following order of toxicity to marine life of cationic metals:

mercury > silver > copper > zinc > nickel > lead
cadmium > arsenic > chromium > tin > iron > manganese
aluminum > beryllium > lithium

For mercury we learned that methyl mercury, an organic complex, is highly toxic and readily soluble in body fats.

Five tragedies of human poisoning due to toxic metals have been documented in Japan: mercury poisonings in Minamata Bay (1953–1960), in Niigata (1964–1965), and in Goshonoura (1973); cadmium poisoning at Fuchu (1947–1965); and chromium poisoning in Tokyo (1975). In each case these metals in sediments were transferred through the marine food web and subsequently were consumed by people. Therefore, it is important to learn the mechanisms of metal transfer from sediments to marine life.

Any significant increase in toxic metals in seawater, sediments, and marine life presents us an early warning signal. A group of international scientists has initiated a global mussel watch. Along the east and west coasts of the United States numerous sampling stations were set up in 1976. Dr. John W. Farrington of the University of Massachusetts, Boston reported in 1983 that the sentinel program is working well. For instance, cadmium was taken up by the mussels along the central California coast during the period of coastal upwelling of deep-sea water during the summer months; its magnitude was about four micrograms per gram of dry mussel tissue (parts per million) while at other mussel sampling stations its magnitude was about one-half of that.

Ocean-dumping of mercury, cadmium, and their compounds is prohibited by an international agreement (London Dumping Convention), while metallic substances requiring special care are arsenic, lead, copper, zinc, and their compounds. As an example of global concern, mercury pollution is discussed below.

Mercury is dangerous for humans. Sensitive human adults show symptoms of mercury poisoning at a weekly intake level of 1.3–2.9 milligrams. The Minamata victims consumed about 14 milligrams of mercury per week. Mercury levels tolerated by the United Nations World Health Organization (WHO) and the United Nations Food and Agriculture Organization (FAO) in human food is a weekly consumption of up to 0.2 milligrams of methyl mercury or 0.3 milligrams of total mercury; these consumption rates have a safety factor of tenfold. Professor Dr. Sebastian A. Gerlach of the Institut für Meereskunde, Kiel, informs us that an average citizen in the Federal Republic of Germany in 1980 had a weekly intake of 0.2 milligrams of mercury (20% from grain, 15% from beer and other alcoholic beverages; 10% from milk and butter, 8% from fish and the remainder from other sources). The mercury concentration of fish is generally on the order of 0.1 milligram per kilogram, mostly as methyl mercury. Therefore, weekly consumption of two kilograms of fish alone would be the upper limit of tolerable mercury intake.

Let us examine the mercury content of the ocean. Total mercury in seawater is about seven million tons. Each year rainfall over the ocean adds about 4,000 tons and rivers add 1,000 tons. The industrial production of mercury in 1968 was about 9,000 tons. If one-half of the human-made mercury entered the ocean, its magnitude would be comparable to the natural addition of rainfall and rivers combined. Sedimentation onto the sea floor may remove 1,000 tons of mercury per year. Thus, this rough calculation shows that the net addition per annum of mercury by nature and humans is on the order of 0.1% of the total seawater mercury content. As such, oceanic fish away from the coast will not show an increase in mercury concentration as a result of human input. However, pockets of local pollution near industrial activities may show mercury contamination in sediments and biota. For instance, in the Santa Barbara Basin off California, the mean mercury concentration in sediments increased twofold in recent years when compared with that of pre-industrialization (Figure 12-4).

An important concern we have about mercury is that mercury concentrates as it goes upward through the marine food web. Studies by Willard Bascom and his coworkers at the Southern California Coastal Water Research Project, Long Beach (1982), yielded fundamental data on bioaccumulation. Henry A. Schaefer and colleagues at the Project showed that mercury concentration increases with the advancement of

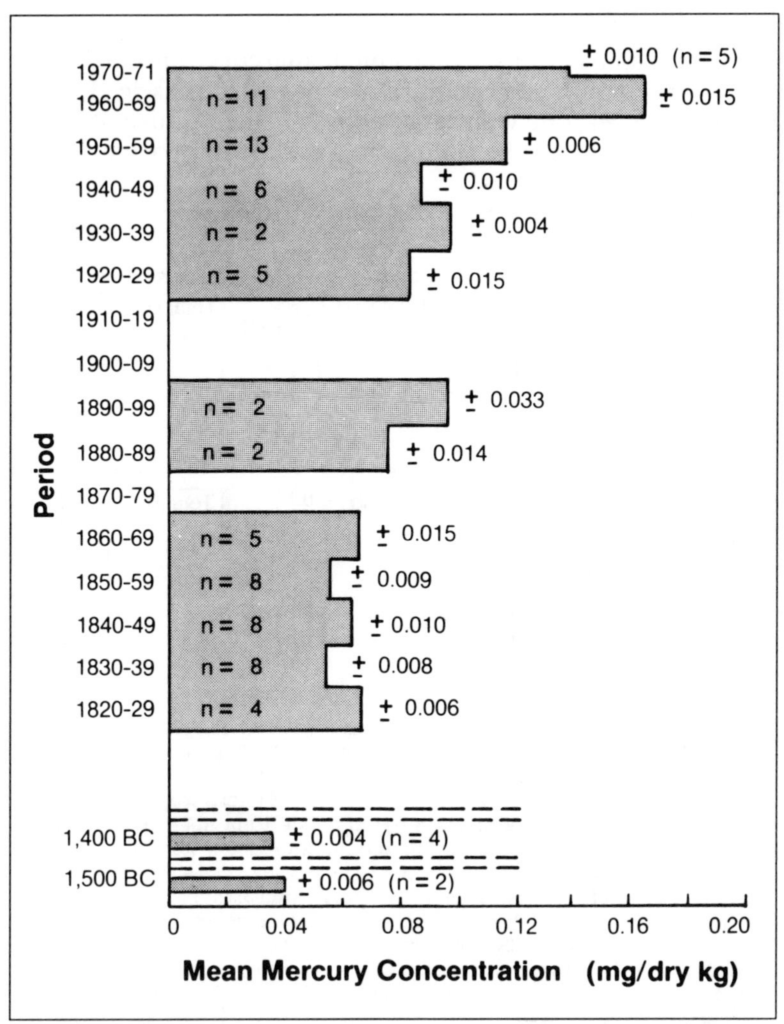

Figure 12-4: Mercury concentration in the sediment of Santa Barbara Basin off California. The water depth was 580 meters where the samples were taken. The term n expresses the number of samples analyzed; the numbers after the ± symbol represent the range of analytical variance (standard deviation). (From D.R. Young *et al.*, 1973, "Mercury concentrations in dated varved marine sediment collected off Southern California," *Nature* 244: 273–274. Reprinted by permission of Macmillan Journals Limited.)

trophic level. Top level predators, such as the silk shark, mako shark, and spiny dogfish, had similar mercury concentrations of 1.5 milligrams in a kilogram of wet edible muscle (Figure 12–5). However, one white shark contained an unusually high value of 8 milligrams per kilogram, possibly due to age.

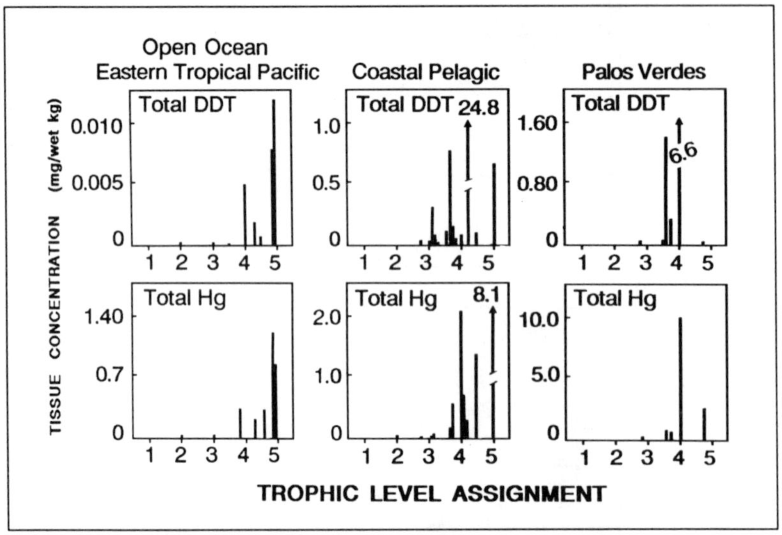

Figure 12-5: Two examples of contaminants (DDT and mercury) that increase with increasing trophic level. Species and their trophic level assignment are as follows: For the open ocean (Eastern tropical Pacific), yellowfin tuna 4.82; silk shark 4.81; skipjack tuna 4.44; frigate tuna 3.92; squid 3.52; flying fish 3.00; zooplankton 2.00. For the Coastal Pelagic, white shark 5.02; mako shark 4.40; sea lion 4.02; swordfish 3.97; thresher shark 3.82; bonita 3.80; barracuda 3.74; Pacific mackerel 3.54; market squid 3.52; Pacific hake 3.09; jack mackerel 3.04; sardine 3.01; basking shark 3.00; blue whale 3.00; anchovy 2.82. For Palos Verdes, near Los Angeles, scorpion fish 4.53; spiny dogfish 4.16; Dover sole 3.52; white croaker 3.36; ridgeback prawn 3.33; mysids and decapods 2.78.

Since we, too, are top-level predators it is possible to bioconcentrate mercury in our bodies. To reduce it, it is prudent to eat fish with low mercury content. Candidate "fish" may be large zooplankton such as krill, anchovies, sardines, and squid; their mercury concentrations reported by Schaefer and his coworkers are in the range of 0.02 to 0.05 milligrams per kilogram. Conversely, bonita, barracuda, yellowfin tuna, and frigate tuna contain mercury in the range of 0.2 to 0.3 milligrams per kilogram.

Synthetic Organic Substances

Our technological society has been synthesizing many new organic substances for our civilization. Wastes generated by the production and use of these substances are adding foreign matter to the sea which did not exist there before. Scientists have coined a new name for these alien substances which affect the marine ecosystem; they are called xenobiotic substances.

A particular concern we have is that some harmful xenobiotic substances may accumulate in the ocean since nature may not be capable of degrading them to innocuous substances. Conversely, since the ocean did not have these xenobiotic substances before humans introduced them, they have become human-made tracers for understanding the dynamics of the ocean (Figure 12-6). For instance, chlorinated hydrocarbons such as DDT and PCBs were found in the biota (plankton, sponges, and fish) of Antarctica, though at low levels, on the order of one part per billion (wet weight) or less. The eggs of Antarctic penguins contained them also.

Among various xenobiotic substances, chlorinated hydrocarbon compounds, such as DDT, PCBs, hexachlorobenzene (HCB) and Mirex, have been studied in the marine environment because they have caused ecological damage. For years to come, even if the manufacturing of these chemicals is halted, I believe there should be continued research on them for two reasons: the xenobiotic substances already produced will enter the ocean in any case, and to learn the processes associated with the recovery phase of the ocean from the insults.

It took 20 years to learn how DDT and its residues caused egg shell thinning and consequent reproductive failure in seabirds. The brown pelican population of Anacapa Island off California had drastically decreased from 1969 to 1972. Professor R. Risebrough of University of California attributed it to egg shell thinning due to DDT residues, mainly DDE (dichloro-diphenyl-ethylene). With DDE concentrations of about 5 and 70 parts per million in the pelican eggs, the egg shells became thinner by 15 and 35%, respectively (Figure 12-7). As a consequence of

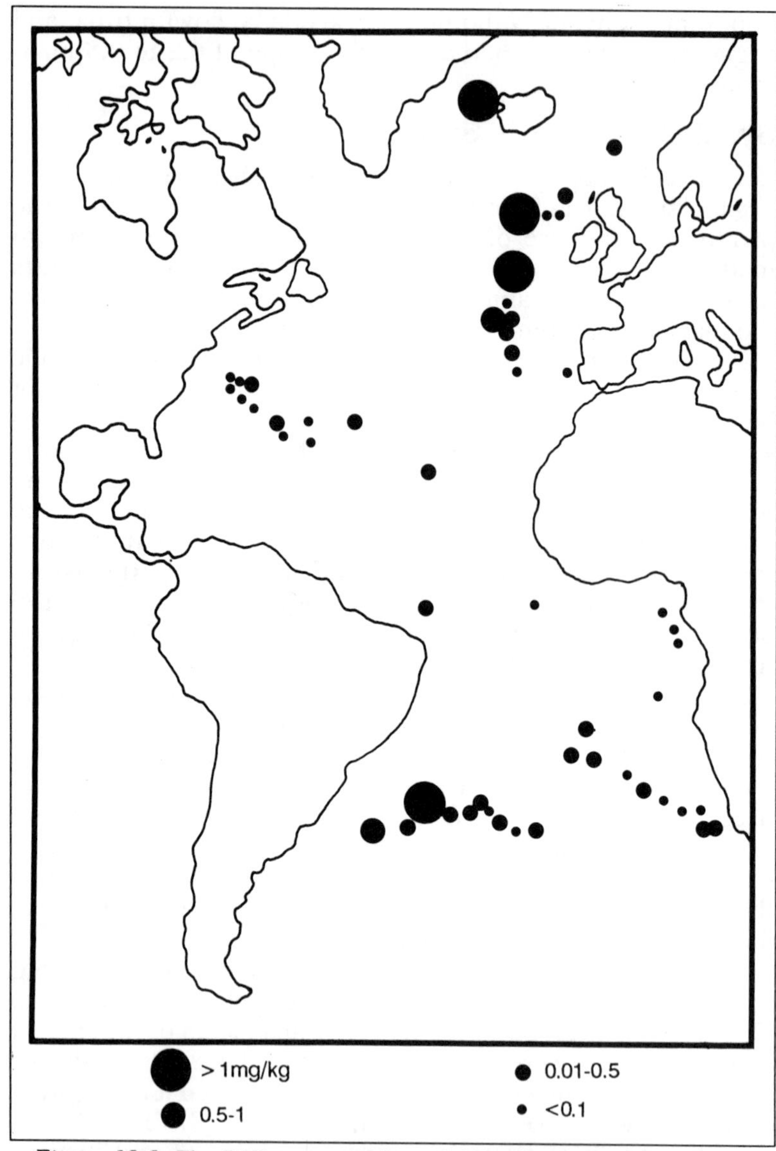

Figure 12-6: The PCB concentration of Atlantic zooplankton, 1970-1972. (From Harvey *et al.*, 1974, "Observations on the distribution of chlorinated hydrocarbons in Atlantic Ocean organisms," *Journal of Marine Research* 38: 103–118. Reprinted by permission of the authors.)

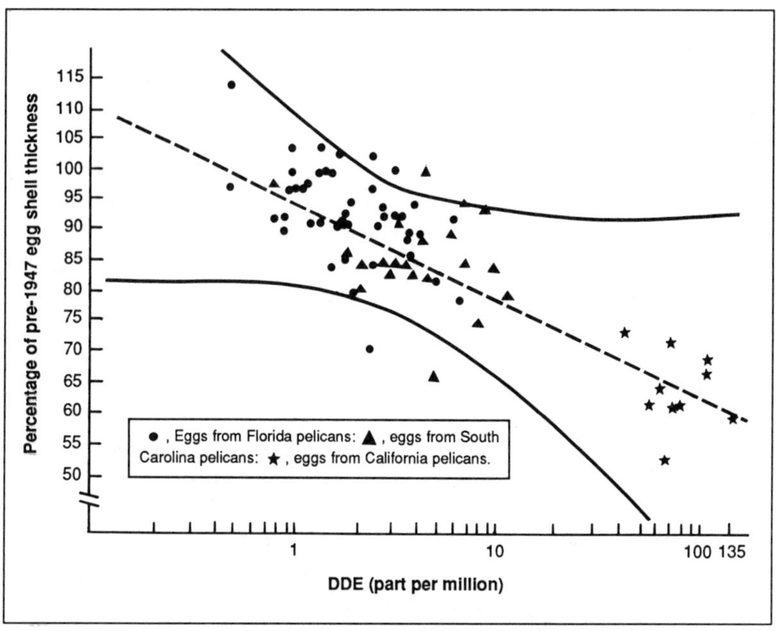

Figure 12-7: The relationship between egg shell thinning and DDE in eighty brown pelican eggs. Solid lines represent 95% confidence limits. (From L.J. Blus *et al.* 1972, "Logarithmic relationship of DDE residues to eggshell thinning," *Nature* 235: 376–377. Reprinted by permission of Macmillan Journals Limited.)

the thinning, the eggs broke quite easily. The source of the DDT residue was a factory in Los Angeles that discharged about 200–500 kilograms of the residues everyday to the sea through the Los Angeles sewage system. These DDT residues were bioaccumulated by fish (Figure 12-5) and the fish were eaten by the pelicans.

The DDT story gives us lessons. The first is that damage to the marine ecosystem can be manifested in the avian domain. The second is that seabirds can be our sentinels to monitor the extent of marine pollution.

The World Health Organization established an acceptable daily intake limit of 0.3 milligrams of DDT per person. The edible muscle of cod caught at George's Bank in the Atlantic Ocean in 1971 contained about 0.01 milligrams per kilogram (wet weight) of DDT and its residues. Therefore, DDT does not pose any significant danger to people who enjoy George's Bank cod. In fact, no harmful effect of DDT upon human health has been observed. In experiments, however, concentrations of one part per billion of DDT in seawater showed a toxic effect on Sargasso Sea diatoms (phytoplankton).

The worldwide production of DDT increased from the 1940s to the 1960s. An estimate given is that almost three million tons were produced and used up by the year 1974. Recently, the use of DDT has been prohibited in some countries; the United States banned its use in 1972. In recent years concentrations of DDT and other pesticides in marine organisms and seabirds have been decreasing (Figure 12-8). Interestingly, the population of brown pelicans in California is increasing. The DDT story demonstrates that we people collectively can reverse marine pollution. The knowledge we persistently have gained in research, monitoring, and vigilance with an open mind has helped to eliminate a cause of pollution. A program of action without a program of learning is often ineffective and costly. I personally salute the dedicated marine pollution scientists for their 20-year effort to understand the dynamics of DDT in the marine environment.

Polychlorinated biphenyls (PCBs), like DDT, are associated with the lipid part of marine life; they inhibit the growth of diatoms at as low a level as 25 parts per billion, and kill juvenile shrimp at five parts per billion in seawater (75% mortality in 20 days). In 1968 over 5,000 people in Japan experienced PCB poisoning with chloracne-like skin eruption; four deaths were attributed to PCBs. The WHO reports that skin effects may occur in man at a daily ingestion of 4.2 milligrams of PCBs; there have been no acceptable standards established by the WHO for daily intake of PCBs by humans.

PCBs are highly stable, non-flammable, and have low water solubility, low volatility and do not conduct electricity, but can sustain an electric field. Therefore, they were used in electric transformers and capacitors. They were also used as hydraulic fluids in high temperature systems. Other uses of PCBs included plasticizers, lubricating and cutting oils, sealants, and as components in paints and printing inks. Therefore, PCBs were broadly distributed and from those sources the ocean has received them (See Figure 12-6).

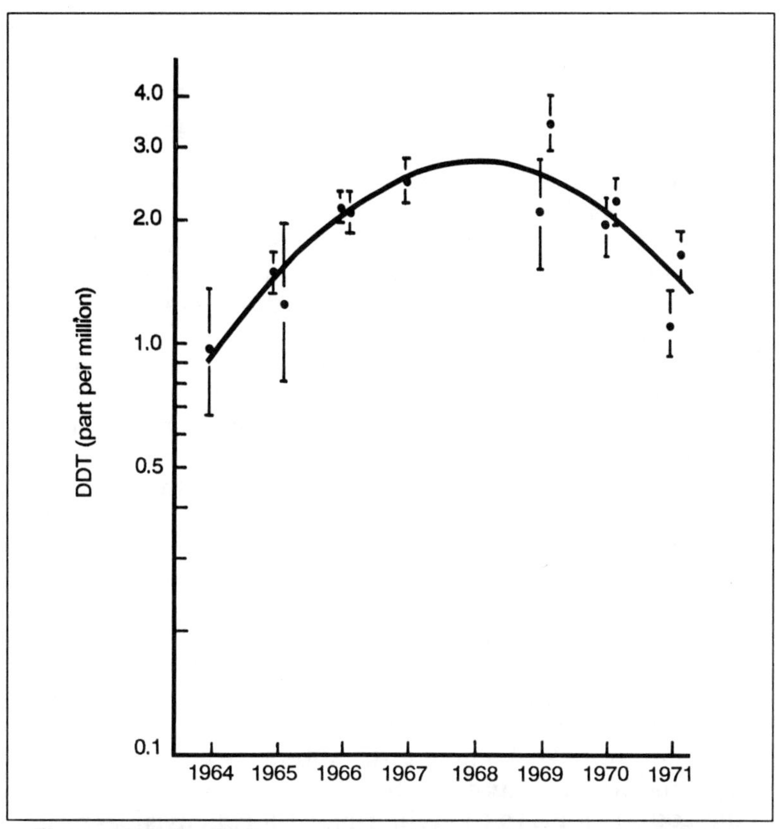

Figure 12-8: The DDT concentration in the eggs of shag at two British bird colonies located at Farne, Northumberland (shown by points) and Isle of May, Scotland (shown by triangles). (Adapted from J.C. Coulson *et al.*, 1972, "Changes in organochlorine contamination of the marine environment of Eastern Britain monitored by shag eggs," *Nature* 236: 454–456. Reprinted by permission of Macmillan Journals Limited.)

About one million tons of PCBs have been produced worldwide; the United States has produced about 50% of them. In 1971 PCB production (in tons) for several industrialized countries was as follows: United States 18,000; the Federal Republic of Germany 8,000; France 7,600; Japan 6,800; the United Kingdom 5,000; Italy 1,500; Spain 1,500. The PCBs, too, appear to be transported primarily from the continents through the atmosphere to the ocean. Recognizing the threat of PCBs to environmental resources, many countries began reduction or cessation of PCB manufacturing. In 1977 the United States ceased production. At present spent PCBs are collected and incinerated at sea.

In most cases PCB concentrations in fish are less than those in meat of 0.3 milligrams per kilogram. For instance, the edible muscle of cod from George's Bank in 1971 had a PCB concentration of 0.04 milligrams per kilogram. In some locations, though, PCBs are concentrated in the sediments and may make their way into the food web. For instance, in 1978, the United States mussel watch program detected heavy PCB pollution in New Bedford Harbor, Massachusetts. The PCB concentration in mussels was as high as 31 milligrams per kilogram; this was six times greater than the United States Food and Drug Administration's recommended upper limit of five milligrams per kilogram in seafood for human consumption. This discovery occurred a year after the cessation of PCB production in the United States. The PCBs had entered the harbor area via two electrical component manufacturing companies nearby. How we will remove the New Bedford Harbor PCBs safely is a challenging operation; we should learn from it.

From the PCB and other studies we must be able to recognize substances which are not suitable for our environment. The criteria we want to know are the substance's toxicity, persistence, and accumulation in the environment (ecosystem). In some countries PCTs (polychlorinated terphenyls) were produced as flame-retardants and in the production of paints and glues. Japan alone produced 2,700 tons of PCTs from 1954 to 1972. When PCTs became known to be toxic, persistent, and accumulating, Japan and the United States ceased their production in 1972. I advocate the toxicity-persistence-accumulation criteria test be applied by manufacturers before a new chemical emerges in our society.

Petroleum Hydrocarbon Compounds

On March 16, 1978, the supertanker *Amoco Cadiz* was grounded off the coast of northern Brittany, France, and its entire 200,000 ton cargo of Arabian crude oil and bunker fuel was released into the marine environment. Seabirds, mussels, and other marine life were affected catastrophically. Other similar events are reported occasionally including offshore oil well blowouts. For instance, the Ixtoc-I platform spill of 1979 in the Bay of Campeche released 400,000 tons of crude oil into the Gulf of Mexico. Major sources of petroleum hydrocarbon compounds enter the ocean unobtrusively in small quantities scattered over many shipping lanes (Figure 12-9) and along coastal zones. In 1975, the United States National Academy of Sciences compiled the major input sources per year as follows:

Transportation activities	2,100,000	tons
River runoff	1,600,000	
Atmospheric fallout	600,000	
Natural seeps	600,000	
Urban runoff	300,000	
Industrial wastes	300,000	
Municipal wastes	300,000	
Coastal refineries	200,000	
Offshore oil production	100,000	
TOTAL:	6,100,000	tons

The above estimate includes 200,000 tons per year for tanker accidents within the transportation activities input of 2.1 million tons.

Transportation-related input of petroleum is one-third of the total. It results primarily from operational discharges. For instance, after discharging cargo oil, tankers on a return voyage used to fill the cargo tanks with seawater to maintain stability. The resulting mixture of ballast seawater and residual cargo oil was pumped overboard before arriving into port. Efforts have been made to reduce the oil content in ballast and tank-cleaning water.

Crude oils are complex mixtures of thousands of hydrocarbons (organic compounds composed solely of carbon and hydrogen atoms), trace amounts of metals such as nickel, vanadium and iron, plus porphyrins, and various organic compounds containing sulfur, nitrogen, and oxygen. The deleterious effects of petroleum to living organisms include mechanical clogging and blanketing, irritation to mucous membranes and respiratory surfaces, and the chemical toxicity of aromatic hydrocarbons which contain one or more highly toxic benzene rings and other harmful substances. Many scientists,

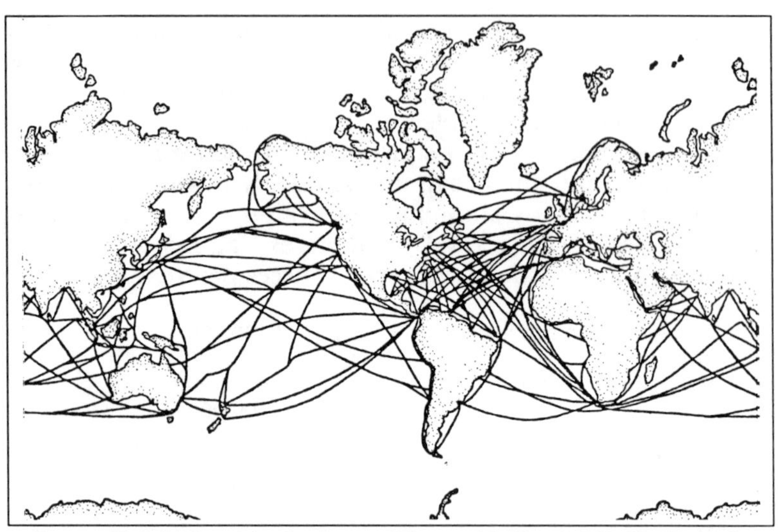

Figure 12-9: Major shipping lanes of the world. (From G.R. Heath *et al.*, 1983, "Why consider subseabed disposal of high-level nuclear wastes?" In: *Wastes in the Ocean, Volume* 3: 303–325. Reprinted by permission of John Wiley & Sons, Inc.)

therefore, have been focusing on the fate and effects of aromatic hydrocarbons and their metabolic and degradation products. For instance, Dr. Donald C. Malins who worked at the Northwest and Alaska Fisheries Center, Seattle, inquires into the putative relationship between the formation of metabolic products coming from aromatic hydrocarbons and the onset of tumor growth and other cellular aberrations in marine organisms. We also know now that aromatic hydrocarbons are soluble in lipids and can accumulate in fish. In the marine environment they settle to the sea floor by attaching to sinking particles.

People, too, can be affected. Shortly after the 1978 *Amoco Cadiz* oil spill, nearby residents experienced headaches, dizziness, nausea and sensations of inebriation, vomiting and abdominal pains. Skin irritation occurred among people who had direct contact with the oil. Carcinogenic aromatic hydrocarbons may be accumulated by food organisms and consumed by humans.

Petroleum discharged into the ocean may go through various processes that include evaporation, dissolution into seawater, emulsification, agglomeration, dispersion, sedimentation, photochemical oxidation, and microbial biodegradation. Mousse and tar balls, too, are produced. Eventually, biodegradation, however slow, will self-purify the marine environment.

An overriding problem facing the scientists who study petroleum pollution is its immense analytical difficulty. Even sophisticated analytical techniques can only measure a small part of the complex sample. Aromatic hydrocarbons with one to five benzene rings can be determined, but not the polar components of crude and refined oils. Also, since petroleum hydrocarbons are often associated with DDT, PCBs, and other toxic substances, their combined effects of synergism and antagonism must be studied to know what actually happens in nature.

Radioactive Pollution

By far the largest radioactivity input to the ocean by man is atmospheric fallout generated by over 1,200 military nuclear tests. In 1961 and 1962 about 50 billion billion becquerels (disintegrations per second) of radioactive hydrogen, hydrogen-3, entered the atmosphere as heavy water via atmospheric hydrogen bomb tests. It then exceeded the natural radioactive hydrogen inventory by about 100-fold. Since hydrogen-3 loses one-half of its radioactivity after 12 years, we have less than one-fourth of the original radioactivity at present. Professor Göte Östlund of the University of Miami and his colleagues have measured the hydrogen-3 distribution in the ocean (Figure 12-10). In 1972 they discovered that hydrogen-3 penetrated down to almost 5 kilometers in depth in the high latitudes of the North Atlantic Ocean. They also found it spread over 3,000 kilometers horizontally at depths between 1 and 2 kilometers.

Other military wastes exist in the ocean. Some examples resulted from the accidental sinking of two nuclear submarines and the loss of airborne nuclear weapons near Palomares, Spain, and Thule, Greenland. In addition, satellites with nuclear power generators have fallen to the earth. They all may constitute an appreciable, yet not quantifiable, source of human-made radioactivity in the ocean.

On the other hand, radioactive waste disposal by our society has been minimal in comparison with the fallout. The United

States disposed of 107,000 canisters of low-level wastes from 1946 to 1970 with a total radioactivity at the time of disposal of 4.3 million becquerels. The recently active dumpsite in the ocean, up to 1982, is in the northeastern Atlantic Ocean (Figure 12-11); during the 12-year period of 1967 to 1979 this site received 22 million billion becquerels of radioactivity. However, the site has not received any wastes since 1983. Nuclear

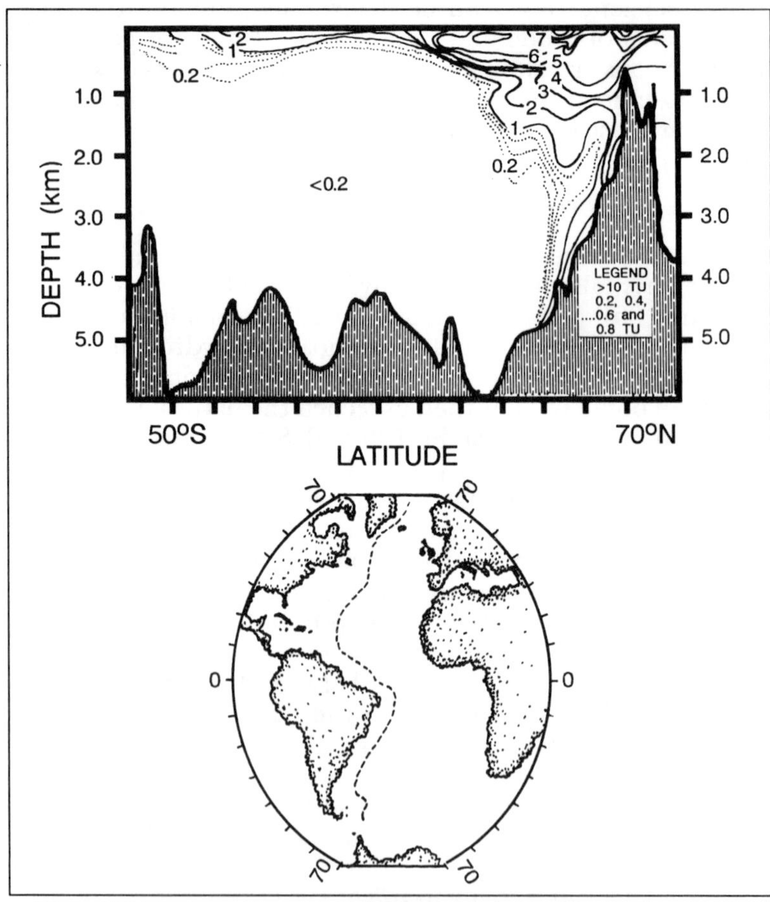

Figure 12-10: The radioactive hydrogen-3 intrusion into the deep North Atlantic water, 1972. The unit TU is the atom ratio of hydrogen-3 over hydrogen-1 (nonradioactive) multiplied a billion billion times (10^{18}). The lower figure shows the sampling line.

fuel reprocessing plants discharge a similar magnitude of radioactivity into coastal waters as effluent.

Contrary to our perception, the ocean is naturally radioactive. One liter of seawater gives off 11 becquerels of potassium-40 radioactivity. Therefore, the entire ocean possesses 15,000

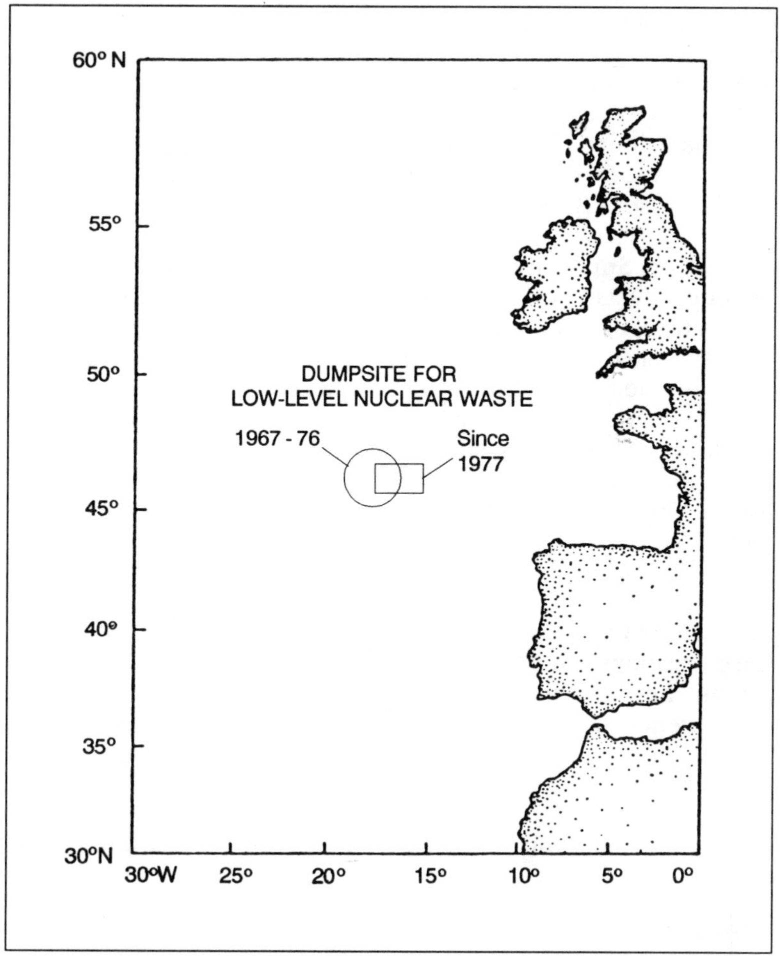

Figure 12-11: Northeastern Atlantic dumpsite for low-level radioactive wastes. It is used by several European member countries of the Nuclear Energy Agency within the Organization for Economic Cooperation and Development.

billion billion becquerels of radioactivity in seawater. In addition, deep-sea sediments are reported to have 500 to 20,000 becquerels of radioactivity in one kilogram, while manganese nodules scattered over the sea floor have 6,000 to 200,000 becquerels in one kilogram. Human bodies, too, are radioactive; a living human adult contains carbon-14 and potassium-40, each of which emits 4,000 disintegrations per *second*. Therefore, when we study radioactive pollution, we should consider the level of natural radioactivity as a baseline.

What are the effects of radiation? Acute and massive exposure experienced by atomic bomb victims resulted in vomiting, convulsions, tremor, loss of muscular control, and subsequent death. Lesser effects included diarrhea, fever, damage to bone marrow, infection, loss of hair, hemorrhage, fatigue, and loss of appetite. Small amounts of radiation may initiate cancer and genetic defects in offspring. For instance, leukemia among the 1945 atomic bomb survivors was first discovered three years after the blast and peaked in 1953. On the other hand, minor damage due to radiation can be repaired within our bodies.

The ocean has not received any high-level radioactive wastes. These wastes are being stored on land. According to *U.S. News & World Report*, August 15, 1983, we have accumulated 8,000 tons of high-level wastes in temporary underwater storage on land.

We have been in the nuclear age for 45 years. Regardless of an individual's view on the future of nuclear technology for power and weaponry, nuclear wastes are already with us. We have a responsibility to future generations to find a safe disposal method and safe disposal sites as well as to develop an adequate contingency plan which will be utilized if our chosen plan becomes unusable. What is safe? Could the ocean be a safer place than land since we live on land? Are we rational enough to transcend our "NIMBY" (not in my backyard) syndrome for the sake of humanity? Our choices need careful consideration and consensus.

Solid Wastes

Over six million tons of solid wastes are inadvertently scattered over the sea every year. Most of this garbage affects coastal areas. Some of it sinks and some floats. It consists of paper, metal, textiles, glass, plastic, rubber, and wood. While this kind of waste does not pose a direct hazard to us most of the time, ship's propellers may become entangled by nylon ropes and plastic sheets. For this and other reasons, throwing garbage overboard from ships is discouraged. In fact, there is an international agreement by the London Dumping Convention to prohibit the disposal into the sea of "persistent plastics and other persistent synthetic materials, for example, netting and ropes, which may float or may remain in suspension in the sea in such a manner as to interfere materially with fishing, navigation or other legitimate uses of the sea."

Recently, Professor Dana R. Kester of the University of Rhode Island visited a preserved wildlife island off the coast of New Zealand. While impressed with the birds walking between his legs without fear, Professor Kester noticed the blight of marine litter, plastic artifacts scattered along the shore. Amenities uplift the human spirit; litter-strewn beaches do not.

In addition to visual litter along a shoreline, plastic is distributed widely in the ocean. In the Sargasso Sea, 3,500 plastic pellets were netted on the average within an area of a square kilometer. Some plastic particles are ingested by organisms and they have been found in the intestines of many marine animals. We collectively must minimize the release of these persistent wastes to the environment. Professor Kester advocates that we strive to make synthetic products that are environmentally degradable by micro-organisms, chemical decomposition, or ultraviolet photolysis.

Sewage and Industrial Wastes

On December 30, 1675, Governor Edmund Andros, second English Governor of the Colony of New York, forbade any person to "cast any dung, dirt, refuse of the city or anything to fill up the harbor or among the neighbors under penalty of forty shillings." Today more than 18 million people live around New York's coastal ocean; this region houses a giant industrial complex and shelters the world's busiest port. The population of this area made a huge jump from less than 20 thousand in the middle of the 17th century to today's population. Along with it, the coastal water has received and been stressed by the inhabitants' wastes, both domestic and industrial. Municipal sewage, storm water runoff, and industrial discharges find their way to the coastal ocean. Unfortunately, the New York story is being reenacted at other megalopoli. Dr. Donald F. Squires of the University of Connecticut recommends that we learn about the interactions of man and the ocean from the extensive information already available in this region.

Sewage contains organic matter, nutrients such as phosphate, bacteria, viruses, and parasitic worms. Some of these bacteria and viruses are pathogenic. When industrial effluents are connected to sewage systems, as is done in New York, a wide range of chemical substances, including oil and metals, enter the sewage system and subsequently the sea. Sewage may reach the sea untreated via storm sewers or may be treated before discharge. The treatment can be mechanical settling of particles (solids), or additional biological or chemical treatment to break down organic matter. Treatment produces effluent and sludge; they are disposed of separately. The sludge from New York used to be ocean-dumped at a site 20 kilometers from the coast; now it is dumped at 170 kilometers offshore.

The worldwide quantity of sewage wastes disposed of into the ocean is proportional to the population size of the coastal inhabitants. When possible, people use sewage for agricultural and aquacultural enhancement. The total sewage sludge the United States produced in 1979 was about six million dry tons. Since an average solid content may be about 3%, the wet sludge coming from sewage treatment plants is about 200 million wet tons per year. Of that number, approximately six million wet tons of sludge were ocean-dumped off New York and about three million wet tons were discharged into the seas through outfalls at Boston and Los Angeles. Therefore, about 5% of the sewage sludge of the United States goes into the ocean. For the United Kingdom, a higher percentage of sewage sludge may enter the sea, since oceanic sludge disposal is practiced widely there. In 1979, about 17 million wet tons of sewage sludge were ocean-dumped by the London Dumping Convention member states.

Transmission of infectious diseases is a major concern of sewage waste disposal for both effluent and sludge. Gastrointestinal illness can occur as a result of bathing in and consuming shellfish from contaminated sites. During the past decade several outbreaks of infectious hepatitis, shigellosis, and cholera have been attributed to either swimming in, or eating shellfish taken from, sewage contaminated areas in the United States. Chlorination of the sewage effluent has been attempted to eliminate the disease organisms, but carcinogenic chlorine containing compounds may be produced in seawater inadvertently.

The oceanic disposal of industrial wastes is well documented by the International Maritime Organization (IMO) which is the secretariat to the London Dumping Convention. For the year 1979 approximately 18 million tons of these wastes were disposed of in the ocean. Some industrial wastes also enter the sea as effluent. However, stringent regulations are enforced to prevent any toxic effluent from entering the environment.

Ocean-disposed industrial wastes include mixtures of hydrochloric or sulfuric acids with iron generated in the production of titanium dioxide pigment, diluted acids, fish offal, ammunition, copper-containing dilute acid, coal ash and colliery wastes, drums and canisters, construction debris, various pharmaceutical wastes, and sodium sulfate solutions. In the Gulf of Mexico metallic sodium was ocean-dumped in the 1950s and 1970s.

I raise concern over the wisdom of connecting domestic sewer systems with industrial effluent systems. Such mixed wastes may be treated ineffectively at sewage treatment plants. In addition, much of our domestic waste can be utilized as fertilizer, especially for silviculture and land reclamation in abandoned open-pit mines. Usable wastes should be reused rather than mixed with toxic wastes. In June 1983 I visited a small sewage treatment plant amidst an apartment complex of 90 thousand inhabitants in Shanghai. The plant separates sewage into effluent and sludge. The effluent is discharged directly into a river nearby and the sludge is carted off to farmland around Shanghai. Thus, the sewage from 90 thousand people is treated at the source and the usable part recycled before it gets mixed with the industrial waste of Shanghai.

Dredged Material

Dredging of waterways and harbors is a continual process, performed in order to maintain marine transportation. Dredge spoil problems occur when coastal and harbor sediments are contaminated by pollutants from municipal, shipping and industrial sources. Conversely, non-polluted dredged material is a resource to nourish beaches, reclaim land, and for construction (sand).

The global magnitude of dredge operations is difficult to estimate. The International Association of Ports and Harbors (IAPH) asked 70 countries to report their dredging activities; it obtained 108 responses from 37 countries. Based on them Professor Kester estimated that 350 million tons of maintenance dredging and 230 million tons of new construction dredging had occurred in 1979. It appears, therefore, that global dredging activities are on the order of one billion tons per year. Contaminated dredged material may be a small fraction of this total.

The substances of concern are toxic metals, synthetic organic substances such as DDT and PCBs, and oil and grease; their concentrations vary depending on the proximity to and magnitude of the original contaminant sources. Pathogens in the sediments are also of human health concern. One of the most important considerations is the remobilization of toxic substances and pathogens from the sediment phase to the water phase above and to the biota. Scientists now know that zinc is released from sediments at low pH (around 5) when seawater still contains some dissolved oxygen; they also know now that copper is released easily when the dissolved oxygen content is very high.

What should we do with highly polluted dredged materials? In order to isolate them, capping experiments have been conducted in the United States and Canada. Contaminated sediments are capped with clean sediments to immobilize toxic substances. There is an artificial island, Crany Island, near Norfolk, Virginia, constructed to isolate the spoils. Here attempts are made to immobilize and isolate them simultaneously.

Discussion

When the dissolved oxygen content of seawater decreased off New York, and fish and shellfish died *en masse* in 1976, marine pollution was blamed as the cause (Figure 12-12). Scientific studies showed later that an extensive bloom of a phytoplankton (dinoflagellate, *Ceratium tripos*) was its cause. Though the unusual weather and oceanic conditions helped the phytoplankton to bloom and decay, thus causing depletion

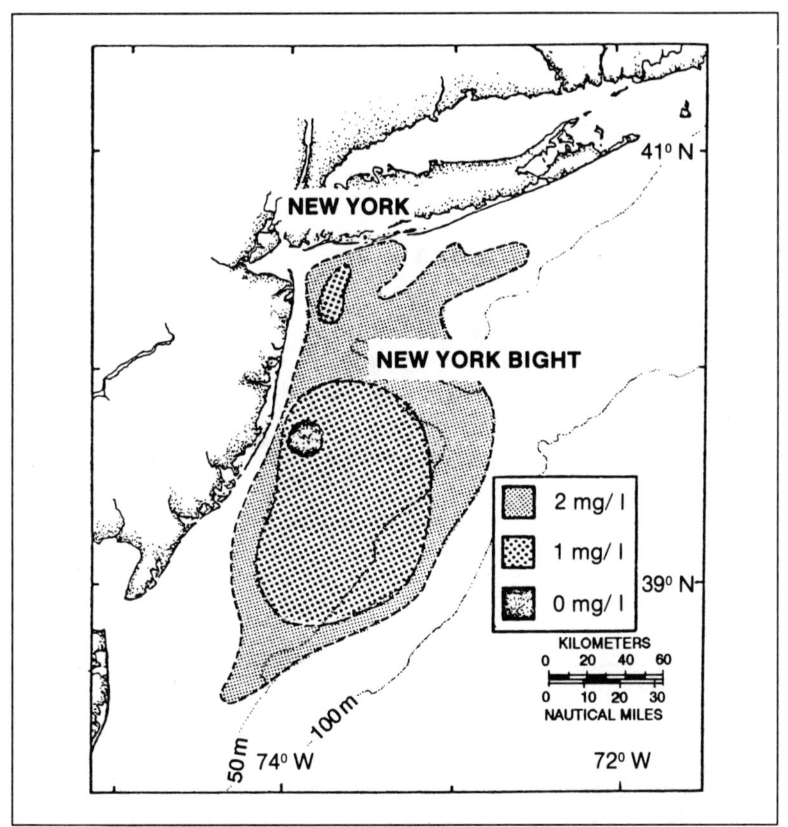

Figure 12-12: Dissolved oxygen distribution in bottom waters of the New York Bight, summer 1976.

of oxygen from the bottom water, an important question still remains: To what *degree* had human polluting activities contributed to the anoxic event? Are we the catalyzing agent? If so, human actions can be amplified with resultant mass killing of marine organisms.

In the Baltic Sea, where bottom water is not replenished readily with oxygen from the atmosphere or the North Sea, the bottom-dwelling organisms live precariously. In recent years conditions have worsened. For instance, in 1900 in the near bottom water of Landsport Deep, oxygen was measured at 2.5 milliliters per liter of seawater; by 1950 it had decreased to 1.5

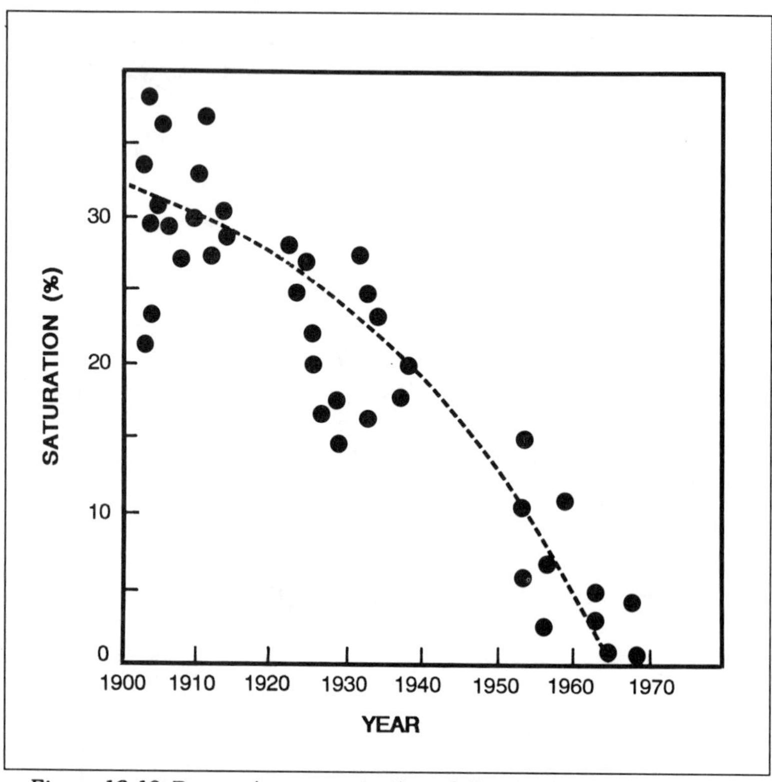

Figure 12-13: Decreasing concentration of dissolved oxygen, expressed as percent saturation with respect to the atmospheric oxygen content, at the Baltic Sea bottom in recent years. (From S.H. Fonselius, 1970, "Stagnant Sea," *Environment* 12: 2–11, 40–48. Reprinted by permission of Helen Dwight Reid Educational Foundation.)

milliliters; recently on occasions no oxygen has been detected and bottom organisms have been found dead (Figure 12-13). The Baltic marine scientists are asking the same question the scientists of the New York area ask: What fraction of the anoxic event is due to human-originated pollution? The Baltic scientists know now that eutrophication is occurring in the surface waters; dead organic substances then rain over the bottom and deplete the oxygen in water.

The oxygen stories of New York and the Baltic are not isolated events. They will happen again and at other places. For instance the Chesapeake Bay was sometimes reported to have low oxygen content. Fundamental oceanographic knowledge of the regions which are likely to be polluted must be known. It takes time to know the oceanic processes in each region. Competent students must be trained.

Realizing the importance of obtaining long-term solutions to marine pollution problems, which include education and training, Professors Kester and Duedall of the United States in 1977 began laying the groundwork to convene international ocean disposal symposia. This symposium series meets once every 18 months in order to discuss among scientists the new findings associated with marine waste disposal practices. The participants live in a dormitory; students from many countries are supported to attend; discussions continue late into the evenings; new contributions are published as books. Thus, in addition to forging the scientific frontier ahead, the several quality seeds (students) nurtured through this cradle may be a cost-competitive strategy to cope with future marine pollution problems. Rome was not built in a day.

People in and around heavily stressed coastal regions and semi-confined water bodies must solve their pollution problems by themselves. The knowledge thus gained will help us protect the integrity of the vast ocean. Each of us, by our cumulative actions, may help keep the ocean alive for generations to come.

In the future we will use the ocean more for our welfare. Ocean thermal energy conversion (OTEC) may produce the energy needed by us in the future. Ocean mining of manganese nodules will give us valuable metals such as copper, cobalt, nickel, manganese, and others. All the planners of these new undertakings must consider the total activities from the front end to the back end (waste management). The only nest we have is the earth.

References

Bascom, W., ed. 1982. *Coastal Water Research Project, Biennial Report 1981–1982*. Southern California Coastal Water Research Project, Long Beach, California, 294 pages.

Champ, M., and P.K. Park. 1982. *Global Marine Pollution Bibliography: Ocean Dumping of Municipal and Industrial Wastes*. New York: Plenum Press, 399 pages.

Duedall, I.W., B.H. Ketchum, P.K. Park, and D.R. Kester, eds. 1983. *Wastes in the Ocean, Volume 1, Industrial and Sewage Wastes in the Ocean*. New York: Wiley-Interscience, 431 pages.

Farrington, J.W. 1983. "Bivalves of sentinels of coastal Chemical pollution: The mussel (and oyster) watch." *Oceanus* 26, No. 2, Summer 1983, pages 18–29.

Gerlach, S.A. 1981. *Marine Pollution, Diagnosis and Therapy*. New York: Springer-Verlag, 218 pages.

Goldberg, E.D. 1976. *The Health of the Oceans*. Paris: the UNESCO Press, 172 pages.

IMO/FAU/UNESCO/WMO/WHO/IAEA/UN/UNEP Joint Group of Experts on the scientific Aspects of Marine Pollution (GESAMP). 1982. *The Review of the Health of the Oceans. Reports and Studies No. 15*. Paris: United Nations Educational Scientific and Cultural Organization, 108 pages.

IMO/FAU/UNESCO/WMO/WHO/IAEA/UN/UNEP Joint Group of Experts on the Scientific Aspects of Marine Pollution (GESAMP). 1983. *An Oceanographic Model for the Dispersion of Wastes Disposed of in the Deep Sea. Reports and Studies No. 19*. Vienna: International Atomic Energy Agency, 1982 pages.

Kester, D.R., B.H. Ketchum, I.W. Duedall, and P.K. Park, eds. 1983. *Wastes in the Ocean, Volume 2, Dredged-Material Disposal in the Ocean*. New York: Wiley-Interscience, 299 pages.

Ketchum, B.H., D.R. Kester, and P.K. Park, eds. 1981. *Ocean Dumping of Industrial Wastes*. New York: Plenum Press, 525 pages.

Malins, D.C., ed. *Effects of Petroleum on Arctic and Subarctic Marine Environments and Organisms, Volume I. Nature and Fate of Petroleum*. New York: Academic Press, 321 pages.

Malins, D.C., ed. *Effects of Petroleum on Arctic and Subarctic Marine Environments and Organisms, Volume 2, Biological Effects*. New York: Academic Press, 500 pages.

Myers, E.P., and C.G. Gunnerson. 1976. *Hydrocarbons in the Ocean*. Boulder, Colorado: MESA Special Report, National Oceanic and Atmospheric Administration, Environmental Research Laboratories, 42 pages.

Park, P.K., D.R. Kester, I.W. Duedall, B.H. Ketchum, eds. 1983. *Wastes in the Ocean, Volume 3, Radioactive Wastes and the Ocean*. New York: Wiley-Interscience, 522 pages.

Squires, D.F. 1983. *The Ocean Dumping Quandry: Waste Disposal in the New York Bight*. Albany: State University of New York Press, 226 pages.

Waldichuk, M. 1977. *Global Marine Pollution: An Overview Paris*: Intergovernmental Oceanographic Commission Technical Series 18, United Nations Educational Scientific and Cultural Organization, 96 pages.

Wolfe, D.A., ed. 1977. *Fate and Effects of Petroleum Hydrocarbons in Marine Organisms and Ecosystem*. Oxford: Pergamon Press, 478 pages.

THIRTEEN

COMMENTARY on P. KILHO PARK'S PAPER

Michael G. Norton

The subject of this paper reminds me of a magazine cover I saw while in Canada in 1969. It was a gravestone on which was inscribed "The Oceans—born c. 2 billion years BC, died 1977." The article inside was entitled "Death of the Oceans" and it postulated the collapse of primary production as a result of increasing quantities of human-made toxic substances entering the ocean. This was symptomatic at that time of a widespread perception of impending global marine pollution problems. These concerns were addressed at the Stockholm Conference on the Global Environment and subsequently led to international agreements, such as the London Dumping Convention, to control marine pollution. The oceans have certainly lasted beyond 1977 and still look as if they have a few years to go—it is thus interesting to see how our present perspective differs from those alarmist ones of 15-20 years ago.

Dr. Kilho Park drew our attention to the dynamics of the ocean systems which are capable of assimilating and processing vast quantities of material of natural origin from land run-off and atmospheric fallout. I would like also to emphasize the characteristics of the living ecosystems which are such an essential component of life on earth. Marine ecosystems have been riding out the changes in the earth's climate through billions of years and were the origin of the terrestrial life whose impacts we are now discussing. Present day ecosystems offer a range of species capable of exploiting the various combinations of substrate, temperature, depth, food sources, and so on that the ocean can offer. Boundaries are ill-defined in the ocean, thus response to change can be rapid. Life will start to colonize a new surface (whether a rock or a ship's hull) overnight; disturbed sediment after a storm will be quickly resettled, since the systems have evolved to adapt to such changes. Individual species may increase or decrease, but overall productivity will tend to be determined by the exploitable energy and food inputs to the system. The discovery of novel life forms along the deep sea volcanic vents shows how adaptable marine organisms can be in exploiting any source of energy (light has been replaced by thermal and inorganic chemical energy sources.)

The natural marine ecosystems are therefore not the fragile systems envisaged in the article mentioned at the beginning of these comments, but are resilient and adaptable assemblages of organisms able to exploit a wide variety of ecological niches. Studies of fisheries have shown that while individual species of fish may go through spectacular population growth and collapse depending on success of spawning, predation, temperature, currents, etc., overall biomass will be much more stable except, of course, where man intervenes by overfishing. The system is also highly dependent as Dr. Park has shown on the input of materials from the land via rivers for many of the essentials in maintaining a productive system (organic detritus, nutrients, etc.). Our activities thus impinge on a system which is both flexible and opportunistic and which is accustomed to responding to a wide range of different environmental conditions and inputs.

On a global scale Dr. Park has shown how the ocean receives much material from natural sources. Some of this is life-supporting, some life-threatening. The system exploits the former and has mechanisms for mitigating the effects of the latter, as can be seen by the fact that disaster does not follow from the input of large quantities of metals which are toxic in other circumstances. How do our efforts look when compared against this natural flux? We have made a mess of a number of estuaries, through discharges of sewage and industrial wastes, which have resulted in the depletion of estuarine species or species with an essential part of their life cycle in estuaries. Some coastal areas have also been affected—more by contamination than by depletion of species. Oil spills still make the headlines occasionally and marine litter is widespread. So let us look to see how far we have made these and other problems into problems of global scale.

The London Convention was concluded to place controls on waste disposal in all the world's oceans. At that time different wastes were assigned to prohibited, strictly controlled, and permitted categories according to the perceived threat they posed to the marine environment and to humans. It is instructive to see how these lists look today and how many problems have subsequently arisen that the authors of the convention did not foresee.

Substances listed in Annex I comprised those perceived as posing the greatest threat to the oceans. Mercury and cadmium were listed, due to the fatal consequences of localized pollution in Japan. Subsequent experience has shown that these metals can continue to pose public health problems in local areas where fish and shellfish accumulate high concentrations, but that more widespread effects have not been detected, nor are any anticipated in view of the fact that anthropogenic inputs are still relatively small compared with natural fluxes. Effects have been limited to bays or inshore coastal waters and have proved controllable by restrictions on discharges to the water body concerned.

Oil pollution—particularly from catastrophic loss of cargo—has proved to be devastating locally and, in some cases, effects at a range of over 100 miles can be found. Additionally, Dr. Park has pointed out that it is possible to find residues of degraded oil throughout most of the ocean regions used for transport. Nevertheless, the major effects have been limited to the immediate vicinity of spills, and the ocean has proved capable of dispersing and degrading large quantities of oil via natural processes—often more effectively than with dispersants. Therefore, while oil continues to pose a major threat locally, it is not currently a threat to the global marine environment.

The convention included organohalogen compounds due to their persistence and tendency to accumulate in marine organisms. These are representative of classes of compounds discussed by Dr. Park (xenobiotics) for which natural degradative and removal processes have not necessarily evolved in the oceans. Contamination by these substances has been shown to be global at the level of analytical detectability and regional at levels which could be of concern from the viewpoint of protecting public health and the well-being of some marine species, particularly mammals. These are the nearest we have come to having a global pollution problem, and further vigilance and controls on these types of compound are warranted.

A further material prohibited in the convention is high-level radioactive waste. Dr. Park has shown how global contamination has already occurred as a result of atmospheric tests and how the oceans already contain a large amount of natural activity. Nevertheless, controversy surrounds the current use of deep water sites in the North Atlantic for *low-level* waste even though the present quantities disposed are very small compared to natural background levels and fallout sources. This may reflect concern that the long half-life of some radioisotopes may raise the possibility of regional if not global contamination if substantial changes in disposal practice were made. Careful prediction and modelling are thus needed before it can be determined how far the role of the ocean in radioactive waste disposal can be expanded. In this case, however, there may be a further option of using the natural stability of the deep ocean sediment sinks as a means of isolating materials with long half-lives from the living marine environment and humans.

Outside of Annex I, the convention also mentions substances requiring "special care," including metals such as zinc, arsenic, copper and lead, and toxic substances such as cyanide. None of these has subsequently proved to present more than local pollution problems. In the case of the metals, all are common within the marine environment and effective sequestering and removal processes exist to make field effects significantly less harmful than might be predicted from laboratory experiments.

The remaining substances mentioned in the convention include those which are part of the natural marine environment but which may still exert an adverse effect via large inputs or due to the location of an input. Organic carbon, particularly associated with sewage, has had a detrimental effect on bays and estuaries and some localized areas in the coastal zone. So too has the removal of silts via dredging and the disposal of spoil. Nutrients have enhanced the primary productivity of some areas where other factors (such as light) are not limiting, and have also caused changes in species composition of the plankton leading to eutrophication or hypertrophication in some areas. Although primarily restricted to enclosed water bodies, such effects have been detected on a regional basis in some coastal waters and have had very damaging consequences, as illustrated by the 1976 anoxia event off the eastern coast of the United States described by Dr. Park.

My conclusion from the above synthesis is that, on evidence to date, controls to avoid serious degradation of coastal zones will ensure adequate protection of the global oceans for most substances naturally present in the oceans. Xenobiotic substances may also be primarily controlled on this basis, but may require an additional degree of control in view of their potential for persistence and transport over long distances. What we must do, therefore, is ensure that controls on coastal pollution are enacted as nations increase their coastal population and industries, and that measures already taken by many countries remain effective. We also need to continue the global programs which have been set up by such organizations as the United Nations Environment Program and the International Council for the Exploration of the Seas to monitor the distribution of metals and persistent synthetic compounds, as well as to continue to support the controls on direct inputs to the ocean under the London Dumping Convention and other international agreements.

So where are the global ocean problems of tomorrow? Not necessarily in the traditional sense of marine pollution, as I share Dr. Park's faith that humanity will continue with its success at controlling traditional sources of marine pollution so as to avoid global problems. We are, however, as Dr. Park has pointed out, whether knowingly or unknowingly increasingly relying on the ocean to provide global stability as humans increasingly widen the scale of their impact on land and in the atmosphere. The oceans comprise two-thirds of the earth's surface, and the interaction between their surface and the atmosphere is becoming one of the key factors in limiting the build-up of atmospheric pollutants. It allows acidic aerosols, both human-made and naturally occurring, to be neutralized, and can be both a source and sink for atmospheric carbon dioxide. Its behavior as concentrations of CO_2 rise will be crucial in determining the global effects of greenhouse gases.

Thus, even though the oceans appear to be surviving threats perceived decades ago, it is even more important to understand global oceanic processes and their buffering capacity as they are increasingly called upon to moderate the effects of our activities.

FOURTEEN

WORLD WATER PROBLEMS: DESERTIFICATION

Gerald Stanhill

Human life is profoundly dependent on the circulation of water at the earth's surface and is very sensitive to even small deviations in the hydrological cycle. In technologically advanced societies such deviations affect comfort and communication; in less industrialized societies they also affect the food supply and health of the population.

Over the centuries man has learned to accept and adapt to the large spatial and temporal variations which occur in his water supply. Indeed, the two characteristic types of extreme hydrological variations have been named after successful examples of human adaption taken from biblical episodes—e.g., Noah's evasive solution for a unique hydrological event, and Joseph's storage strategy for dealing with a sustained deviation from the norm.

Although water has always posed a major global environmental problem to humans, it is the new aspects of this problem —those posed by the growth in human population—and the new possibilities for solutions—those made available by science and technology—that will be considered in this chapter. The new aspect of the problem considered is the likelihood that human activity is inadvertently affecting the hydrological cycle.

Among the solutions explored are the possibilities of exploiting new water sources, of reusing old ones, and of reducing water requirements, particularly in agriculture, the major consumer of water.

The desertification process may be considered as an extreme example of a global water problem illustrating the interaction between human and natural factors in causing the phenomenon. Humans and nature, separately and in combination, have transformed large areas of useful and ecologically significant but fragile landscapes into desert areas, drastically reducing both their human carrying capacity and ecological diversity. Recent theories provide a biofeedback mechanism to explain how human overexploitation of fragile semi-arid landscapes could initiate or reinforce climatic changes, strengthening and extending the desertification problem.

Adequacy of Water Supply in Different World Regions

The annual fluxes of precipitation and evapotranspiration over the land surfaces of the globe are shown in Figure 14-1 as depths per unit land area. Difference between the two fluxes, i.e., the potential water supply available for human exploitation, are also shown.

The fluxes, taken from Baumgartner and Reichel (1975), show minima for all three quantities at the Poles. Maximum values of both precipitation and evapotranspiration occur in the equatorial region (15°S to 15°N). The zones of maximum annual runoff lie to the south of the equator. This region is mostly oceanic and the land areas are sparsely populated.

The equatorial zone of high hydrological activity is bounded to the north and south by the major arid and semi-arid regions. These regions are centered around the 30° north and south latitudes, although in the northern hemisphere the vast desert and semi-desert regions of central Asia extend considerably to the north up to latitudes greater than 40°.

The maximum atmospheric demand for water occurs in the mid-latitude belts of low hydrological activity. The reason for the high potential evapotranspiration in these regions is connected with the low levels of precipitation and the associated near absence of cloud cover. For this reason maximum fluxes

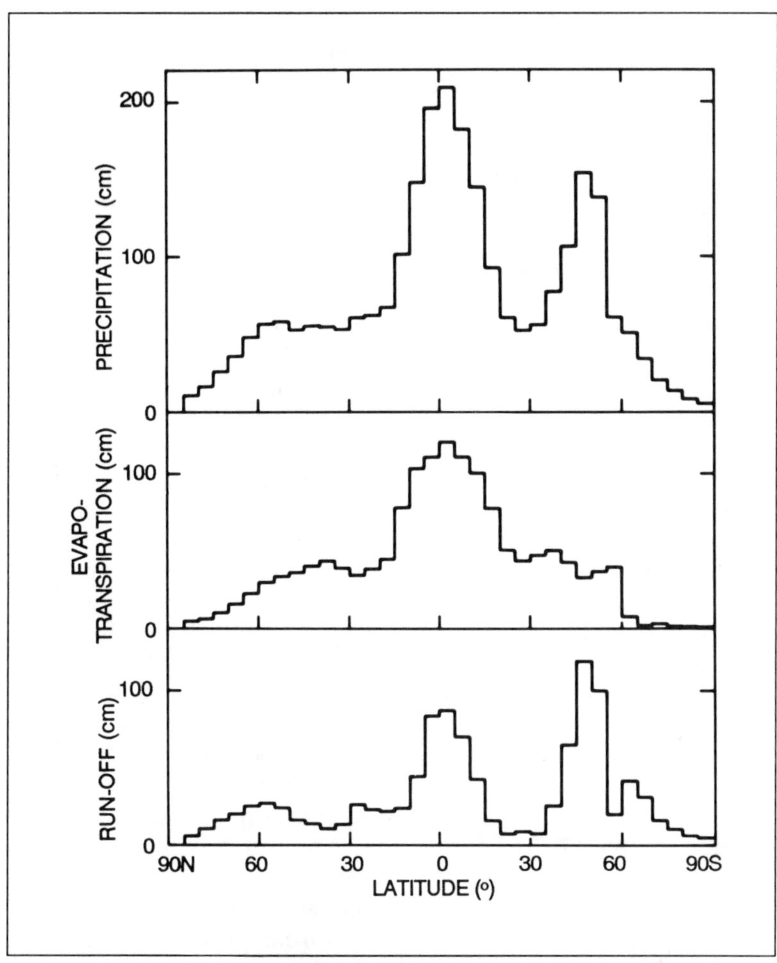

Figure 14-1: The distribution of precipitation and evapotranspiration, and their difference—runoff—over the land surfaces of the globe. All values are in cm water depth equivalent per year and are taken from Baumgartner and Reichel (1975).

of global solar radiation are recorded in these mid-latitudes rather than at the equator as astronomical considerations would suggest. Maximum annual totals of insolation measured on the earth's land surface, over 8 GJm^{-2}, were measured in the mid-latitudes of the Southern Hemisphere and the highest mean for any 10° latitude zone and 7.27 GJm^{-2}, was recorded between 20° and 30°S. By comparison, in the equatorial meridians between 10°N and 10°S, annual insolation averaged 6.70 GJm^{-2}. (Stanhill 1983). At the same time the low levels of actual evaporation ensure that most of the net radiation energy available at the surface is expended in heating the air. This heating increases the atmospheric water deficit and rates of air movement. Thus, atmospheric transport and heat balance characteristics combine to enhance the rates of potential water demand.

The inadequacy of water supply in the mid-latitudes of the globe is made more serious by the high interannual variability of precipitation. It is commonly three or more times greater than in temperate or equatorial regions and in many cases the high interannual variability rather than the absolute shortage makes this region particularly prone to desertification.

An example of this is found in the desertification problems of the Sahel region of sub-Saharan Africa. In this area the boundaries between bioclimatic zones have been correlated with the mean annual precipitation. The isohyet of 750 mm/yr corresponds to the border separating the semi-humid from the semi-arid zone. The 325 mm/yr isohyet to the border between the semi-arid and arid zone, and the 100 mm/yr isohyet to the border between the arid and hyper-arid or true desert zone. Elsewhere in Africa, the isohyets corresponding to the same ecological zone boundaries differ considerably from those appropriate to the Sahel. However, the climatic aridity index developed by the Meteorological Applications group of the World Meteorological Organization was found to be of wide general application. This index consists of the total annual precipitation divided by the total annual potential water demand calculated by Penman's combined heat balance and aerodynamic method. Values of 0.5, 0.2 and 0.06 for this index coincide with the borders separating, respectively, the semi-humid from the semi-arid, the semi-arid from the arid, and the arid from the hyper-arid zones.

Using long-term rainfall records, the standard deviation of the above index was calculated for 17 stations along three transects across sub-Saharan Africa. The results, presented in Table 14-1, were used to map the areas within which the borders of the regions can be expected to lie, assuming that annual rainfall follows a normal distribution (Figure 14-2). In practice, the distribution of annual rainfall in arid zones is usually found to be somewhat skewed towards the lower than average values and may also display symmetrical non-normality, i.e., kurtosis. Because of this the size of the border areas given in Figure 14-2 are probably somewhat underestimated.

Table 14-1: Interannual Hydrological Variability in Sub-Saharan Africa

Bioclimatic Zone	Station	Coordinates N	Coordinates E	No.of Years Record	Aridity Index* (Mean)	Aridity Index* (Stand. Deviation)	Coefficient of Variation %
Arid	Abu-hamed	19°32'	33°20'	24	.006	.008	133
	Bilma	18°41'	12°55'	20	.017	.013	75
	Atbara	17°42'	33°58'	24	.042	.027	64
	Tessalit	20°12'	0°59'	20	.070	.031	44
	Khartoum	15°36'	33°33'	24	.108	.040	37
	El-dueim	13°59'	32°20'	17	.162	.048	30
	Gao	16°16'	0°03'	30	.179	.047	26
	Menaka	15°22'	2°13'	20	.192	.048	25
Semi-arid	Kosti	13°10'	32°40'	24	.259	.056	22
	Soroa	13°14'	11°59'	20	.287	.063	22
	Tillabery	14°12'	1°27'	15	.373	.062	17
	Niameya	13°29'	2°10'	27	.418	.106	25
	Maduguri	11°51'	13°05'	54	.443	.090	20
Semi-humid	Malakal	9°33'	31°39'	24	.500	.079	16
	Garoua	9°20'	13°23'	20	.663	.097	15
	Kandi	11°08'	2°56'	30	.692	.115	17
	Nattingou	10°19'	1°23'	20	.984	.180	18

* Aridity index equals annual potential evapotranspiration divided by annual precipitation. Source: unpublished data of the author.

The map shows that the northern border separating the hyper-arid from the arid zone will, for two years out of three, be found within a 200- to 100-km-wide-region. The width of the zone is least in central Africa. The border separating the arid from the semi-arid zone, probably the zone at greatest risk of desertification, falls within a region which is 225 km wide in western Africa, but only 80 km wide in the upper Nile Valley. The border separating the semi-arid from the semi-humid zone lies, in two years out of three, within a region ranging in width between 225 km in the Nile Valley to 175 km at Dahomey-Mali.

Thus, areas at climatic risk of desertification, and more generally those with inadequate water supply, cannot be precisely delineated other than in a stochastic sense. Variation with time is considerable, both in total area and in position. How large these areas are, and how they are interconnected, are important but inadequately studied aspects of the global water problem.

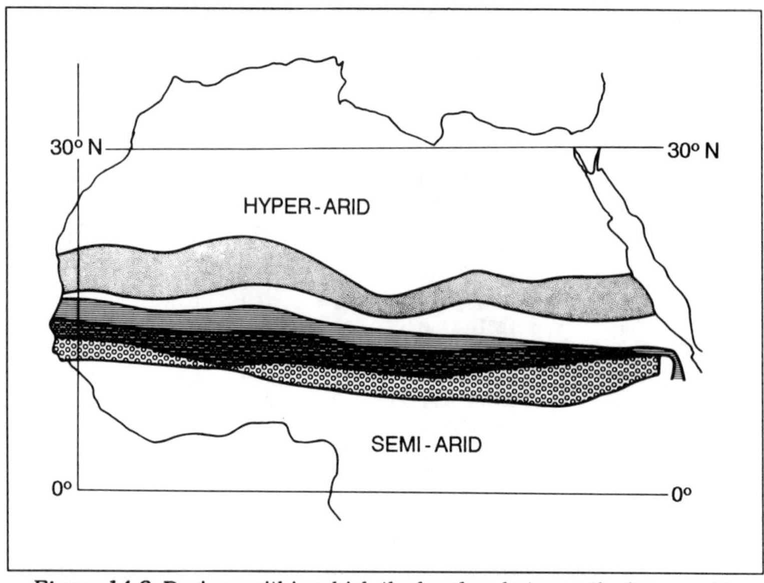

Figure 14-2: Regions within which the borders between the hyper-arid, arid, semi-arid and semi-humid ecological zones can be expected to fall during two out of three years in sub-Saharan Africa.

Border region between hyper-arid and arid zones; annual precipitation divided by annual evapotranspiration = 0.06.

Border region between arid and semi-arid zones; annual precipitation divided by annual evapotranspiration = 0.20.

Border region between semi-arid and semi-humid zones; annual precipitation divided by annual evapotranspiration = 0.50.

The Sahel drought of 1970–1972 is illustrative in this respect. During this period all three borders of the bioclimatic zones of the Sahel moved in a southerly direction and considerably expanded the total area prone to desertification. However, for

Africa north of the Equator as a whole, the expansion of the arid areas south of the Saharan desert was compensated for by wetter conditions north of the Sahara and in the 0–15°N zone (Figure 14-3).

The redistribution of rainfall during the early 1970s which led to the Sahel and Ethiopian famines was, however, not confined to Africa (Lamb 1982) (Figure 14-3). It is significant

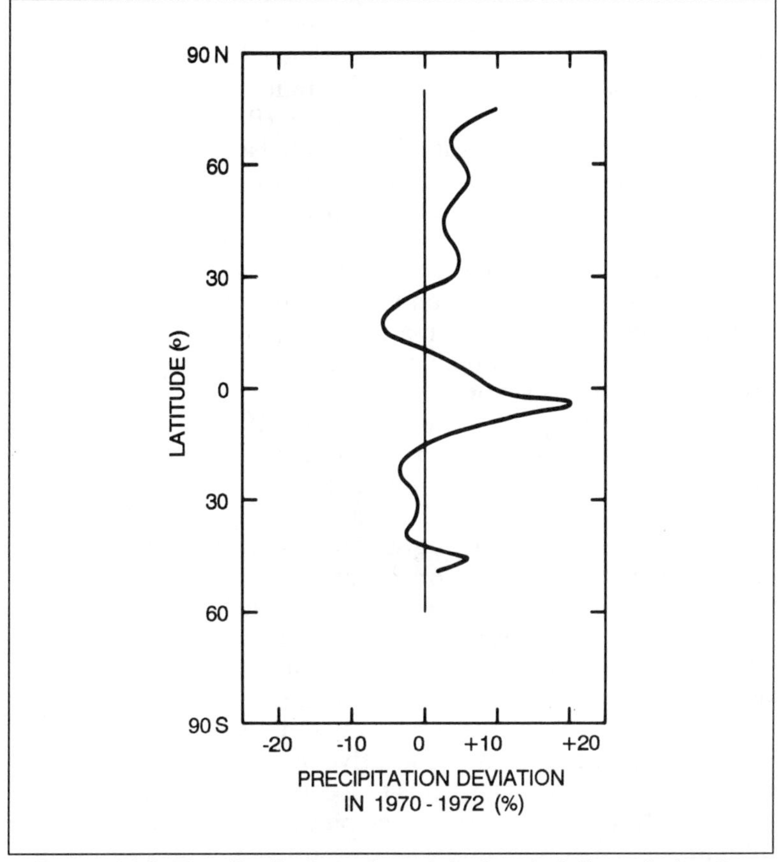

Figure 14-3: The distribution of precipitation deviations during the period 1970–1972 averaged meridionally over the earth's surface. All values are in % deviations from the 1931–1960 30-year average, and are taken from Lamb who analyzed the data from several hundred measurement stations.

that there were much larger positive deviations in precipitation south of the equator. These anomalies attracted little attention and evidently had a minor impact compared with the major effects of the smaller, negative deviations occurring in the more fragile mid-latitudes to the north.

Desertification and Climatic Trends

Recent analyses of the 1970s Sahel disaster belittle the role of changing climate and attribute the famine to human mismanagement of land resources (Garcia 1981). These analyses show that the increasing aridity experienced in the region since the 1950s falls within statistical expectations and so cannot be construed as evidence of climate change (Bunting *et al.* 1976).

Nevertheless, desertification in the Sahel and similar episodes in this century in the drought-prone midwestern states of the United States and in the southern grain-growing provinces of the Soviet Union, show a recurring pattern of interaction between climate and humans. During periods of benign climate coupled with advances in technology, agriculture is intensified and expands in climatically marginal regions. Population densities increase to levels which can no longer be supported when rainfall deficiencies reappear.

It has been suggested (Charney 1975, 1976) that the desertification process includes an important positive feedback mechanism. Denudation of semi-arid zone vegetative cover due to overgrazing or cropping inhibits rainfall and thereby reinforces or even initiates drought. The mechanism proposed to explain this biodesertification process is that the reduction in vegetative cover leads to increased radiative loss to space, both from the denuded surface and from the air column above it. This increased radiation loss is due primarily to the higher short wave reflectivity of bared land, and, to a lesser extent, to their higher surface temperatures and hence greater long-wave emission. The net radiative cooling accompanying reduced vegetation cover leads to a cooling, and subsidence in the air column. Convective precipitation is thereby strongly inhibited.

Using realistic values of short-wave reflectivity for overgrazed and protected semi-arid areas, (Otterman *et al.* 1975) in computer simulation studies shows that such surface changes could reduce annual rainfall in the Sahel by 20% (Ellsaesser *et al.* 1976). This reduction is similar to that recorded during 1970–1972.

Unfortunately, for reasons of scale, this theory explaining desertification as a climatic enhancement of human-initiated changes in land productivity is not amenable to experimental verification and must be considered as an interesting possibility rather than a demonstrable physical explanation.

On a larger global scale, the possibility exists that we are inadvertently altering the climate. The major concern is a potential warming of the atmosphere which, it is predicted, will be especially pronounced in the upper latitudes. It is also predicted that this warming will exceed the "noise" level of random temperature changes before the end of this century.

The apparent major cause of this global warming trend is the 30% increase in the CO_2 concentration of the atmosphere which appears to have occurred during the last century. Two human activities are thought to have been of approximately equal importance in increasing the CO_2 level. One is the release of carbon held in the terrestrial biomass by deforestation and the conversion of shrub and grassland to cultivated, agricultural areas. The second way man has contributed to the CO_2 increase is by the release of the carbon fixed in fossil fuels during the carboniferous era (Clark 1982).

The climatic changes expected to result from a CO_2 induced warming are many and important and will almost certainly include changes in the distribution pattern of precipitation over the globe's land surfaces. Details of these changes and some of their implications for food production and other human activities are presented in another contribution to this volume (Kellogg 1989) and will not be considered further here.

Before leaving the question of CO_2-induced changes on water supply and demand, one important and complex physiological mechanism should be mentioned. This is the effect of increased CO_2 concentration on the size of stomatal apertures through which water vapor diffuses through leaf surfaces into the atmosphere. An increased CO_2 concentration leads to closure of stomatal apertures and hence an increase in water vapor diffusion resistance (von Keulen *et al.* 1980). In regions where precipitation exceeds the climatically limited water loss, a marked decrease in water loss can be expected; in regions where the potential demand exceeds the water supply, i.e., those requiring irrigation, some compensating decrease in the rate of actual water loss and hence improvement in the water balance may occur.

Changes in the Global Water Balance and Circulation

As the volume of water on the globe (1.4×10^9 km^3) is essentially constant, significant changes in the current water balance or in the rate of its circulation can only come about through a change in the relatively small fraction of the total volume which is now in active circulation. Table 14-2 shows that the annual flux through the earth's atmosphere is only 496 x 10^3 km^3/yr, i.e., less than 0.03% of the total water volume takes part in the annual circulation. Thus, the average residence time for water on the globe is 2,800 years.

Table 14-2: Global Water Balance and Circulation

		Land	Oceans	Atmosphere
Total Volume	(10^3km^3)	64 (+24 polar ice)	1,370,000	13
Annual Flux	precipitation	111	385	
	evaporation	71	425	496
	runoff to ocean	40		

Source: Baumgartner and Reichel 1975.

Over the ocean surface the volume of water in active circulation through the atmosphere is limited by the area of the surface exposed and the solar energy available for its vaporization. Any change in evaporation rate requires either an extraterrestrial change in the amount of solar radiation reaching the earth's atmosphere or, more probably, a change in the amount of solar energy transmitted through the atmosphere, i.e., in the degree of cloud cover or in the composition of the atmosphere.

Over land surfaces the volume of water in circulation is dependent primarily on the precipitation. This is controlled by the pattern of atmospheric circulation and to a lesser extent by the precipitation efficiency—the fraction of the atmospheric water content which precipitates as rain or snow. This fraction is itself affected by the aerosol composition of the atmosphere and hence influenced by human activity.

A rather weak positive feedback mechanism exists for water circulation over land surfaces in that a minor fraction of the precipitation is, after evaporation, reprecipitated over the same region. There is, however, one region of great significance to the biosphere where this mechanism may be important. This is the Amazon basin, where as much as half of the rainfall is believed to originate from water transpired within the region (Salati *et al.* 1979). As this is the earth's major area of tropical forest, in which a significant proportion of the total carbon in the biosphere is both stored and exchanged, any changes in vegetative cover in this area which affected the rate of transpiration could, in turn, affect precipitation and carbon exchange.

Measurements of changes that have taken place in the volume of water storage are available on a global scale for only one component of the water balance—the oceans. A recent review of measured changes in sea levels (Barnett 1983) has shown that during the period between 1903 and 1969, this has increased at the rate of 15 cm per century. This rate of increase corresponds to the addition of 542 km^3 water volume to the oceanic pool each year—a 1.3% increase in the current runoff from the land to the oceans—39.7 x 10^3 km^3/yr (see Table 14-2).

The reasons for this increase are not known. Suggested causes include thermal expansion, following surface heating, melting of the polar icecaps and a decrease in the liquid water stored on and under the land surfaces of the globe. Moreover, it should be noted that the increase in land runoff calculated from the rise in the ocean levels is less than the $\pm$ 2 x 10^3 km^3/yr error associated with global estimates of this land to ocean runoff term.

Nevertheless, it is instructive to consider the extent to which our current activities could be responsible for this storage change. Lvovich (1970) has estimated that in 1965 man withdrew 2848 km^3/yr of water for various purposes, of which 1051 km^3/yr was available in return flow; the difference of 1797 km^3/yr was presumably lost to the atmosphere. It can be seen from the estimates presented in Table 14-3 that irrigation was by far the largest sink for water use by humans, 2,300 km^3/yr

was estimated to have been withdrawn for this purpose in 1965, of which only 600 km³ were available for reuse. Thus, 1,700 out of the 1797 km³/yr of water used by man which was not available for reuse was consumed in irrigated agriculture.

Table 14-3: Estimated World Water Use by Humans, in 1965 and 2000 (km³/yr)

Year	1965		2000	
Use	Withdrawal	Consumption	Withdrawal	Consumption
Irrigation	2,300	1,700	4,250	3,850
Energy	250	15	4,500	270
Industry	200	40	3,000	600
Urban	98	42	950	190
TOTAL	2,848	1,797	12,700	4,910

Source: Lvovich, 1970.

It would be reasonable to expect this large and increasing water loss to the atmosphere attributable to human activities, to reduce runoff from the land surfaces to the oceans. Keller (1970) showed that such a change has occurred in one local study of changes in the water balance of an industrialized temperate region. Nevertheless, the rise in sea level indicates that on a global scale land runoff has increased, indicating that the net effect of human activity, if responsible for this change, has been to decrease rather than increase evaporation from land surfaces.

The critical question concerning this issue is the extent to which irrigation, the major human water use, has caused a net increase in evapotranspiration compared with that from the natural vegetation which agriculture has replaced. In desert areas this is almost certainly the case, but in tropical areas, where the majority of irrigated land is situated, evapotranspiration from annual irrigated crops may, in fact, be less than that from the perennial natural vegetation which it has replaced.

If, globally, the net effect of irrigation has in fact been to increase water loss to the atmosphere, then this must have been more than offset by a concurrent but opposing change in the volume of water lost to the atmosphere from non-irrigated lands. This could well have come about from changes in land use, for it is well established that deforestation reduces evapotranspiration and increases runoff, as does the conversion of permanent grassland to cultivated, arable land use (Shachori and Michaeli 1965).

A study of the effects of the large changes in land use which have taken place during this century on the regional and global water balance appears to be urgently needed to resolve this issue. Any such study demands a global approach even if it is only the water supply to a restricted land area which is of interest. This is because ultimately the water supply depends on precipitation from the atmosphere which, to a major degree, is derived from water evaporating outside the area in which it falls. For the world's land surfaces as a whole, two-thirds of the water precipitated originates outside the same continental land surface on which it falls (Table 14-4). As the global water balance given in Table 14-2 shows there is a net annual flux of 40 x 10^3 km^3 of water from the oceans to the land surfaces via the atmosphere, the oceans clearly form a major source of this water supply.

Only in the case of the largest of the land masses, e.g., Eurasia, or for the special conditions of the Amazon basin (relatively slight air motion, tropical moist air mass), does the

Table 14-4: Atmospheric Water Balance of the World's Land Surfaces (10^3 km^3/yr)

Area	Advected Water Vapor	Precipitation Total	External Origin	Local Origin	(%)	Evaporation
Africa	25.0	20.0	15.0	6.0	28	16.0
Asia	21.0	26.0	17.0	8.0	32	16.0
Australia	6.0	4.0	3.0	0.65	17	3.0
Europe	9.0	6.0	3.0	1.0	19	3.0
North America	15.0	14.0	10.0	3.0	25	9.0
*Sellers 1969	14.0	12.0	8.8	3.2	27	10.0
South America	30.0	24.0	18.0	6.0	24	15.0
Arizona	0.309	0.036	0.034	0.002	6	0.036
European Russia	5.4	2.4	2.1	0.3	11	1.4
U.S.S.R.	12.0	12.0	9.0	3.0	26	7.0
Eurasia	23.0	32.0	21.0	11.0	34	20.0

Source: Kalinin *et al.* 1976. *Sellers 1969; estimates for Arizona, North America and European Russia combined.

local contribution to precipitation reach 50%. Thus, any major changes in the circulation of the atmosphere, or in the rate of evaporation (primarily from the oceans but to a less extent also from land areas) can be expected to exert a major influence on the size and distribution of the world's water supply. The speed with which such changes could be expected to occur is a function of the residence time of water in the atmosphere, globally averaging less than 10 days (see Table 14-2). Given the rate of atmospheric circulation the length of the path over which the water cycle occurs can be calculated, averaged globally, this path length exceeds 1,000 km.

Human activities which could affect both the atmospheric circulation and the rates of evaporation have already been mentioned. To summarize very briefly both urban and agricultural activites: the former primarily acts through its release of CO_2 by combustion of fossil fuels but also through the addition of aerosols, while the latter acts primarily through changes in land use via their effects on increasing the flux of CO_2 to the atmosphere and reducing the flux of water.

Perhaps the most catastrophic effect we could have on the world's water balance would result from a very large scale and prolonged pollution of the oceans with an evaporation-reducing film. Even a monomolecular layer of oil could have a significant effect on the world water balance and climate if a sufficient area of ocean surface were covered for a sufficient length of time. Fortunately, the area involved would have to be *very large*. A 1% reduction in the water flux from the ocean to the atmosphere would, for example, require a 10% effective film covering some 30,000 km^2 of the Pacific Ocean. Nevertheless in view of the vast amounts of petroleum hydrocarbon compounds and other pollutants now reaching the oceans and documented in another contribution to this volume (see Park, chapter 12), there is no place for complacency.

Adjusting to Shortages and Variations in Water Supply

Although, for the reasons just given, the study of human and climate influence on the world water problem demands a global approach, the practical problems of water shortages and uncertainties in supply require a much more restricted framework of local requirements and possibilities. The appropriate strategy for a given locality depends on the type of use, as this will determine what solutions will be economically feasible. However, this strategy is also strongly influenced by the general level of the local economy. For example, a farmer in Western Europe can afford to pay for a piped and treated water supply for his cattle, whereas the nomadic pastoralist in the Sahel cannot afford to pay for the minimal construction costs to secure watering points for his family. Thus, the problem of water supply is, as with food and energy, one that must be considered within a realistic framework of economic constraints. The relevant context, emphasized in the following discussion, is the extent to which technical and scientific progress is likely to reduce the costs of existing solutions, and to develop new and more economic solutions at a level appropriate even in the developing areas of the world.

Water Conservation

Conservation of existing water resources, rather than the development of new ones, is certainly the preferred strategy for solving the water problems of developing territories and is particularly relevant for the major consumer—irrigated agriculture. Currently, irrigation efficiency is extremely low; commonly one-quarter or even less of the water allocated for irrigation is productively consumed by crop evapotranspiration. The remaining water is, however, not merely wasted but is extremely damaging. Entire irrigation districts have been abandoned because of damage attributable to low irrigation efficiency.

The process is a simple one—surplus water application causes a rise in ground water table; drainage is impeded, crop root development is restricted, and soil salinity is induced. These factors combine to reduce crop growth and inhibit evapotranspiration, further reducing irrigation efficiency as well as yields. These problems can be ascribed to faulty design and management, which fail to ensure that the water conveyance to the fields and its distribution therein are in phase with crop water requirements, both in time and in space.

Although the reasons for the low water use efficiency of much of present-day irrigation practice are well established and solutions are readily available, in practice improvements are difficult. This is not only due to the large capital sums needed to implement efficient drainage on currently cultivated land, but also to the complex social and educational problems involved in improved irrigation practice. To arrange an even distribution of water on an individual farmer's fields at the correct time and in the correct amounts over an entire irrigation district, demands a level of community organization and cooperation that is often not available.

In the United States the water use efficiency of some existing irrigation schemes has been doubled so that over 80% of the irrigation water is used productively. This was accomplished without replacing surface application methods with energy- and capital-intensive high-pressure systems such as center-pivot sprinklers or drip irrigation. The techniques used to achieve this increased water use efficiency included laser land levelling, automatic pulsed water application and runoff recovery (Jensen 1981).

Additional substantial savings in water application are possible through modification of the cropping practice. Substitution of the currently cultivated crops or varieties by those with a shorter growing season or by crops which can be grown during seasons of lower climatic water demand can ensure substantial savings which exceed any reduction in yield. This is possible because the amount of crop growth per unit water transpired is inversely proportional to the absolute, climatically determined water requirement (Stanhill 1985). An example of this is seen in the results of a series of cotton irrigation experiments in which the same water application resulted in a 50% greater yield per unit of water in the cooler and more humid coastal plain of Israel than in the hotter and drier Jordan rift valley (Shalhevet and Bielorai 1978).

Finally, substantial water savings are possible in arid and semi-arid irrigation farming by applying the precise amount of water needed for the leaching requirement, i.e., that part of the application needed to wash out the soluble salts concentrated in the soil solution by evapotranspiration. This leaching forms an essential part of irrigation practice in arid and semi-arid districts and is often used to justify the excessive water applications given in such regions. Relatively simple methods of calculation are now available to determine the minimum amount of water needed to prevent soil salinity building up to toxic levels (Shainberg and Oster 1978). In many cases the amounts needed are only one-third of those previously applied on the basis of rule of thumb methods of calculation.

Another major aspect of water conservation through which significant savings can be made at relatively low cost concerns land use practice on water catchment areas. Land use practices which reduce evapotranspiration or runoff losses will increase the water supply available for a given level of precipitation.

An indication of the different levels of water yield to be expected for the two major groups of ground cover can be obtained from Figure 14-4. Water yields from a catchment area receiving 700 mm of precipitation a year could be increased by as much as 1,400 m^3 ha^{-1} yr^{-1} by converting a perennial woody, deep-rooted vegetation cover to a shallow-rooted, short-season grass cover. However, the effects of land use changes are complex and may sometimes be counterproductive. Increased runoff may be so marked as to lead to flooding and soil erosion. Catchment areas are increasingly being managed in a multi-purpose fashion and in such cases considerations other than water yield, such as recreational value, timber production, or grazing, must also be borne in mind. Despite such limitations, there is little doubt that in some areas a substantial, sustained, and economic increase in water yield from catchments can be attained by a land use management system carefully designed to meet local requirements and circumstances.

Another aspect of water conservation is the reduction of evaporation losses from water storage and conveyances. In many cases natural and artificial water channels and reservoirs are bordered and sometimes actually covered by perennial vegetation, this not only unproductively transpires substantial volumes of water but interferes with water flow and shipping. In some areas of the world, the volume of water so lost is vast. For example, in the Sudd swamp region of Sudan, which covers more than 8,000 km^2, Migahid, on the basis of two years of measurement from papyrus vegetation and from an open water surface, estimated evapotranspiration to be 19 km^3/yr, a rate of water loss which exceeds that from an open water surface exposed to the same climate (Migahid 1952).

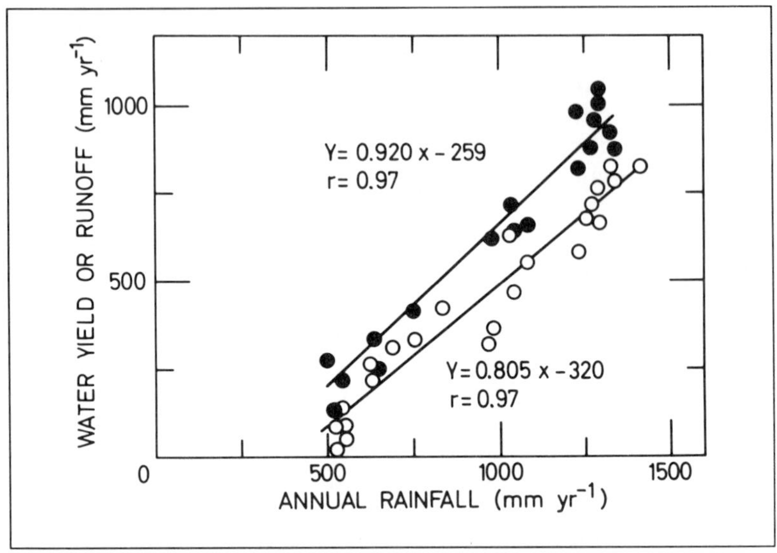

Figure 14-4: Relationship between annual water yield and precipitation for two land-use practices: Each point represents the result of a paired catchment study reviewed by Shachori and Michaeli (1965). The lines refer to the equations fitted by a linear regression model.

*open O Forest, woodland or maquis scrub cover.

*filled ● Grass or bare ground cover.

To date, no simple economic method of controlling such vegetation is available. Perhaps, in view of the many and as yet not fully understood effects that such changes may have, it is fortunate that any large-scale operations of this type are unlikely to be feasible before their full implications are better understood.

The above remarks also apply to the much studied question of evaporation reduction from open water surfaces. Here again, very substantial losses occur if relatively shallow reservoirs are created in arid areas. Lake Nasser, the 500 km^2 reservoir formed by the Aswan High Dam, provides an excellent example. It has been calculated to lose 13.7 km^3 of water a year by evaporation (Omar and El-Bakry 1981), a volume exceeding the amount that it has made available for irrigation. A review of evaporation reduction studies carried out on open water surfaces (Frenkiel 1965) indicates considerable uncertainty in the efficacy of treatments now available. It has even been suggested that the deforestation involved in producing paper to report the results of such research has to date been more effective in reducing water loss (Stanhill 1970)!

Finally, and potentially most important, the possibility of water conservation through reducing plant transpiration should be mentioned. A review of the possibilities (Stanhill 1986) shows that two research approaches can be distinguished. In the first, chemicals have been used to reduce stomatal apertures and so increase diffusive resistance to water vapor loss. To date, three main difficulties have prevented significant progress with this approach. One is the lack of an inexpensive, non-toxic, long-lasting, and effective stomatal closure agent. The second difficulty is that of effectively applying the agents that do exist to reach the stomata-bearing undersurfaces of leaves, especially in dense tree canopies. Third, and most restrictive, are the apparently inevitable negative side effects resulting from stomatal closure. These include the reduction of carbon dioxide influx, photosynthesis, dry matter production, growth and ultimately yield that follows the increase of diffusion resistance, and also the increase in leaf surface temperature and hence vapor concentration gradient that inevitably follows a reduction in transpiration.

The second approach for reducing plant water loss to the atmosphere is to increase the short-wave reflectivity of crop stands, typically between 0.15 and 0.25, and so reduce the solar energy available for evapotranspiration. As this energy is reflected to outer space, the negative feedback effects associated with transpiration reduction by stomatal closure are avoided. However, a major difficulty—the need for an inexpensive and effective non-toxic material easily applied and long-lasting—is common to the two approaches.

Field experiments with reflectant materials applied to crops growing under semi-arid conditions have failed to demonstrate any significant water conservation, but do show small but significant increases in yield and water use efficiency (Stanhill *et al.* 1976). In the field studies referred to, the yield increase was two kg sorghum grain per kg kaolin clay reflectant applied to the crop canopy before the seed head emerged. This yield response may be compared with that to irrigation water, which was 1.7 g grain per kg water evapotranspired.

The two major factors controlling plant water loss—solar energy absorption by the plant canopy and the diffusive resistance of its stomates—also control carbon dioxide fixation and hence yield. This explains the very high correlation between water loss and dry matter production. It is this strong coupling between H_2O efflux and CO_2 influx that makes the goal of genetic selection or engineering of new crop varieties with low water requirements and high yields such a difficult one. However, in view of the major role that irrigation farming plays in our water requirements, and the limiting role of water in food production, the modest resources currently devoted to this goal might well be increased.

New Sources of Water

The Liquid Phase

Although several methods of making sea water suitable for use have been demonstrated on a commercial scale, their cost makes the sea an uneconomic source of fresh water even for the most essential and minor of our requirements—domestic needs. Where no other water sources are available, such as at coastal desert or island sites, the use of desalination installations has been restricted to providing the essential water needs of tourists or service personnel.

The high cost of desalinated water is due to the capital costs of the material capable of withstanding sea water corrosion and the high energy requirements needed to separate salt from water. Any substantial reduction in desalination costs are, therefore, dependent on major advances in heat exchange and corrosion technology and/or the development of new sources of cheap energy.

Occasionally claims are made that untreated sea water can be used directly for irrigation. However, the range of soils and crops which can survive irrigation with sea water is so restricted, the yields obtained so low, and the dangers of soil and aquifer salination so great, that the use of sea water in conventional agriculture is not currently of any significance. Nor does the state of research into the salt tolerance of crop plants suggest that a radical change in this situation is likely in the foreseeable future.

In contrast, some promising results are reported from research on the culture of plants and animals native to marine environments. Even in such cases, the high cost of the installations needed to cultivate and harvest these crops has restricted land-based mariculture (sea water farming) to high-value products, such as algae used for health food and carotene B production and sea foods such as oysters and shrimps.

The situation with regard to the use of brackish water is much more promising and large volumes of this once neglected water resource are now being used in agriculture without any desalination pretreatment. Special care is needed to avoid the dangers of damage to the physical structure of the soil and to prevent salt accumulation to toxic levels, but both these dangers can be avoided. The techniques necessary are: the use of soil ameliorants—especially gypsum; correct cultivation practices to avoid damage to soil structure; the addition of the correct quantities of surplus water for leaching; and selection of salt-tolerant crops to avoid salt accumulation in the soil (Shainberg and Oster 1978).

Some yield reductions are to be expected following irrigation with brackish water; their size depends on the degree of salinity, to a lesser extent on the nature of the salts, and the degree of salt tolerance to the crop cultivated. The results of much research on this topic are generalized in Figure 14-5, based on the review of Maas and Hoffman (1977). This shows, for example, that irrigation with brackish water of moderately high salinity—a conductance of 8 deci Siemens/meter—would not cause any reduction in the yields of tolerant crops such as barley or cotton, whereas a moderately sensitive crop such as sugar cane would not produce any yield.

It should be noted that the major advances that have permitted the large-scale use of brackish water in sustained crop reduction have resulted from agronomic research in crop, soil, and water management practice rather than from an increased physiological understanding of salt tolerance or by the introduction of more salt-resistant varieties. Current progress in brackish water management such as the use of trickle or microjet application methods to control automatically the timing and placing of the irrigation water suggest that the limits of improvement obtainable through advanced management

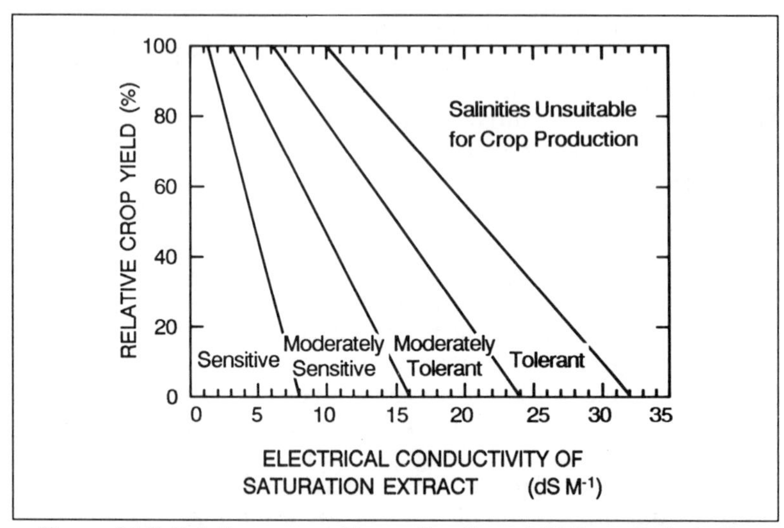

Figure 14-5: Relative crop yield as a function of (1) the salinity in the root zone from which two-thirds of the water uptake occurs and (2) the sensitivity of the crop type. Based on review by Maas and Hoffman (1977). Note for the great majority of soils the concentration of total dissolved solids (TDS ppm) is related to the electrical conductivity of the soil solution, and hence irrigation water (EC measured in dS m^{-1} or mmho cm^{-1}) by the relationship TDS = 640 EC.

practice have not yet been reached. Unfortunately, the cost and sophistication of these irrigation methods make them unsuitable for irrigated agriculture in many parts of the world.

The economic limitation to the use of desalinated sea water also applies to the reuse of domestic and industrial waste water. The capital and running costs of even the minimal treatment needed to render these sources suitable for agricultural application make reclaimed water too expensive for either the rural users or urban and industrial suppliers in all except developed countries with advanced economies.

In developed areas two options are available with different costs and applications. Where the supply of fresh water is limiting, complete tertiary treatment can be used to render used water suitable for reuse in every application, including domestic. Although the cost of such treatment is high, there is no need for a separate storage and conveyance system. These are needed for the second option, suitable where the main purpose of treatment is to prevent pollution and environmental deterioration by waste waters. Cheaper secondary treatment is then sufficient to render the waste water suitable for agricultural use.

Care is, of course, needed to avoid health hazards and the accumulation of salts and toxic metals in crops, soils, and aquifer. But there are a number of examples of successful sustained agricultures based on such waste water. Indeed, the sewage water of Paris has been so used for over a hundred years.

The Vapor Phase

Although the atmosphere is the smallest of the global water compartments, considered dynamically it is the only source of water to the land surfaces. Moreover, the residence time of water in its vapor phase is so short that any increase in the speed of the hydrological cycle through the atmosphere would have a major effect on water supply. The potential for achieving such an acceleration is based on the rather low efficiency of the precipitation process which is, in many circumstances, limited by the number of condensation nuclei available to initiate precipitation. In such circumstances the addition of suitable artificial nuclei, e.g., from vaporized silver iodide, can lead to significant increases in precipitation.

As an example of a successful and statistically significant result, the second Israeli randomized cloud seeding experiment is quoted (Gagin and Neumann 1981). Over a period of six rainfall seasons, the average increase in precipitation achieved was 13% at a significance level of $p = 0.028$; over the major target—the catchment area of the National Water Carrier—the increase was 18%, $p = .017$. Much larger increases were found when the cloud top temperatures were low (-15° to -21°C) and the daily rainfall amounts moderate (less than 15 mm).

The aim of the above experiment was to improve the hydrological situation, i.e., increase the amount of precipitation contributing to the surface and underground water storage. In other situations the desired effects are more immediate and local, for example, as a form of supplementary irrigation or to improve snow conditions in resort areas.

Whatever the aim, the cost of additional water made available by a successful cloud seeding program is very low and far less than that involved in developing most new sources or reclaiming waste water. A review of cloud seeding from the point of view of water resource management has appeared which includes the topics of seeding modes and instrumentation as well as the social, legal, environmental, economic, and scientific aspects of the subject (ASCE 1983).

A new approach to increasing precipitation which is described in this volume may have considerable potential for some arid coastal regions (Assaf, this volume). It is based on the mixing of the upper 100 m of sea surface to increase its surface temperature and evaporation during winter and so enhance the water vapor content of and precipitation from the passing air.

Control and Reduction of Desertification

The human factor is generally regarded as the decisive one in desertification, since, in areas at climatic risk, human overexploitation of the natural vegetation and overcultivation of the soils will cause desertification even if precipitation and water supply remain unchanged. Under natural conditions when significant fluctuations in rainfall do occur, overexploitation during years of below-average rainfall will initiate deterioration of the natural resources—even if the level of exploitation does not exceed the long-term average carrying capacity of the region.

In this sense, theoretically, the solution to the desertification problem is a simple one—the reduction of the human and animal populations to the lowest level sustainable under the minimum, statistically expected rainfall regime. Preferably this will be achieved by resettling surplus populations in underexploited regions or, if this option does not exist, by importing alternative sources of animal feed, human food, and fuel, generally in that order of priority.

As the areas at risk of desertification include the poorest countries of the world, the practical possibilities of resettlement or import of alternative supplies is limited. In such circumstances so is the contribution that science and technology can make. Possibly, in the few areas where this is likely to succeed, the most important contribution is precipitation enhancement by cloud seeding. In other situations it is the development of strategies that will permit the maximum sustainable carrying capacity to be maintained for the statistically most probable rainfall regime.

Estimates of this maximum rate of exploitation, together with its areal, seasonal, and annual variation, can be made considerably more accurate by the timely provision of adequate information on the resource status, including that of rainfall; here remote sensing can make a large contribution. The optimum management practice also lends itself to the use of sophisticated new techniques of simulation modelling and systems analysis. However, the major problem in controlling and reducing desertification is in the implementation of a rational program for the exploitation of natural resources by a pastoralist society. The problem is extremely complex, and depressingly few successful examples are available.

In order to conclude on a more hopeful note, some of the techniques used successfully in the long-desertified Negev region of southern Israel will be described briefly. The region extends from the so-called "drought line" (corresponding to an annual rainfall of 230 mm concentrated in four mid-winter months) below which no drought compensation is available for farmers, to the extreme south and east of the region where rainfall is completely negligible (~ 30 mm a year).

In the coastal section and at the eastern border of this extreme desert region, land use is suitable for either tourism or intensive cultivation of high value crops. In both cases the favorable climate is the major natural resource and efficient modern methods are employed to make maximum use of the extremely limited amounts of underground water. The income from both tourism and the high value, out-of-season export crops produced is sufficient to pay for the import of basic foods. The system was rapidly and successfully adopted by the Bedouin inhabitants of the Sinai coastal region, and led them to reduce their livestock flocks to the minimal level needed for an insurance system.

Land use in the major part of the desert region has sought to maintain a maximum degree of vegetation cover to reduce soil erosion and flash floods. In the northern part of the region, down to the 200 mm/yr isohyet, winter cereal cropping and improved natural pastures for lamb production have been economically successful. In the drier regions, attempts to reduce grazing of natural vegetation down to sustainable levels has proved difficult. A recent suggestion is that this limitation should be imposed by strict control of animal watering points, the capacity of which could be matched to that of the vegetation.

Perhaps the most exciting system of land use in the region is runoff farming—a combination of extensive and intensive systems of land use developed in the region by the Nabateans over 1,000 years ago. In this system, water runoff from a catchment area is encouraged by various forms of land treatment and utilized in contoured strips, flood plains or valley bottoms. In areas with an annual rainfall of 200 mm, experiments for the large-scale cultivation of wheat and other crops in strips separated by treated soil show promise. Farther south, in the highland areas with only half this rainfall, several reconstructed Nabatean farms have produced a wide variety of crops, including orchard fruits, some for more than a decade (Evenari *et al.* 1971).

These few concrete examples of what has been done in a desert landscape emphasize the positive aspect of our role in the desertification process, for they show that it *is* a reversible one, dependent above all on human will, ingenuity, and enterprise.

References

Assaf, G. 1989. "Enhancement of precipitation via sea mixing." This volume.

Barnett, T.P. 1983. "Recent changes in sea level and their possible causes." *Climatic change* 5: 15–38.

Baumgartner, A. and E. Reichel. 1975. *The World Water Balance*. Amsterdam: Elsevier Scientific Publishing Co.

Bunting, A.H., M.D. Dennett, J. Elston, and J.R. Milford. 1976. "Rainfall trends in the West African Sahel." *Q. Jl. Roy. Met. Soc.* 102: 59–64.

Charney, J.G. 1975. "Dynamics of deserts and drought in the Sahel." *Q. Jl. Roy. Met. Soc.* 101: 193–202 and 1976 Reply in *Q. Jl. R. Met. Soc.* 102: 468.

Clark, W.C., ed. 1982. *Carbon Dioxide Review 1982*. Oxford: Clarendon Press.

Ellsaesser, H.W., M.C. MacCracken, G.L. Potter, and F.M. Luther. 1976. "An additional model test of positive feedback from high desert albedo." *Q. Jl. Roy. Met. Soc.* 102: 655–666.

Evenari, M., L. Shanan, and N.H. Tadmor. 1971. *The Negev*: the Challenge of the Desert. Cambridge, MA: Harvard University Press.

Frenkiel, J. 1965. "Evaporation reduction." *Arid Zone Research* XXVII. Paris: UNESCO.

Gagin, A. and J. Neumann. 1981. "The second Israeli randomized cloud seeding experiment: Evaluation of the results." *J. appl. Met.* 20: 1301–1311.

Garcia, R. 1981. *Drought and Man: The 1972 Case History, Volume 1: Nature Pleads Not Guilty*. New York: Pergamon Press Inc.

"Guidelines for cloud seeding to augment precipitation." 1983. *J. Irrig. Drainage Div.* ASCE 109: 112–183.

Jensen, M.E., ed. 1981. *Design and operation of farm irrigation systems*. ASAE Monograph No. 3, St. Joseph, MI: American Society of Agricultural Engineers.

Kalinin, G.P. and V.P. Bykov. 1976. "Water resources." In *Encyclopedia Britannica,* Chicago, IL: Encyclopedia Britannica, Inc.

Keller, R. 1970. "Symposium of World Water Balance." International Association of Scientific Hydrology. Publication No. 93: 300–314.

Kellogg, W.W. 1989. "Carbon dioxide and climate change: implications for humankind's future." This volume.

Lamb, H.H. 1982. *Climate, History and the Modern World*. London: Methuen.

Lvovich, M.E. 1970. "World water balance: General report." Symposium on the World Water Balance, University of Reading, England. Publication No. 93: 401–405.

Maas, E.V. and G.J. Hoffman. 1977. "Crop salt tolerance—current assessment." *J. Irrig. Drainage Div.* ASCE 103(IR2): 115–134.

Migahid, A.M. 1952. *Further observations on the flow and loss of water in Sudd swamps of the Upper Nile*. Cairo: Cairo University Press.

Omar, M.H. and M.M. El-Bakry. 1981. "Estimation of evaporation from the lake of the Aswan High Dam (Lake Nasser) based on measurements over the lake." *Agricultural Meteorology* 23: 293–308.

Otterman, J., Y. Waisel, and E. Rosenberg. 1975. "Western Negev and Sinai ecosystems: Comparative study of vegetation, albedo and temperatures." *Agro-Ecosystems* 2: 47–59.

Park, P.K. 1989. "Worldwide pollution of the oceans." This volume.

Salati, E., A.A. Dall'Olio, E. Matsai, and J.R. Gat. 1979. "Recycling of water in the Amazon basin: An isotopic study." *Water Resour. Res.* 15: 1250–1258.

Sellers, W.D. 1969. *Physical Climatology*. Chicago, IL: The University of Chicago Press.

Shachori, A.Y. and A. Michaeli. 1965. "Water yields of forest, maquis and grass covers in semi-arid regions: A literature review." Methodology of plant eco-physiology. *Proceedings of the Montpellier Symposium. Arid Zone Research XXV*: 467–477. Paris: UNESCO.

Shainberg, I. and J.O. Oster. 1978. *Quality of Irrigation Water*. IIIC Publication No. 2. Bet Dagan, Israel: International Irrigation Information Center.

Shalhevet, J. and H. Bielorai. 1978. "Crop Water requirements in relation to climate and soil." *Soil Sci.* 125: 240–247.

Stanhill, G. 1970. "A new method of reducing evaporation," *J. Irreproducible Res.* 18: 55–57.

Stanhill, G. 1983. "The distribution of global solar radiation over the land surfaces of the earth." *Solar Energy* 31: 95–104.

Stanhill, G. 1985 "The water resource for agriculture," *Phil. Trans R. Soc. Land.* B 310:161-173.

Stanhill, G. 1986 "Water Use Efficiency." *Advances w. Agrmomg.* 39:53–85.

Stanhill, G., S. Moreshet, and M. Fuchs. 1976. "Effect of increasing foliage and soil reflectivity on the yield and water use efficiency of grain sorghum. *Agron. J.* 68: 329–332.

van Keulen, H., H.M. van Laar, W. Louwerse, and J. Goudriaan. 1980. "Physiological aspects of increased CO_2 concentration." *Experientia* 36: 786–792.

FIFTEEN

COMMENTARY on GERALD STANHILL'S PAPER

Avraham Melamed

Desertification, much induced by humans themselves, and the lot of the population inhabiting the gradually desertified arid and semi-arid areas may generally be viewed pessimistically. This is to a great extent a result of the present state of general development and lack of the minimal know-how and capital resources required to halt the process of desertification.

Nevertheless, much could be achieved in reversing the trend by applying rather simple technologies, e.g., developing shallow groundwater resources for watering cattle and small-scale irrigation, soil conservation methods and afforestation (applied successfully in Israel under similar climatic conditions).

Yet, there are great prospects for development of many of the presently desertified regions when in some future era the vast natural resources with which those regions are blessed are combined with the application of proper technologies. Some of these technologies are already applied elsewhere and some may be expected to be available in the future: For example, there are great rivers in the climatic regions of concern that are practically still unused for irrigation. The most important ones are the Shari and Logone rivers in Chad. The flow of these two rivers is equivalent to the present share of Egypt of the Nile

River waters, where this source of water provides the main economic basis for a population some 10 times greater than that of the Shari and Logone basins.

Just south of the Nile, Shari and Logone river basins the climate is humid, predominantly a tropical rainforest area, drained by the vast basin of the Congo River (Figure 15-1). In these areas, similar to those of the Amazon of South America, water yield averages some 400 mm per annum and the total flow of the Congo River (half the flow of the Amazon) is almost 20 times the flow of the Nile.

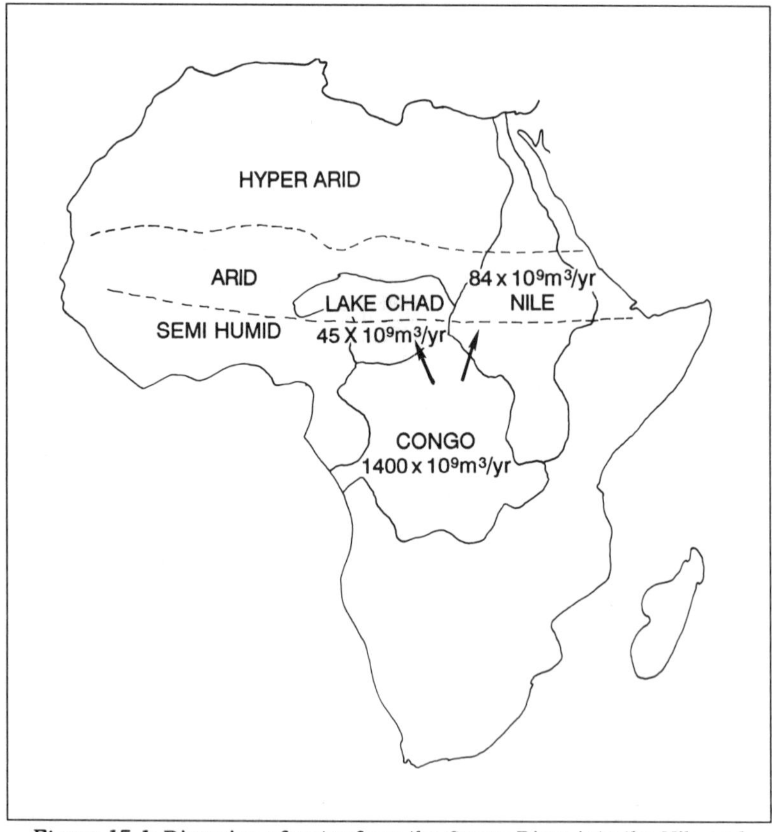

Figure 15-1: Diversion of water from the Congo River into the Nile and Lake Chad basins.

Topographic barriers between the Congo basin and the basins adjacent to it in the north may be overcome by present-day hydrotechnical technologies already applied elsewhere (e.g., in the United States and the Soviet Union). Diversion of only 5% of the Congo basin water would thus augment the water resources of the arid belt by an amount equal to the entire flow of the Nile!

In many arid regions the ratio between groundwater reserves and natural replenishment of water by rainfall is far greater than in more humid areas. Therefore, the one-time reserves of groundwater underlying the vast arid belt of Africa, in particular in Nubian Sandstone, are extremely significant as a great resource for the future. This resource, tapped now on only a very minor scale, may for a whole century yield quantities of water equal to the Nile's entire flow.

One may expect that future technologies would provide the means to affect climatic conditions. For example, the first steps we are taking today to materialize concepts of artificial cloud generation may bring about great changes in the natural water balance of arid regions. Reduction of evapotranspiration from irrigated areas shaded by such clouds along with enhanced runoff and water yield from large areas covered by natural vegetation may result in changes in the agricultural productivity of such areas.

Last but not least, is the great natural resource with which the arid zones of the globe are endowed—solar energy. In the future, with means available for its economic exploitation, large energy resources may facilitate the management of water resources in the pumping and conveyance of water as well as in improving its quality.

The peoples of the arid regions may look forward to a much brighter future when part or all of the above resources and technologies are applied.

With assistance received from outside in technology and capital, combined with an education effort on their own part, these goals may be achieved sooner.

PART FOUR:

HUMAN INVOLVEMENT

SIXTEEN

ENHANCEMENT of PRECIPITATION by SEA MIXING

Gad Assaf

The borders of the deserts in Israel and Egypt coincide with the southern deserts of the Mediterranean Sea. The desert borders here are narrow: for example, in the Israeli coastal area the annual precipitation changes from 100 mm near the Egyptian border to 400 mm only 50 km north of the border. The precipitation in this area develops in winter, when cold continental air from Europe intrudes over the warm sea. The sharp desert boundaries indicate that the cloud convection decays rapidly when the air masses leave the sea.

The decay of precipitation is less pronounced going windward towards the east. Here orographic effects make the precipitation pattern more complicated. Nevertheless, on the average, annual precipitation declines from 500 mm in coastal zones to 200 mm only 150 km inland.

The Mediterranean heat storage is a major element which induces precipitation in the Levant area. There is little doubt that the warm sea in winter feeds the energy and the vapor to the winter storms. Anati (1977) examined the heat content of the sea and found that more than half of summer insolation is accumulated in the upper sea. The heat, which is accumulated during the summer, is removed to the atmosphere during the winter. In fact, the evaporation heat and buoyancy fluxes from the Mediterranean are larger during the winter than in the summer (Table 16-1).

Table 16-1: Characteristics of Physical Parameters over the Eastern Mediterranean

	Summer	Winter	Storms
Solar Insolation I (w/m^2)	260	100	50
Heat Storage Variation ΔQ (w/m^2)	180	-180	-500
Mechanical (wind) Mixing Energy M (w/m^2)	3 x 10^{-3}	5 x 10^{-3}	5 x 10^{-2}
Buoyancy Flux in the sea B_s (m^2/sec^3)	-6 x 10^{-8}	10^{-7}	2.5 x 10^{-7}
Buoyancy Energy Flux (in the sea) b_s* (w/m^2)	-3 x 10^{-3}	10^{-2}	2.5 x 10^{-2}
Atmospheric Buoyancy Flux B_a (m^2/sec^3)	3 x 10^{-4}	2 x 10^{-3}	10^{-2}
Atmospheric Buoyancy Energy Flux b_a* (w/m^2)	0.3	5	50

*See Appendix 16-A for detailed discussion.
Note: The subscripts s and a are used for sea and atmospheric conditions.

The sea absorbs some 95% of the incidental solar radiation and practically all of the atmospheric long-wave radiation. The outgoing energy consists of latent heat (vapor flux), long-wave radiation, and sensible heat fluxes. These fluxes all increase with surface temperature. The equilibrium temperature T_e

keeps the energy fluxes from the surface at the same level as the incident fluxes.

In summer the sea temperature is below the equilibrium level and more heat comes into the sea than goes out. The surface temperature depletion depends on the rate of wind mixing. An intensive wind mixing entrains cold water from below, which keeps the sea surface at a lower temperature and enhances the heat accumulation in and above the seasonal thermocline (Figure 16-1). In winter, the heat storage keeps the sea above the equilibrium temperature and enhances the heat fluxes from the sea to the atmosphere. The winter heat flux from the Mediterranean is about 300 w/m^2, two-thirds (200 w/m^2) of which is contributed from the heat storage depletion.

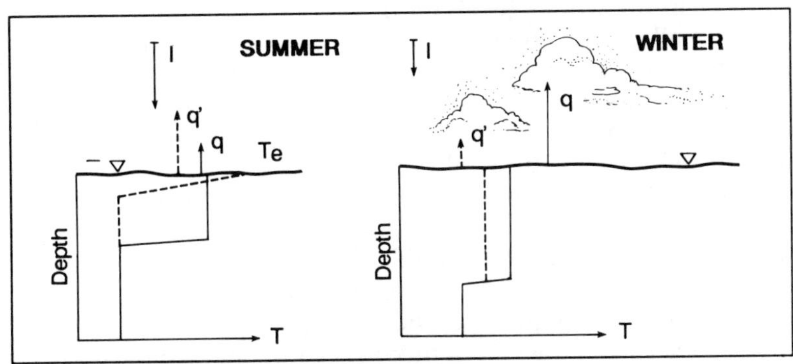

Figure 16-1: Schematics of the heat fluxes.
I In-going (solar and atmosphere fluxes)
q' Outgoing heat from the sea surface without sea mixing
q Outgoing heat from the sea with summer mixing
Te Equilibrium temperature

The evaporation rate and heat fluxes from the Eastern Mediterranean (EM) in winter exceed the summer fluxes (see Table 16-1, Figure 16-1). The winter fluxes are enhanced by intrusion of cold continental air masses into the EM. The cold atmosphere enhances heat fluxes and at the same time induces convective instability which extends above the condensation level. The condensation converts latent heat into sensible heat which feeds its energy to the cloud convection. The latent heat dominates the heat fluxes from the EM and it affects the atmosphere through condensation.

Tzevtkov and Assaf (1982) examined the variations in heat storage of the sea and the precipitation pattern in Israel. They found a fair correlation ($r = 0.56$) between the heat storage in early winter and the winter precipitation. A remarkable correlation ($r = 0.9$) was found between heat storage *depletion* and winter precipitation. This indicates that, even in years when heat storage characteristics work against atmospheric features, the actual heat removed from the sea is proportional to the actual precipitation; for example, in cases where cold winters are associated with small summer heat storage or when warm winters are associated with large heat storage.

Haney and Davis (1976) performed numerical experiments on the seasonal thermocline model. They found that with wind mixing the surface temperature is lower in summer and warmer in the winter by about 0.5°C.

Physically, wind mixing converts kinetic into potential energy by entrainment of dense water from lower layers. The conversion rate of wind into potential energy is exceedingly small: in the open sea, it is of an order of magnitude of few mW/m^2 (see Table 16-1). This should be compared with a rate of 200 W/m^2, which is the rate at which the heat storage is released from the sea to the atmosphere in winter. In other words, the winter heat fluxes which are associated with summer wind mixing are larger by five orders of magnitude than the energy flux which generates the mixing itself. This finding led to the idea that artificial sea mixing can be utilized to increase the seasonal heat storage over the EM.

Assaf and Bronicki (1980) explored the technological aspects of artificial sea mixing and consequent increase of heat storage and winter precipitation.

Mixing Energy

Assume that someone who reads this chapter decides to do something about the weather and mix the sea in the summer. Our reader rents a boat, a cable, and a container which is open on top and has a hole in the bottom. The reader lowers the container to the deep layer of the EM where the density is $\rho = 1028\ kg/m^3$ and starts to pull the container upwards (Figure 16-2).

As long as the container is submerged in dense water the force exerted on the cable will just balance the extra weight of

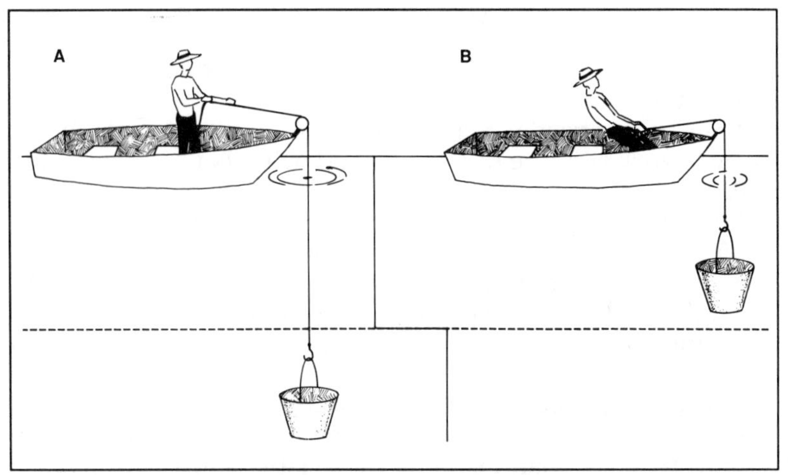

Figure 16-2: The effort associated with sea mixing.
A: As long as the container is within the dense water, the effort is negligible.
B: Large effort is needed to raise the dense water within the light water above.

the walls and the friction, which we assume to be negligible.

When the container is lifted to the upper layer, where the density is $\rho = 1026$ kg/m^3, our reader will have to pull the cable with a force of $g\Delta\rho V$, where:

$\Delta\rho = 2$ kg/m^3 = the density difference
V = volume of the dense water in the container
g = the gravity acceleration

The dense water in the container will spread from the hole at the bottom and will be replaced with light water from above. This will reduce the force exerted to pull the container. We assume that the replacement rate of the dense water will be uniform, so that near the surface all the dense water will be mixed throughout the upper layer and our reader will be ready to allow the container to sink for the next mixing cycle.

The average force exerted while the container was lifted is half of the maximum force, i.e., $g\,V\,\Delta\rho\,/\,2$ and the energy invested by our reader in one cycle is:

$$M = h\,g\,V\,\Delta\rho/2$$

For: V = 10 m^3, $\Delta\rho$ = 2 kg/m^3 and h = 50 m one obtains:
M = 5000 Joule

Thus, if our reader makes one cycle in 100 seconds, he or she will work at a rate of 50 watts and will mix the sea at a rate of 0.1 m^3/sec.

To increase the volume of the upper layer by 10%, from 50 to 55 m depth, over the entire 600,000 km^2 one needs some three million devoted persons who will work day and night throughout the summer period. The power exerted by the three million is 150 MW, which is the power of a medium to small power turbine.

There are probably more practical means to mix the sea, some of them will be discussed in the next section. Before we go into this, we shall examine the relationships between sea mixing and sea temperature (in Appendix 15-B). Natural summer mixing reduces the Mediterranean surface temperature by 2.5°C from its equilibrium value. This induces accumulation of heat at a rate of 200 W/m^2 and it consumes only 3.7 mW/m^2 of mechanical energy. Thus one may say that the ratio of heat accumulation rate to mixing rate is given by:

$$\frac{2\,c_p}{g\,\alpha\,h} = 40{,}000$$

where: h = the depth of the mixed layer, α =thermal expansion coefficient.

Such an efficiency calls for artificial mixing as a practical means to enhance heat storage.

Artificial Mixing

Long surface waves dissipate on the Mediterranean coasts at a rate of 7 kW/m. There are numerous ideas how to convert wave energy to power. One of the ideas, due to the late John Isaacs of the Scripps Institute, was to introduce deep pipes on floats with one-way valves that would allow the water to flow only upward (Figure 16-3). These pipes can be converted to effective sea mixers by extending the pipes to the depth of 80 m and allowing the cold water to be distributed in the upper sea (Figure 16-4). A 2 m diameter pipe which floats on a 4 m diameter buoy will induce sea mixing at a rate of 4 kW.

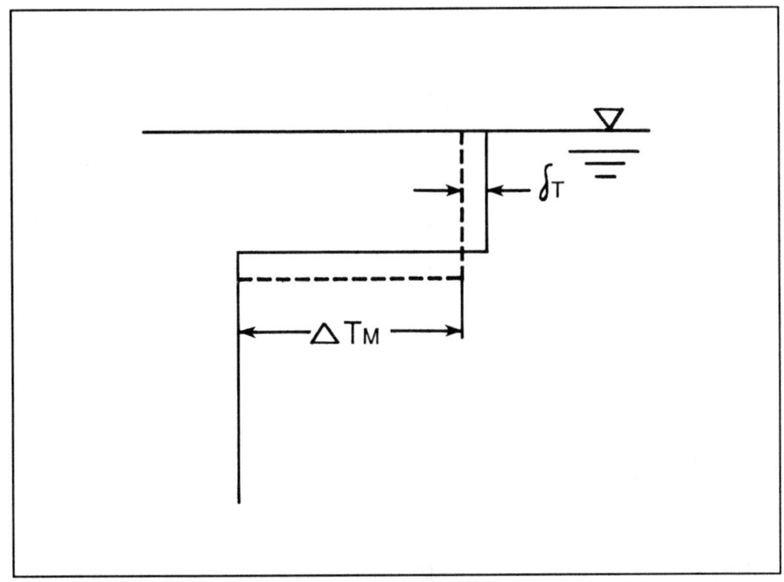

Figure 16-3: Schematic of two-layer one-dimension-sea mixing.

Ten thousand such mixers distributed over an area of 3 x 3 square degrees (330 km x 330 km) would amplify seasonal heat storage by 10%. As a result December surface temperature would be elevated by 0.5°C (Haney and Davis 1976). From the work of Tzvetkov and Assaf (1982) one may see that this yields an excess precipitation of 70 mm in Jerusalem and 30 mm in Amman. Conservatively we assume excess precipitation of 20 mm over an area of 10^5 km^2, or a total of 2 x 10^9 m^3 of excess overland precipitation. Each mixer will thus yield an annual output of 200,000 m^3 of overland precipitation.

The total amount of energy for mixing in an area of 10^5 km^2 will be 10^8 kWh/yr (40 MW for 2500 hours). Thus each m^3 of excess overland precipitation consumes 0.05 kWh of mixing energy. It is interesting to compare this with the performance of a desalination plant which consumes 10 kWh/m^3. It is possible to consider not only wave mixers but also intensive mixing platforms which would pump deep cold water to be mixed with the warm water above. As the hydraulic head is very small (0.2 m) and the flow rate is large, intensive mixing should be

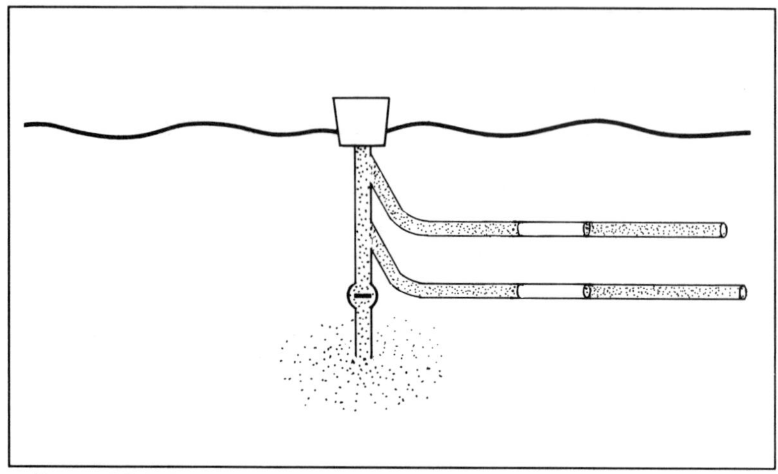

Figure 16-4: One version of wave mixer.

associated with new pumping technology. Air lift pumps would seem to be the proper means. But one may also consider jet pumps, where a small volume of water will induce a larger volume flow rate to be mixed.

It should be noted that one may pump cold water to be mixed with warmer water or warmer water may be pumped to deeper levels to be mixed with cold water below. We envision that each platform will induce mixing at a rate of, say, 4 MW, i.e., it will pump some 2000 m^3/sec across an hydraulic head of 0.2 m. The pipe will be between 50 m to 100 m long and 15 m in diameter. Each platform will hold 3 such pipes that will inject the water as plumes with 1 m diameter or so.

The actual mixing would be developed as an entrainment into these plumes. Ten platforms with a total power generation of 10 MW and mixing generation rate of 4 MW each (i.e., 40% pumping efficiency), functioning for 3000 hours a year, would be needed to enhance overland precipitation by 2×10^9 m^3, i.e., 0.15 kWh/m^3 overland precipitation.

A detailed engineering study will be needed to evaluate the most economical means for sea mixing. Personally, I believe that a wave mixer will win the bid.

Appendix A

The 10-fold increase of buoyancy convective energy fluxes in storms (see Table 16-1) is mostly related to condensation and cloud formation.

The total flux of kinetic energy is related to the buoyancy flux B as:

(i) $$b^* = \rho B h \quad (\mathrm{w/m^2})$$

Eventually this flux of energy is dissipated in the system. Before it dissipates, the energy can be pumped into different scales of atmospheric motions. In three-dimensional (3D) isotropic turbulence, the energy usually cascades into smaller scales and eventually dissipates into Kolmogorov microscales.

Levich and Tzvetkov (1982) argued that initial energy input is at the scale of the cumulus clouds, which is a few km in size. They assumed that some of the energy cascades into two-dimensional (2D) scales. It was shown by Kraichnan and Montgomery (1980) that the 2D turbulence is characterized by two inertial ranges, where entropy is propagated into the small-scale eddies and energy cascades into the large-scale eddies.

Levich and Tzvetkov found that the build-up rate of energy in the cyclone is comparable with the conversion rate of heat to convective energy in the clouds.

Appendix B

Surface Temperature and Mixing

Let us approximate the seasonal thermocline as a two-layer fluid (see Figure 16-2). At $T = T_e$ the surface temperature is in equilibrium and the heat balance can be written as:

(i) $$\rho_a \cdot c_{pa} \cdot c_b \cdot u\,(1 + R)\,(T_e - T_a) + IR\,(T_e) = Q_a \;(1)$$

where a denotes air parameters, ρ = density, c_{pa} = specific heat of air, c_b = the bulk exchange coefficient, R = inverse Bowen ratio (R = LH/SH), SH = sensible heat flux, LH = latent heat flux, u = air velocity, IR = infrared radiation balance, Q_a = atmospheric and solar radiation which is absorbed by the sea surface.

Assuming that natural or artificial sea mixing is now turned on (the effect of mixing is displayed by the dashed line in Figure

16-2), entrainment will increase the volume and reduce the surface temperature by δT.

The heat fluxes from the surface (Q_M) will be reduced by:

$$\text{(ii)} \quad \Delta Q_M = \rho_a \cdot c_{pa} \cdot c_b \cdot u\,(1 + R) + \frac{\partial\,(IR)}{\partial T}\,\delta T.$$

Eventually the surface temperature will be stabilized at a new equilibrium T_M, which can be estimated by equating with the cooling rate of the surface due to entrainment of cold water:

(iii)

$$\rho \cdot c_p \cdot W_e \cdot \Delta T_M = \rho_a \cdot c_{pa} \cdot c_D \cdot u\,(1 + R) + \frac{\partial\,(IR)}{\partial T}\,\delta T$$

where W_e = the entrainment velocity, ρ = water density, c_p = specific heat of water, $\Delta T_M = T_M - T_D$, T_D = the deep water temperature, c_D = the bulk coefficient. Equation (iii) can be solved for $\delta T/\Delta T_M$.

For summer conditions in the Mediterranean we may approximate: R=8, u=6m/sec, $\partial(IR)/\partial T$ = 5 W/m^{2}-° C and W_e = 5x10^{-6}m/sec. For these parameters we obtain:
$\partial T/\Delta T_M$= 0.25; i.e., without mixing, the temperature of the Mediterranean surface will be elevated by:
∂T = 2.5, $\Delta T_M \approx$ 10°C and the accumulated heat during the summer months is estimated as:

$$\rho \cdot c_p \cdot h \cdot \Delta T_M = 2 \times 10^9\, J/\text{m}^2 \text{ or } 550 \text{ kWh/m}^2,$$

which is equivalent for heating at a rate of 200 W/m^2, where

$$h = W_e \cdot t = 50\, m,\ \Delta T_M = 10°\text{C, and } t = 10^7 \text{ sec.}$$

The mechanical energy which is involved in this mixing process can be approximated as:

$$ME = g \cdot \rho \cdot \alpha \cdot \Delta T_M \cdot W_e \cdot h/2 \approx 3.7 \times 10^{-3}\, W/\text{m}^2,$$

where: $\alpha = \frac{1}{\rho} \cdot \frac{d\rho}{d\text{T}}$

Thus, the natual summer mixing reduces the Mediterranean surface temperature by 2.5°C from its equilibrium value. This induces accumulation of heat at a rate of 200 W/m^2 and it consumes only 3.7 mW/m^2 of mechanical energy. Thus one may say that the ratio of the heat accumulation rate to the mixing rate is given by:

$$\frac{\Delta Q}{ME} = \frac{2\,c_p}{g\,\alpha\,h} = 40{,}000$$

Such an efficiency calls for artificial mixing as a practical means to enhance heat storage.

References

Anati, A. 1977. "Topics in the Physics of Mediterranean Seas." Ph.D. Thesis, Weizmann Institute of Science.

Assaf, G. and L. Bronicki. 1980. "Method and means for weather modification." United States Patent Application.

Haney, R.L. and R.W. Davis. 1976. "The role of surface mixing in the seasonal variation of the Ocean Thermal Structure." *J. of Phys. Ocean.* 6: 504–510.

Kraichnan, R.H. and D. Montgomery. 1980. *Report on Progress in Physics 43*, No. 5: 547–619.

Levich, E. and E. Tzvetkov. 1982. "Hydrodynamic model for a single rain situation in the eastern Mediterranean." Proceedings of Charney's Workshop on Weather Modification, Cesarea.

Tzvetkov, E. and G. Assaf. 1982. "The Mediterranean heat storage and Israeli precipitation." *Water Resources Research* 18: 1036–1040.

SEVENTEEN

EFFECTS of ARTIFICIAL SEA MIXING

OCEANOGRAPHIC AND METEOROLOGICAL PROCESSES

Colette Serruya

The idea of artificial mixing of the Eastern Mediterranean (EM) as a practical means of increasing heat storage and rainfall on the Middle East is based on the relationship found by Tzvetkov and Assaf (1982) between these two parameters. The technology of mixing consists in entraining deep cold water to the surface. The profitability of the enterprise is guaranteed by the very high ratio (~40,000) existing between the amount of energy stored as heat accumulation and the amount of energy required for the mixing necessary to allow this heat accumulation. This very attractive technique, although designed for increasing overland precipitation, will affect many other parameters and processes. The purpose of this chapter is to discuss the most prominent effects of artificial sea mixing on oceanographic and meteorological processes.

Sea Mixing and Nutrient Recirculation

The EM is a stratified water body having a "permanent thermocline" around 300 m and a "seasonal thermocline" approximately at 100 m in summer. During the winter mixing, the upper 200 m are fairly homothermal at a temperature of 16 to 17°C.

The oxygen profiles of the EM are very informative of the mixing conditions, since a slight decrease in mixing rate is always accompanied by a drastic decrease of oxygen, a substance which is incessantly consumed by the oxidative processes. Deep oxygen profiles measured by Oren (1971) show two important features: (1) a drastic decrease of concentration below 300 m in all seasons, confirming the presence of a permanent thermocline; and (2) the presence of at least 3 mgl^{-1} oxygen in deep water, indicating that a slow but constant exchange of water exists between the upper layers and the water masses below the permanent thermocline.

An inverse pattern is found for the main nutrients: nitrogen, phosphorus, and silica. In contrast with oxygen, these substances accumulate with depth. The low productivity of the EM is related to the poor recirculation of these nutrients. The high photosynthetic potential of the area, due to the highly favorable radiation situation, does not materialize into high productivity because of the lack of phosphorus, nitrogen and silica.

Deep profiles measured by Oren (1971) down to 1,800 m show a net increase of phosphorus at 200 m (0.2–0.3 μgl^{-1} in comparison with 0.05 μgl^{-1} at the surface). The concentration of phosphorus continues to increase with depth (0.4 μgl^{-1} at 400 m and 1.0 μgl^{-1} at 1,000 m).

The artificial mixing of the upper 300 m of the sea could enhance photosynthetic activity and algal production by two mechanisms:

(1) The redistribution of a given amount of nutrients stored in the layer between 200-300 m at a given time. A very crude calculation shows that by mixing an area of 300 km x 300 km we can recirculate as much as 30 x 10^{12} mg phosphorus stored in the 200–300 m layer. The redistribution of this phosphorus into the upper 300 m of the considered area would increase the phosphorus concentration by 60% were the recirculated phosphorus to remain in the considered mixing area. Taking into

account losses by advection, the local increase would be much less but its effect would be felt on a wider area.

(2) The recirculation at a faster rate of the dead organic matter sinking as a constant rain to the 200–300 m layer. This permanent recirculation will accelerate the mineralization of the organic phosphorus (P) and nitrogen (N) into inorganic forms assimilable by algae. In other words the turnover time of phosphorus and nitrogen would be shortened, and a given amount of P and N atoms would allow the formation of more biological material.

These processes should lead to a significant increase of primary productivity, which is the first link of any terrestrial or marine food chain—and more specifically, of fish production. The many EM countries depending on sea fisheries for their protein supply should see this increase of productivity as a very positive by-product of artificial mixing. It is also a non-negligible fringe benefit for the project from the purely economic point of view.

Sea Mixing and Surface Temperature

The Levant summer climate is dominated by the breeze regime: landward breezes during daytime and seaward breezes at night. This characteristic regime is due to the very different thermal capacity of seas and continents. During daytime the land is more rapidly heated by the sun than the sea. Consequently the air mass above land masses expands rapidly and rises, initiating a landward flow of cooler marine air. At night, the continental air rapidly becomes cooler than the marine air mass and flows seaward.

Since the air flow is governed by the difference in temperature of the land and sea surfaces, it is clear that artificial mixing which would decrease the temperature of the sea surface would affect the breeze regime.

Mahrer (1985), utilizing a model developed by Segal *et al.* (1982) to simulate the diurnal variation of the sea and land breezes, investigated the possible effect of the decrease of the Mediterranean sea surface temperature which would result from artificial mixing. Two numerical simulations were carried out, one with the July average of the Mediterranean sea surface temperature (27.5°C) and one with a sea surface temperature of 24.5°C. The effects of the lowering of the sea surface temperature are as follows: (1) The changes in flow pattern mainly concern the coastal area; (2) With a sea surface at lower temperature, a stronger westerly wind will prevail during daytime (an approximate increase of 15% of wind velocity at noon is predicted for the given 3°C difference in sea surface temperature; (3) At night the seaward breeze is reduced since the difference of temperature between land and sea diminishes; (4) A cooler temperature will prevail during daytime.

It is difficult to estimate the effect that a substantial increase in wind velocity would have on human activities (fisheries, tourism) in the coastal areas. A more positive aspect of this modification could be envisaged were the wind energy on the Israeli coast to be utilized on a large scale.

As far as the decrease of daytime air temperature is concerned, this may be considered a fully positive side effect.

Sea Mixing and Winter Climate

The weather modification project has already generated considerable research concerning cloud formation in the EM. Outstanding results in this field have been obtained.

(1) The clouds producing rain over the Levant are not born over the whole EM on the path of displacement of cold air masses, but are rather formed along the African coast and near the Levantine coast (Tzvetkov *et al.* 1982). This is in perfect agreement with the finding by Hecht *et al.* (1985) that, in September 1980, high heat storage was concentrated at the eastern edge of the EM forming a thermal front near the Levantine coast. A similar feature having been reported earlier by Levine and White (1972), it seems that this accumulation of warm water near the Israeli coast represents a permanent feature strongly correlated with the formation of rain clouds in the EM.

(2) Dry years and wet years do not differ in the number of rain events but in the duration of each event (Tzvetkov *et al.* 1982).

These observations indicate that artificial mixing is far from exhaustive, and more research should be done to assess the effect on human communities and economic activities.

References

Hecht, A., Rosentroub, Z. and Bishop, J. 1985. "Temporal and Spatial variations of heat storage in the eastern Mediterranean Sea." *Israel J. Earth Sci.* 34:51–64.

Levine, E.R. and White, W.B. 1972. "Thermal frontal zones in the Mediterranean Sea." *J. Geophys. Res.* 77:1081–1086.

Mahrer, Y. 1985. "A numerical study of the effects of sea surface temperature on the sea and land breeze circulation." *Israel J. Earth Sci.* 34:91–95.

Oren., O.H. 1971. "The Atlantic water in the Levant basin and on the shores of Israel." *Cah. Oceanogr.* 23:291–297.

Segal, M., Mahrer, Y. and Pielke, R.A. 1982. "Application of a numerical mesoscale model for the evaluation of seasonal persistent regional climatological patterns." *J. Appl. Meteorol.* 21:1754–1762.

Tzvetkov, E. and Assaf, G. 1982. "The Mediterranean heat storage and Israeli precipitation." *Water Resour. Res.* 18:1036–1040.

Tzvetkov., E., Assaf, G., and Mannes, A. 1985. "Synoptics and cloud fields connected with rain situations in Israel," *Israel J. Earth Sci.* 34:102–109.

EIGHTEEN

GLOBAL EFFECTS of NUCLEAR EXCHANGE

Robert U. Ayres

It would be normal in a volume such as this to report recent research in which one has played a significant role. As it happens, however, my work on the subject of this chapter was mostly done over 25 years ago. Specifically, I carried out a three-year study on the environmental consequences of nuclear war at the Hudson Institute during the years 1962–1965. It was sponsored by the Office of Civil Defense, then part of the Department of Defense, though later it shifted to the Federal Emergency Management Agency (FEMA). The work was documented in a series of unclassified research reports that were (and probably still are) available through the United States Department of Commerce National Technical Information Service (NTIS).

In November 1983, ABC-TV presented a docu-drama "The Day After," purporting to be a realistic account of the conditions that might exist following a nuclear war. The docu-drama was criticized by some as an unwarranted and misleading attempt to scare the public and influence the nuclear freeze debate. But the most surprising criticism came from the well-known planetary astronomer Carl Sagan of Cornell University, who flatly asserted that the movie far *understated* the severity of the aftereffects. In a televised interview immediately following the TV movie, Sagan stated that a "small" nuclear exchange involving as little as 1000 megatons (TNT equivalent), with its associated fires, might inject enough fine dust and smoke into the atmosphere to intercept most of the incident sunlight for several weeks. Thus the land surface of the northern hemisphere would be in a dense shadow that would cause a sudden and extreme cold wave. He suggested that the land surface temperature could drop by as much as 30°C, depending on the season. This (hypothetical) phenomenon has since come to be known as the "nuclear winter." As it happens, I predicted a similar (though milder) cooling effect in my 1965 study, based on historical experience with large volcanic eruptions. More of this later.

I chanced to be present at a scientific meeting a few days after the ABC-TV presentation, where a small group, including many of the authors represented in this volume, assembled informally at the suggestion of Fred Singer to exchange notes and ideas on the nuclear winter phenomenon. Inevitably, my past interest in the problem emerged, as did some apparent discrepancies between my 1965 predictions and those made by Sagan *et al.* (R.P. Turco, Toon, Ackerman, Pollack and Sagan, summarized in *Science*, 1983. "Global Atmospheric Consequences of Nuclear War." I'll refer to these authors hereafter as TTAPS). It was immediately obvious that the matter would have to be reconsidered—if only to satisfy my own curiosity. Has the available data changed since 1965? Has the state-of-the-art of atmospheric modelling—especially with the help of computers—changed so dramatically? Did I overlook something crucial? Or, is this one of those surprisingly common situations involving long and complicated chains of reasoning where reasonable people making apparently reasonable assumptions and using the same data end up coming to different conclusions?

Considering that no atmospheric nuclear tests have been carried out by the United States since 1961, I thought it unlikely that the data available to TTAPS would have differed significantly from those available to me in 1963–1965. Nor did it seem likely that the improved sophistication of computerized atmospheric modeling would entirely account for the differences in our quantitative predictions. But I couldn't do much about satisfying my curiosity until the TTAPS papers were published and the still unpublished backup document finally reached me (in January 1984) by a roundabout route. To anticipate the outcome, I now believe the discrepancies between the two independent studies (TTAPS was apparently unaware of the earlier work), to be due to a combination of factors, mainly different assumptions, but not to actual major mistakes of omission or commission on either side. This is an interesting conclusion, since it suggests the enormous sensitivity of complex chains of reasoning—such as the nuclear winter scenario—to very small and seemingly innocent assumptions by the analyst.

The Physical Basis of the Nuclear Winter Scenario

The baseline nuclear war scenario of TTAPS involved 10,400 nuclear bursts ranging in yield from 0.1 to 10 megatons (MT) of TNT equivalent and in altitude from the ground or water surface to beyond the atmosphere. (The baseline TTAPS originally used was different. This was not the only assumption differing without explanation.) These explosions were assumed to raise an average of 0.333 tons of dust into the stratosphere for each ton of explosive power in a ground-burst weapon and 0.1 tons of dust for above surface bursts. The dust (9.6×10^8 tons in all) was assumed to have a log-normal particle size distribution for particle diameters less than 1 micron (10^{-6} meter) with 8.4% of the mass consisting of particles in this size range. The baseline scenario also assumed widespread fires of all kinds including urban/industrial fires (52% of emissions), fire storms (7% of emissions), wild fires (30%) and long-lived fires in exposed peat deposits, coal seams, etc. (7%). These fires generated an assumed total of 2.25×10^8 tons of smoke emissions (soot and fly ash) of which 5% reached the stratosphere. The rest remained in the troposphere for a period of up to several weeks before being washed out, largely as "black rain." The soot and fly ash particles were assumed to be distributed around a very small modal radius, about a tenth of a micron (0.1μ).

Atmospheric Effects of a Nuclear Exchange

The meteorological/climatological effects can be divided into two categories: (1) short-term effects lasting a few weeks, resulting primarily from the dense smoke pall (due to fires), and (2) longer term effects resulting from the stratospheric dust layer.

In both cases TTAPS assumed that the primary effect would be a net cooling, due to the fact that very small particles (i.e., with diameters of less than 1μ) tend to be very effective scatterers of light in the wavelengths characteristic of solar radiation. In fact, long ago, W.J. Humphreys (1940) showed that the intensity of solar light passing through a dusty cloud layer, with index of refraction m = 1.5 but no absorption, falls off as exp(-λx) where x is the path length in cm. of the light through the dusty layer. The attenuation coefficient, λ, in this case is defined by

$$(1) \quad \lambda_r \cong 2\pi r^2 \rho_0 \times 10^{-8}\, cm^{-1} \; (for\ r < 1\,\mu)$$

where ρ_0 is the density of scatterers (all with the same radius r), per cubic centimeter in the layer. It is interesting to note that a 20% attenuation—i.e., 20% reduction in solar flux (insolation) at the earth's surface—would only require a mass of about 3 x 10^6 metric tons of particles, distributed in a uniform cloud layer, if we assume all particles have radii of 0.25 μ, index of refraction 1.5, and a density of 2 gm/cm^3 (corresponding roughly to silica). To cut the insolation level by 40% (instead of 20%) would require the mass of the dust cloud to be a little more than twice as great, or about 7.0 x 10^6 metric tons. Reduced insolation would lead to cooling (over a period of time) of the surface of the earth, as will be discussed later.

Obviously the simple calculation is unrealistic, since particle sizes are not all uniform (monodisperse). However, a realistic (polydisperse) cloud can be assumed to be a superposition of many monodisperse clouds, distributed over the entire range of sizes. Thus, calculation of the effective attenuation of insolation requires integration over the entire (log-normal) particle size distribution. Also, it must be pointed out that while most glassy particles have indices of refraction close to 1.5, a correct scattering calculation should also take into account the absorption coefficient ξ, which is usually defined as the imaginary part of the index of refraction. The effect of absorption is negligible in the UV and visible part of the solar spectrum but becomes very significant in the infrared (IR) region. Absorbed radiation is re-radiated in the IR region at a rate depending on the temperature of the radiating body. TTAPS assumed for convenience that IR absorption and re-radiation by the dust cloud in the infrared would be negligible. However, this assumption deserves closer scrutiny, as will be pointed out later.

TTAPS also considered smoke from fires as a large source of particles which would not reach the stratosphere at all, but (because of their very small sizes) might remain in the troposphere for several weeks. This assumption is perhaps the single major difference between the two studies.

Returning to the basic dust light scattering and attenuation phenomenon, it appears that a 20% to 40% reduction in surface insolation would require no more than 3 to 7 million tons of monodisperse scatterers with radii of 0.25μ. The same effect would require proportionally smaller quantities of smaller radius particles, and conversely. (In other words, for 0.5μ particles, roughly double the tonnage of scatterers is required.) Obviously, if on the order of 10^9 tons of dust and smoke are injected into the stratosphere by the nuclear bursts, the problem is potentially very severe. The critical questions are as follows: (1) How rapidly would the particulates diffuse and spread laterally into a uniform layer? and (2) How rapidly would the particulates be removed from the atmosphere by physical processes?

With regard to question (1), it must be recalled that almost all of the weapons would be exploded within the latitude range 30°–60°(N), constituting around 15% of the earth's surface. Normal tropospheric circulation is primarily west to east, with very little mixing across the equator. Mixing in each hemisphere is primarily due to the so-called jet streams, which hover at the boundary of the temperate zones and the tropics in summer and at the boundary of the temperate zones and the polar zones in winter, as shown in Figure 18-1. Thus the dust and smoke would diffuse (under normal conditions) only quite slowly out of the north temperate zone where it was first injected.

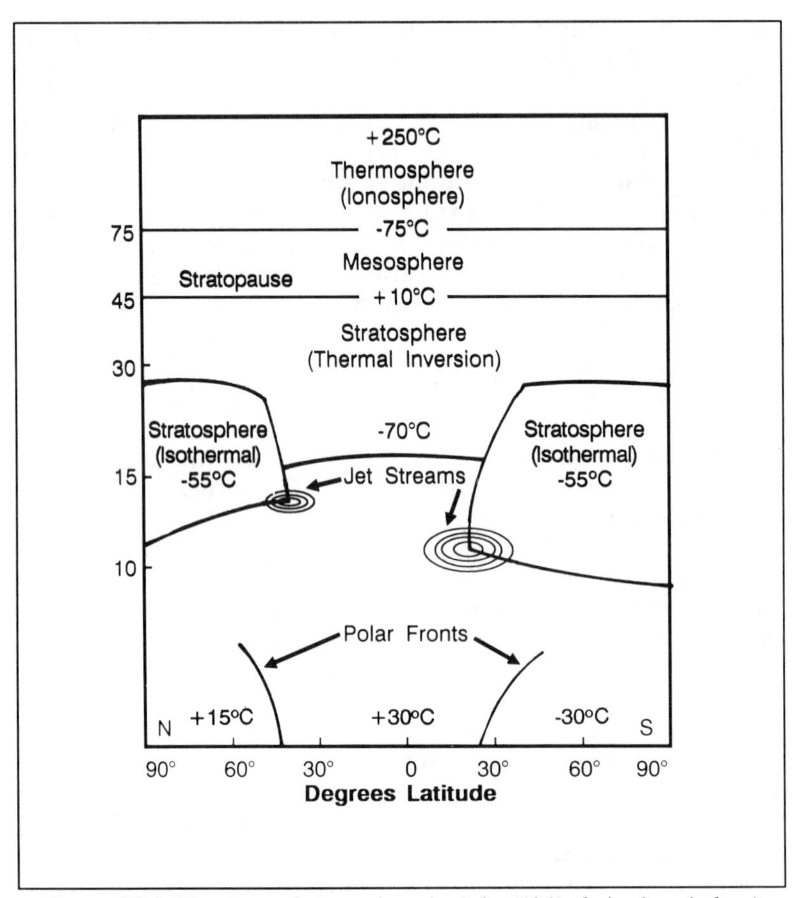

Figure 18-1: Structure of atmosphere in July. (Altitude is given in km.)

In this connection, TTAPS made two alternative simplifying assumptions, namely (1) instantaneous hemispheric mixing and (2) slow horizontal diffusion within the hemisphere. The first assumption is clearly unrealistic and merely provides a limiting case. The rate of diffusion observed in connection with large volcanic eruptions and other such events is of the order of 2×10^{11} cm^2/sec. In the event of an actual war with thousands of individual bursts, it is not unlikely that the mid-latitude zone (45°–60°N) would be largely covered by clouds within a matter

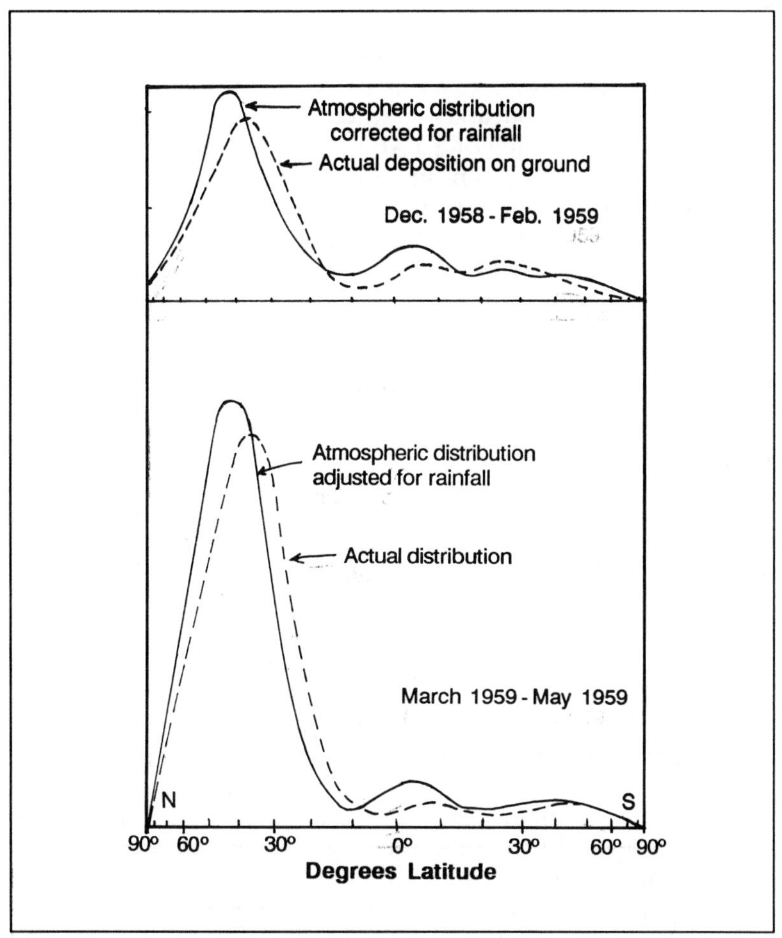

Figure 18-2: Meridional distribution of world-wide fallout (Based on Sr-90 data).

of days. However meridional diffusion over the whole of the northern hemisphere might still require several months (Covey *et al.* 1984). During this period, however, much if not most of the material initially in the clouds will have fallen out. Direct evidence of this fact can be found in the meridional distribution of fallout from past atmospheric nuclear tests, as shown in Figure 18-2.

Dust particles are normally removed from the atmosphere by two basic mechanisms: vertical diffusion (driven by gravity) and rain-scavenging. The latter mechanism is by far the most rapid, where it is applicable—namely in the troposphere in the temperate zones. Much of TTAPS case for a very severe short-term surface cooling phenomenon depends on suppression of the usual rain-scavenging process and the creation of a stabilized thermal inversion. The smoke and soot particles, being carbonaceous, would be effective absorbers of light. Such a cloud would become very hot, thus possibly suppressing thermal instabilities and some vertical mixing processes. While TTAPS acknowledges that fires caused by the Nagasaki and Hiroshima bursts were almost immediately followed by severe rain storms (that were probably induced by the fires themselves), they speculate that a sufficiently dense smoke cloud *might* inhibit rain droplet formation, partly because fresh soot particles tend to be hydrophobic, partly because absorption of heat by the cloud could cause atmospheric temperature inversions, thus inhibiting convection, and partly because suppression of normal evapotranspiration would cause reduced humidity.

This set of arguments by TTAPS is plausible, but it is also highly speculative and quite possibly wrong. In the first place, the combustion processes that generate smoke must also produce significant amounts of water vapor—a mechanism not considered by TTAPS, but pointed out by Singer (1984). For each ton of (dry) hydrocarbon or cellulosic fuel burned, between 0.6 ton and 1 ton of water vapor is generated in addition to the water content of the fuel itself. In fact, for typical fuel-loadings, the air column over the burned area—where the smoke cloud is concentrated—would double or triple its water-vapor content. Moreover, the rising air-column of a large fire automatically entrains surrounding air masses and tends to carry them to higher altitudes where they expand and cool, resulting in condensation of ambient water vapor and (often) precipitation.

In the second place, large fires would seem to increase—not suppress—atmospheric convection. Hence it is difficult to see how large-scale cloud-induced temperature inversions could occur in the first place. Thus, while there is some uncertainty (admittedly, the condensation-precipitation phenomenon is very complex and not fully understood), it seems probable that at least the first few hours and days after a nuclear war would be characterized by violent and widespread storms and a greatly enhanced—not diminished—rate of rain scavenging and particulate deposition.

In any case, the scavenging issue has not been settled and deserves far more careful study. The most reasonable conclusion that can be drawn from TTAPS is that *if* tropospheric scavenging mechanisms are suppressed for any reason after a nuclear war, the attenuation of solar radiation at the surface of the earth might be both severe and protracted. Lower temperatures would certainly follow.

With regard to the removal of dust particles from the stratosphere (i.e., above 10 km or so) there seems no doubt that the so-called Stokes Law diffusion is the only applicable mechanism (See Appendix 18-A).

Figure 18-3, derived from calculations in Appendix 18-A, shows that particles with radii $r \geq 1\mu$ survive less than one year in the stratosphere. On the other hand, small particles with radii $r \leq .25\ \mu$ diffuse so slowly that they can remain in the stratosphere for as long as 10 years or more.

Let me now return to the question of light scattering and attenuation of insolation at ground level. A crude order-of-magnitude estimate was given earlier, but it is possible to do a somewhat better job without excessive computational labor.

Attenuation of Solar Flux at Earth's Surface

If we consider only direct illumination (ignoring contributions from scattered light) the net solar flux W arriving at a point on the surface of the earth, as a function of latitude θ and angle ϕ with respect to the zenith, will be given by an integral over all wavelengths λ of the solar flux arriving at the top of the atmosphere, times the attenuation factor

$$(2) \qquad I = I_o \exp\left[-\gamma(\lambda,t,\theta)\ D \sec\theta \sec\phi\right]$$

In the above expression I stands for the intensity and $\gamma(\lambda,\tau,\theta)$ is the scattering cross-section per unit volume (of the cloud) in units of square microns (μ^2). Here a "unit volume" is a cylinder $1\mu^2$ in cross section and 1 kilometer in altitude. The diurnal

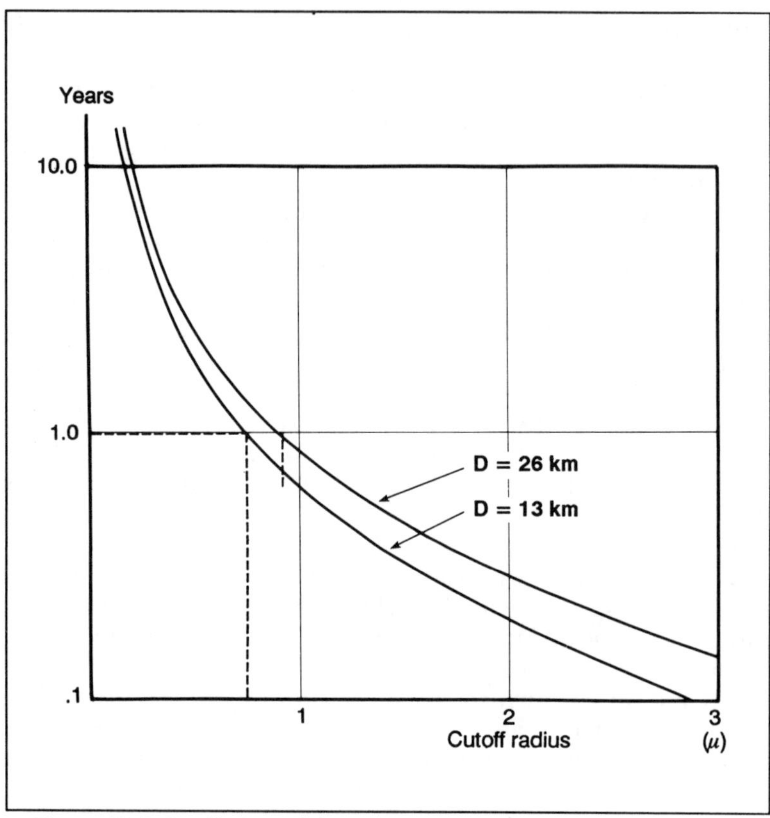

Figure 18-3: Smallest particle radius (r_{cutoff}) surviving in stratosphere after time t as a function of initial cloud thickness D.

variation is averaged out by integrating over ϕ. Details are worked out in Appendix 18-B.

To compute the fractional net change in insolation on the surface of the earth, after a time t, at a latitude θ, one must integrate over all wavelengths λ (or frequencies v) and incidence angles ϕ. Since W has dimensions of energy flux, it is more convenient to integrate over frequency v than wavelength (each photon has energy hv). The only parameter appearing in the integrand is the product ρ_0D.

$$(3) \qquad \frac{\delta W(\theta,t)}{W(\theta,t)} = \frac{\int_0^\infty dv\, S(v) \int_{-\pi/2}^{\pi/2} d\phi \cos\phi\, (I_0 - I)}{\int_0^\infty dv\, S(v) \int_{-\pi/2}^{\pi/2} d\phi \cos\phi\, (I_0)}$$

$$= 1 - \int_0^\infty dv\, S(v) \int_{-\pi/2}^{\pi/2} d\phi \cos\phi \exp\left[-\frac{\gamma(v,t,\theta)}{\rho_0\, g(\theta)} \rho_0 D\, g(\theta) \sec\theta \sec\phi\right]$$

The function S(v) is simply the sun's spectral distribution as a function of frequency. The ratio $\gamma(\lambda,t)/\rho_0 g(\theta)$ is plotted in Figure 18-4 below. Sample results for insolation attenuation at the end of one year are shown in Figure 18-5 for a range of values of $\rho_0 D$. The curves for various values of absorption ξ lie close together, so I have chosen to indicate the envelope as defined by $\xi = 0$ and $\xi = 0.3$. Evidently, the results are fairly insensitive to the absorption coefficient. The fractional net change of insolation as a function of latitude is probably more significant than the overall net change for the earth as a whole. The latter can be obtained, however, by integrating (3) over all latitudes θ:

$$(4) \qquad \frac{\delta W(t)}{W} = \tfrac{1}{2} \int_{-\pi/2}^{\pi/2} \frac{\delta W(\theta,t)}{W(\theta,t)} \cos\theta\, d\theta$$

It remains to show how the product $\rho_0 D$ depends upon the actual quantity of dust in the stratosphere. This is done in Appendix 18-C.

On inserting numerical values it follows that the case $\rho_0 D = 0.1$ resulted from a total original (t = 0) volume of dust particles of 0.25 km^3 or ~.5 x 10^9 metric tons, for the log-normal distribution with $r_o = 0.5\mu$. Note that assuming smaller particles ($r_0 = 0.25$) would result in (roughly) twice the attenuation effect for the same mass of dust. Obviously, the smoke and soot particles from fires would be four to five times as effective as light scatterers as dust from soil, because of their smaller size—as long as they remain suspended in the atmosphere. Since these particles are not likely to be injected into

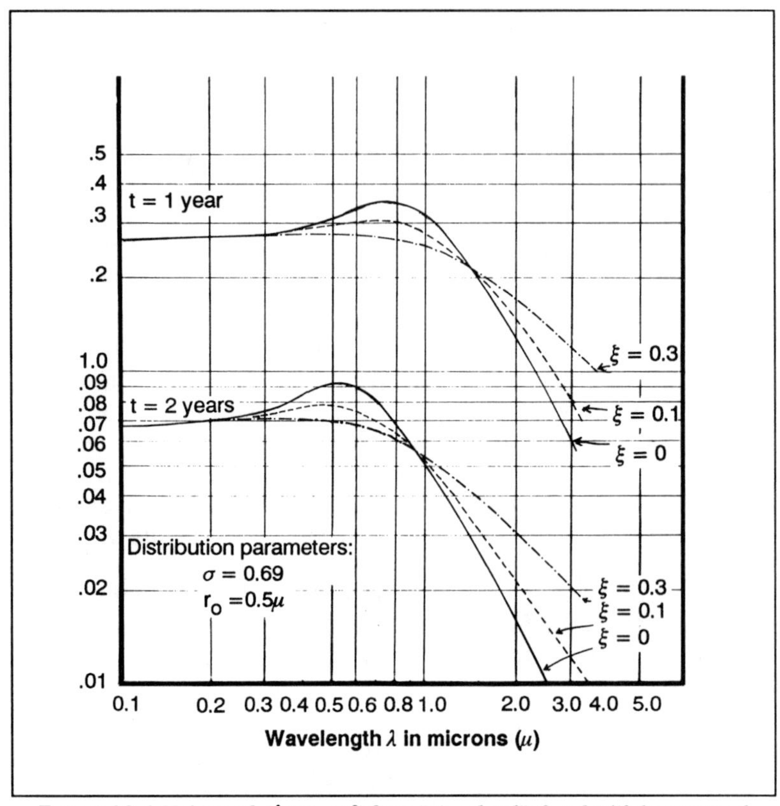

Figure 18-4: Values of $\gamma(\lambda,t)/\rho_0 g(\theta)$ for a "standard" cloud with log-normal particle size and distribution and various absorption coefficients ξ after one year and two years.

the stratosphere, however, they will probably be removed rather quickly by mechanisms other than diffusion.

It should be noted that 10^{11} metric tons of dust—the amount injected into the stratosphere by as little as 100 MT of groundburst nuclear weapons or 10,000 MT of airburst weapons—is not a particularly large quantity on the geological scale. The weight of the atmosphere itself is of the order of ~ 3 x 10^{14} tons. Atmospheric CO_2 accounts for 2.2 x 10^{12} tons. The weight of the hygroscopic materials (salt and various sulfates) picked up and deposited annually is probably ~ 3 x 10^{10} tons (Woodcock

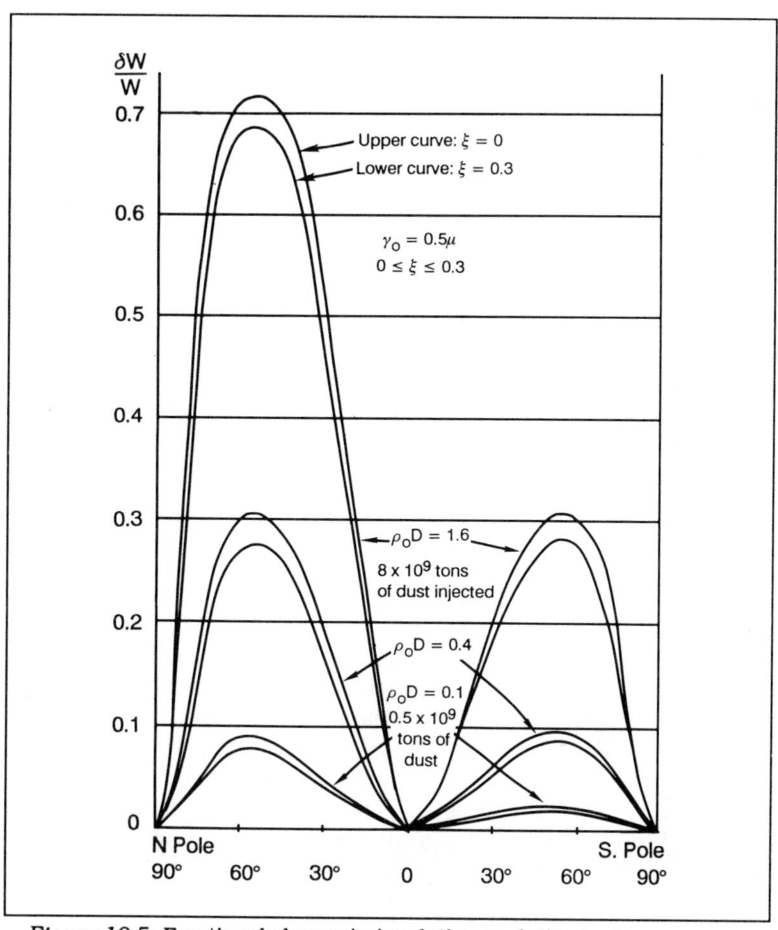

Figure 18-5: Fractional change in insolation vs. latitude, 1 year.

1957). Each year more than 10^8 tons of air pollutants are injected into the atmosphere, in the United States alone, of which more than 10^7 tons are in the form of fly ash and other particulate wastes (e.g., smokes). Pollutants of biological origin may contribute a further 10^9 to 10^{10} tons of pollens, spores, and organic esters and terpenes. A major volcanic eruption of thc cxplosive type may contribute 10^{10} tons or more. For example, the eruption of Krakatoa yielded ~ 5 km^3 or ~ 10^{10}

tons (Royal Society 1888) while micrometeorites from outer space contribute 1.7×10^7 tons per year to the upper atmosphere (Cadle 1966).

Insolation and the Earth's Heat Budget

An adequate and fully detailed analysis of the effect of a given temporary fractional decrease in insolation ($\delta S/S$) on the climate of the earth is certainly beyond the scope of this chapter; (see, however, Covey, this volume). Hence the following, highly schematized picture is presented merely in order to indicate some of the significant interactions. One may divide the major atmospheric energy transfer processes into three categories: (1) short-wave (optical) radiation, (2) long-wave (infrared) radiation, and (3) other processes including convection, turbulent transfer, evapotranspiration and condensation. For convenience these may be labeled SW, LW, and OP.

If the SW radiation arriving at the top of the atmosphere is arbitrarily set at 100 units, then the income and outgo for the atmosphere (as distinguished from the geosphere) are summarized approximately as shown in Table 18-1 compiled by Kondrat'yev (1965) and reproduced below. The table indicates the relative importance of various major energy exchange processes affecting the thermal balance of the earth. Taking the estimates of Budyko, Yudin, and T.G. Berly and (in the first column) as a basis for calculation, the situation can be summarized briefly in terms of aggregate inputs and outputs.

A highly simplified approximation can be expressed as the temperature adjustment equation

$$(5) \qquad R\frac{\partial T_S}{\partial t} = \tfrac{1}{4} S(1-\alpha) - F_{IR}(T_S)$$

where R is the thermal inertia of the earth, T_S is the average surface temperature, S is the incoming solar flux, α is the earth's albedo (normally 0.3) and F_{IR} is the infrared (long wave) flux escaping into space. The IR flux is approximately

$$(6) \qquad F_{IR}(T_S) = a + b\,(T_S - 273°K)$$

where

$$a = 0.289 \text{ cal.cm}^{-2}\text{min}^{-1}$$
$$b = 2.88 \times 10^{-3} \text{ cal.cm}^{-2}\text{ min}^{-1}(°K)^{-1}$$

Table 18-1: Average Annual Thermal Balance of Earth

Components of the thermal balance (%)	A	B	C	D
Shortwave radiation				
Received at the upper boundary of the atmosphere	100	100	100	100
Reflected from clouds into space	27	25	27	30
Reflected into space by atmospheric scattering	7	9	6	8
Absorbed by clouds	12	10	11	
Absorbed by the atmosphere				15
Solar radiation	6			
Radiation reflected by the earth's surface	2	9	3	
Reaches earth's surface;				
As direct solar radiation	30			30
As diffuse radiation	18			17
Absorbed by the earth's surface;				
Direct solar radiation	27	24	11	27
Diffuse radiation	16	23	34	16
Reflected from earth's surface;				
Direct solar radiation	3			3
Diffuse radiation	2			1
Thermal radiation				
Total thermal radiation of the atmosphere, including	151			146
Radiation into space	55	66*	48	50
Atmospheric emission reaching the earth's surface	96	105		96
Thermal emission of the earth's surface including:	116	119		120
Absorbed by the atmosphere	108			112
Radiation into space	8		17	8
Net radiation of the earth's surface	20	14	23	24
Other components of thermal balance				
Turbulent heat transfer from the earth's surface to atmosphere	4	10		-4
Latent heat of condensation (or evaporation)	19	23		23

* Including thermal radiation from the earth's surface.

References: (A) Budyko *et al.* 1952. (B) Houghton 1954. (C) Moller 1950 (D) Baur & Phillips 1944.

It follows that, to first order, a 1% decrease in S causes a decline in the equilibrium average temperature of the earth's surface of 0.65°K. Most of this adjustment occurs on a scale of months. Clearly a better estimate requires the use of radiative-convective atmospheric circulation models. TTAPS initially used a one-dimensional model (altitude only) which led to their prediction of a short-term 30°K temperature drop under the smoke cloud associated with a baseline nuclear war scenario. Their published paper (1983) assumed a similar nuclear war scenario and physical model but with a two-dimensional radiative-convective model with meridional (latitude) circulation and more compensatory negative feedback effects. It predicted a more moderate 15°K drop. Alexsandrov and Stenchikov (1983) obtained similar results with a quasi-three-dimensional (two-layer) model.

Covey, Schneider, and Thompson (1984) have utilized a three-dimensional general circulation model (GCM) and obtained more complex patterns, but qualitative agreement with results of the simpler models. There seems to be general agreement among the climatologists that long-term effects would probably be insufficient to trigger a change in equilibrium, such as a new period of glaciation, for instance. It must be reiterated, however, that precipitation and other smoke removal mechanisms cannot be fully modeled at present.

In more recent work, e.g. Schneider & Thompson (1988), Singer (1988) and Singer (this volume) a further question about the TTAPS model has been raised. Specifically, Singer points out that nuclear explosions would tend to lift water vapor into the stratosphere and that this would trap heat in the lower atmosphere by a variant of the "greenhouse effect." It is unclear whether the greenhouse effect suggested by Singer would compensate for the cooling effect of the Rayleigh scattering from micro-particulates, but these phenomena appear to be of comparable magnitudes.

The results of the simulations demonstrate some important qualitative conclusions, notably that *ceteris paribus* injection of a large quantity of smoke into the troposphere (mainly 45°N–75°N) in summer would dramatically alter the general circulation pattern in both hemispheres. The major features are: (1) higher air temperatures in the upper troposphere of the northern latitudes (45°N–75°N) and significant differences extending as far as 30°S; (2) average surface cooling of 10°C or more in the north temperate zone between latitudes 50° and 70°; and (3) acceleration of the upper level west-to-east winds and reversal of the lower level east-to-west circulation in the northern latitudes, and acceleration of the east-to-west flows at all altitudes in the equatorial regions. Generally speaking, temperature gradients from the tropics to the north polar regions would be increased, and the "storm belt" should be moved to the north.

The biological impacts of nuclear winter are clear, at least qualitatively. If the war took place in the winter, the "freeze" would drive temperatures below the level of tolerance for many perennial species living near the northern edge of their natural range. Many tree species, for instance, would be killed outright or severely weakened and subject to later attack by pests and/or disease during the subsequent growing season. Almost all overwintering birds would die. On the other hand, dormant seeds of most annuals and insect eggs would be unaffected.

On the other hand if the war took place in spring or summer, crops would either be freeze-killed outright or their growth rates would be severely retarded. For example, winter wheat in the Pacific Northwest requires about 1,150 day-degrees centigrade to ripen. A decline in average temperature of 10°C over a 30-day period would produce a deficit of 300 day-degrees by the time ripening normally occurs—enough to delay the harvest a month or more and allow time for insects and disease to take a heavy toll.

Putting it another way, productivity of many crop plants would be far less than normal. In the case of corn, for instance, a 10°C temperature differential extending throughout the season would reduce the yield by nearly 70% (see Figure 18-6). However, the period of severe cold would probably be considerably shorter, and the productivity impact correspondingly less. Actually, most crops in the field would be badly contaminated by radioactive fallout and perhaps unusable for that reason. It is also highly uncertain that crops could be harvested, given the likely breakdown in social infrastructure, transportation and distribution. But these are different, albeit important, problems.

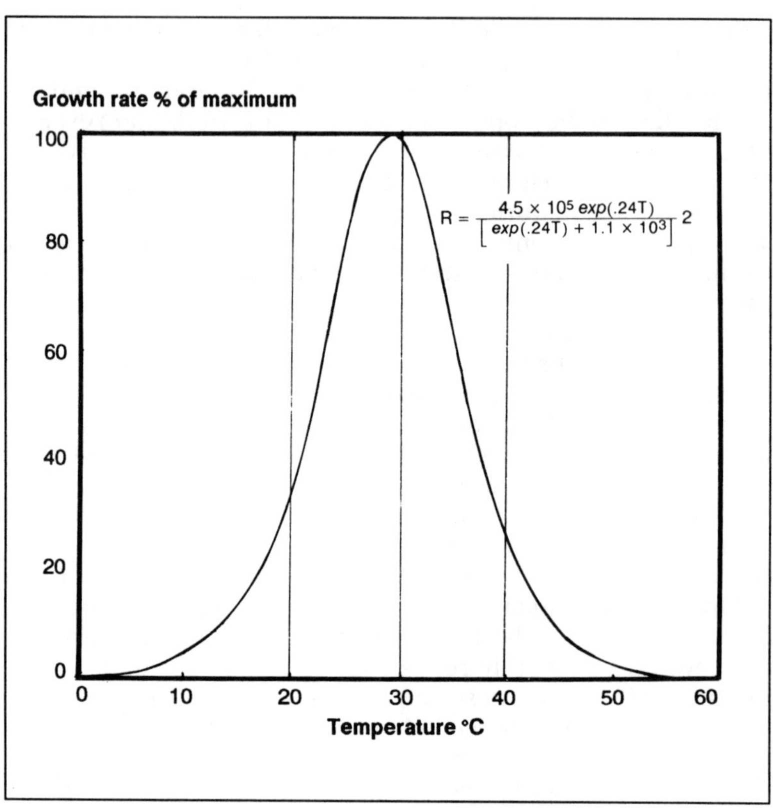

Figure 18-6: Corn Growth as a Function of Temperature.

Given that periods of intense cold may be expected to occur throughout the latitudes where most of the weapons are used, for periods of anywhere from a few days up to a few months (depending upon the nuclear scenario, the location and effectiveness of tropospheric scavenging mechanisms) a critical question is the following: would the global environmental effects be so devastating *per se* that even the survivor of a successful preemptive first strike would be unable to survive the aftermath? I think one can say with reasonable confidence that the nuclear winter and its biological aftereffects—however bad—would not be as severe as the direct damage caused by the use of nuclear weapons and the economic/social collapse that would almost certainly result from a large-scale nuclear exchange. On the other hand the nuclear winter—like the fallout—would unquestionably interfere seriously with both ecological and economic recovery from a nuclear war. It is even conceivable that the combined post-attack biological effects of fallout and severe cold would take a greater human toll in the long run than the immediate destruction itself.

A second critical question is the one emphasized by Sagan. He argued that even a preemptive *one-sided* nuclear war (in which the attacker successfully destroyed all of the other side's weapons) would set in motion a chain of secondary effects that would essentially destroy the "winner" as well as the "loser." It would be nice if this could be demonstrated convincingly, since it would largely eliminate any rational motive for a preemptive first strike and immeasurably strengthen the arguments for a nuclear freeze.

The difficulty is, in a sense, that the transparency of Sagan's motives may have detracted from the effectiveness of his argument. On the one hand, the nuclear hard-liners in an American administration will be ultra-suspicious of any conclusions based on mathematical models that are not completely and fully tested and verified. They will point out, correctly, that it would be doubly disastrous if the United States leadership believed in the Sagan thesis while the Soviet leadership did not.

Furthermore, Sagan and his scientific colleagues (TTAPS and others) may themselves have been subtly influenced by the strategic argument they wanted to make. It is hard to understand, otherwise, why they were so meticulous in documenting their model calculations and so casual about justifying the critical assumptions about the stability of the tropospheric cloud of smoke and soot that accounts for most of the surface cooling.

My own opinion is that the short-term problem (from smoke) would not be as severe or widespread as TTAPS has assumed and that the nuclear winter would *not* be totally devastating to the successful attacker in a one-sided nuclear war. The successful attacker would not be able to avoid significant post-attack consequences—both radiological and meteorological—but they would be far less damaging than the more localized effects of the attack itself (and the immediate post-attack effects) on the victim. The indirect costs to the attacker would probably be enough to deter a rational regime, to be sure. But such a regime is already deterred by the strong likelihood of retaliation. In summary, I am not sure that the nuclear winter fundamentally alters the present nuclear strategic situation—much as I would like to be convinced.

Appendix A

The Stokes-Cunningham equation for the downward drift velocity v(r,Z) in cm/sec for spherical particles of radius r (in cm) at altitude Z (in Km) is as follows:

$$\text{(i)} \qquad v(r,Z) = \frac{2\,gd}{9\eta} r^2 \left[1 + \frac{A}{rp(Z)}\right]$$

where g is the gravitational constant 981 cm/sec^2, d is the density (~2.3 gm/cm^3), η is the viscosity of the air, p(Z) is the barometric pressure as a function of altitude Z, and A is an empirical constant. Normally η is a function of temperature, but in an isothermal stratosphere (-55°C), one has

$$\text{(ii)} \qquad \eta = 1.416 \times 10^{-4}\ \text{gm cm}^{-1}\ \text{sec}^{-1}$$

and A = 4.6332 x 10^{-3}cm^2. In the stratosphere, pressure p(Z) is numerically given (in millimeters of Hg) by

$$\text{(iiia)} \qquad p(Z) = 165 \exp[-0.146\ (Z-10)]$$

or, converting to cm(Hg) for convenience,

$$\text{(iiib)} \qquad p(Z) = 16.5 \exp[-0.146\ (Z-10)]$$

where Z is in kilometers.

Now suppose the initial distribution of the stabilized cloud is uniform of thickness D between lower and upper altitudes Z_1 and Z_2 (Z_2 >10), so D = Z_2 - Z_1 . This cloud can be thought of as a superposition of many monodisperse clouds. Define the drift distance u(t) = X(o) - Z(t), or

$$\text{(iv)} \qquad Z = Z(0) - u$$

whence,

$$\text{(v)} \qquad V(r,t) = \frac{dZ}{dt} = \frac{du}{dt}$$

Then, for a particle of radius r starting at the top of a cloud of thickness D, we obtain

(vi) $$\frac{du}{dt} = 11.2\, r^2 \left\{1 + \frac{0.28\times10^{-4}}{r} \exp\left[0.1469(D - u)\right]\right\}$$

where r is measured in microns (10^{-4} cm) and D, U are in km and t in years. This expression can be integrated in closed form to yield

(vii)

$$u(r,t) = 6.85 \ln\left\{\left[1 + \frac{0.28 \times 10^{-4}}{r} \exp(0.146D)\right] \exp(1.72\, r^2 t) - \frac{0.28 \times 10^{-4}}{r} \exp(0.146D)\right\}$$

Clearly u can never exceed D, since by the time the top layer of the cloud has drifted down that far (the initial cloud thickness) it reaches the bottom of the stratosphere and is quickly removed by condensation and precipitation. Equation (vii) therefore defines the longest time a particle of radius r can remain in the stratosphere. The results are plotted in Figure 18-3.

Appendix B

The scattering cross-section per unit volume—also called the "extinction coefficient"—is defined as follows:

(i) $$\gamma(\lambda,t,\theta) = \pi \rho_0 \int_0^\infty dr\, r^2 Q(r,\lambda,\xi) \int_0^\infty dz\, F(r,z,\theta)$$

$$\cong \pi \rho_0 g(\theta) \int_0^\infty dr\, r^2\, Q(r,\lambda,\xi)\, f_0(r) \left[1 - \frac{u(r,t)}{D}\right]$$

where u(r,t) is given by equation (vii) in Appendix 18-A, and $g(\theta)$ can be approximated as:

(ii)

$$g(\theta) = \left\{ \begin{array}{lll} 3 \sin^2\theta \cos^2\theta & -\pi/2 \le \theta \le 0 & \text{(South Hemisphere)} \\ 12 \sin^2\theta \cos^2\theta & 0 \le \theta \le \pi/2 & \text{(North Hemisphere)} \end{array} \right\}$$

For convenience, I have used the simple log-normal form of particle-size distribution, i.e.,

(iii) $f_0(r) = (2\pi)^{-1/2}(\sigma r)^{-1} \exp\left[-\left(\ln\frac{r}{r_0}\right)^2/2\sigma^2\right]$

where $\sigma = \ln 2 = 0.69$ and r_0 is assumed to be 0.5. This choice differs slightly from TTAPS assumption, but both were estimated from the same original source of data (Russell and Nathans, 1966). The kernel-function Q (r, λ, ξ is derived from the general theory of electromagnetic scattering of spherical particles with an absorption coefficient ξ. It can be approximated by the following (see Van de Hulst, p. 176):

(iv) $$Q(r,\lambda,\xi) = 2 - 4\exp(-R\tan\beta)\left(\frac{\cos\beta}{R}\right)\sin(R-\beta)$$

$$- 4\exp(-R\tan\beta)\left(\frac{\cos\beta}{R}\right)^2\cos(R-2\beta) + 4\left(\frac{\cos\beta}{R}\right)^2\cos 2\beta$$

where

(v) $$R = \frac{4\pi(m-1)r}{\lambda}$$

and

(vi) $$\xi = \tfrac{1}{2}\tan\beta$$

Assuming a (real) index of refraction m = 1.5, (v) reduces to

(vii) $$R \cong 2\pi r/\lambda$$

Most glassy or crystalline substances will not differ substantially from the typical value m = 1.5, but the imaginary part of the index of refraction, ξ (the absorption coefficient) may vary considerably. Hence I have performed all calculations for a range of values of ξ. The extinction coefficient $\gamma(\lambda,t)$ divided by $\rho_0 g(\theta)$ is plotted as a function of wavelength λ in 4 for t = 1 year and t = 2 years in Figure 18.4.

Appendix C

Dimensionally it is evident that $\rho_0 D$ is the normalizing constant proportional to the number of scattering centers in a vertical cylinder of one square micron (or 10^{-12} meter^2) cross-section and D kilometers in altitude. Since we are dealing with a collection of particles with a range of sizes, the average volume of material originally in the cylinder at time t = 0 is

$$\text{(i)}\ \langle V_{cyl} \rangle = \rho_0 D\, g(\theta) \frac{4\pi}{3} \int_0^{\infty} r^3 f_0(r) dr = \rho_0 D\, g(\theta) \frac{4\pi}{3} r_0^3 \exp\left(\frac{9}{2}\sigma^2\right)$$

for all distributions of log-normal form. In the case $r_0 = 0.5(\mu)$ and $\sigma = \ln 2 \cong 0.7$, the average volume per particle works out to be $\sim 5\mu^3$. The volume of dust over the whole earth is

$$\text{(ii)} \qquad \langle V_{Earth} \rangle = \rho_0 D \int_{Earth} dA\, g(\theta) \cdot \frac{4\pi}{3} r_0^3 \exp\left(\frac{9}{2}\sigma^2\right)$$

$$= \rho_0 D \ \ 2\pi R^2 \int g(\theta) \cos\theta d\theta \cdot \frac{4\pi}{3} r_0^3 \exp\left(\frac{9}{2}\sigma^2\right)$$

$$= \rho_0 D\, (4\pi R^2) \cdot \frac{4\pi}{3} r_0^3 \exp\left(\frac{9}{2}\sigma^2\right)$$

where R is the earth's radius.

References

Aleksandrov, V.V. and G.L. Stenchikov. 1983. *Proc. Appl. Math.* Moscow: Computing Center of Academy of Sciences.

Baur and Phillips in Obolenskiii, *Textbook of Meteorology*, Gidrometeoizdat, Leningrad (1944), cited by Kondrat'yev.

Budyko, Yudin and Berlyand, in Alison, Drosdov and Rubenstein, *Textbook of Climatology*, Gidrometeoizdat, Leningrad (1952) cited by Kondrat'yev.

Cadle, R.D. 1966. Chapter 2 in *Particles in the Atmosphere and Space. New York: Reinhold Publishing Corp.*

Covey, C., S.H. Schneider and S.L. Thompson. 1984. "Global atmospheric effects of massive smoke injections from a nuclear war: Results from general circulation model simulations." *Nature.*

Cunningham. 1910. *Proc. Roy. Soc.* 83: 357.

Houghton, G.H. 1954. *Journal of Meteorology 11* No. 1, cited by Kondrat'yev.

Humphreys, W.J. 1940. *Physics of the Air*, 3rd ed.. New York: McGraw-Hill.

Kondrat'yev, K. Ya. 1965. *Radiative Heat Exchange in the Atmosphere.* Translated by O. Tedder, New York: Pergamon Press, p. 254, 346.

Moller, F.J. 1950. *Experientia* 6, No. 10 (1950), cited by Kondrat'yev.

Royal Society, The Report of the Krakatoa Committee of the Royal Society. 1888. "The eruption of Krakatoa and subsequent phenomena." London.

Russell, Irving, M.W. Nathans, *et al.* 1966. Joint OCD-DASA Fallout Phenomena Symposium, April 12–14, 1966, San Francisco: United States Naval Radiological Defense Lab.

Schneider, S.H., S.L. Thompson. 1988. *Nature* 333: 221–227.

Singer, S. Fred. 1984. *Nature* 310: 625

Singer, S. F. 1988. "ReAnalysis of the Nuclear Winter Phenomenon," *Meteorol. Atmos. Phys.* 38: 228–239

Singer, S.F. 1989. This volume.

Turco, R.P., Toon, O.B., Ackerman, T.P., Pollack, J.B., Sagan C. 1983. "Nuclear Winter: Global Consequences of Multiple Nuclear Explosions." *Science* 222: 1283–1292.

Van de Hulst, H.C. 1957. *Light Scattering by Small Particles.* New York: John Wiley & Sons, p. 16, 176 *et seq.*

Woodcock, A.H. 1957. "Salt and rain." *Scientific American*, October 1957.

NINETEEN

NUCLEAR WINTER or NUCLEAR SUMMER?

S. Fred Singer

The Controversies

The hypothesis that a major nuclear war would lead to severe climate consequences affecting the whole earth has become quite controversial. On the one hand, supporters of the theory (Turco *et al.* 1983) argue that weeks and months of darkness and freezing (the "nuclear winter") would destroy agricultural crops and much of the world's biota, causing misery to the surviving human population; some have gone further and speculated that such a climate event would spell the end of the human species on this planet (P. Ehrlich *et al.* 1983).

On the other hand, the theory has been attacked as a scientific hoax (Sparks 1985), with political motives ascribed to its originators, who are most charitably described as political simpletons.

There is some middle ground here, but not much. Careful reading of the scientific paper (Turco *et al.* 1983) reveals that the authors (Turco, Toon, Ackerman, Pollack, and Sagan, or TTAPS) leave open many loopholes that would make the nuclear winter less certain. A subsequent report by the prestigious United States National Academy of Sciences (1985) raises even more uncertainties. This has given rise to a sub-controversy, with Professor Sagan claiming that the NAS report supports the TTAPS work while at least one member of the NAS panel strongly disagrees (Sagan 1984a and Katz 1985).

A separate controversy has developed about the policy implications of the TTAPS study. On the one hand, supporters of the nuclear freeze movement have taken over the theory to prove their point (*Disarmament*, 1984), with the willing cooperation of Professor Sagan (Sagan 1983). On the other hand, the nuclear winter phenomenon has been used to support the case for the Strategic Defense Initiative ("Star Wars") (United States Department of Defense 1985). One interesting point: Soviet use of nuclear weapons is more likely to lead to worldwide climate effects than United States use; also, the Soviet Union is more vulnerable to these effects than the United States (Singer 1984a). European Russia could be wiped out, leaving the Asiatic parts of the USSR as a "greater Afghanistan."

A veritable cottage industry has sprung up around the nuclear winter/nuclear freeze issue, complete with bumper stickers, T-shirts, and, of course, a "grass roots" committee with offices in Washington, D.C. The "bible" of the movement is *The Cold and the Dark*, a book that reprints or rehashes what has been published. It has been gushily reviewed as the "most important work since the Bible and the collected works of Shakespeare."

Even the International Council of Scientific Unions has weighed in, with yet another study that elaborates much of the earlier work. An expensive undertaking, it has been expensively published (SCOPE 1985). According to SCOPE, all of humanity would not be wiped out, merely 2 to 3 billion people. By contrast, a sober analysis of the aftereffects of nuclear war leads to quite different prescriptions on how to avoid a holocaust (R. Ehrlich 1986).

The Scientific Basis

During all this controversy, sight has been lost of the essential scientific issue: Is the nuclear winter phenomenon real? My thesis is simple: even granting many of the artificial assumptions made by TTAPS, and by the subsequent NAS study, a worldwide cooling is unlikely. On the contrary, if we take proper account of the so-called atmospheric greenhouse effects, then a *warming* is a possible outcome. Most likely, the greenhouse effects will roughly cancel the cooling effects, producing little change in global temperature (Singer 1984b).

Before explaining the greenhouse effects, I will review some of the history of nuclear winter and then deal with the assumptions that enter into the theory.

As far as I know, Robert Ayres first studied the optical absorption effects of dust and smoke produced by nuclear bursts (Ayres 1965, 1989). Many years later, Crutzen and Birks (1982) independently hit upon the idea that nuclear explosions could ignite many widespread fires whose smoke could absorb enough of the incoming sunlight to cause darkness and cooling of the surface of the earth—at least in the northern hemisphere. They may have been stimulated by the earlier publications of Luis Alvarez and colleagues who surmised that the extinction of the dinosaurs 65 million years ago was caused by the absorption of sunlight by dust created by the impact of a large asteroid (Alvarez 1987).

Even earlier, Sagan and colleagues (Hanel *et al.* 1972) had observed cooling of the Martian surface during occasional dust storms on Mars. But TTAPS did not extrapolate these findings to the earth until after the appearance of the work of Crutzen and Birks. Since then, of course, they have been following up the subject with a vengeance, coining the term "nuclear winter" and promoting the concept politically as well as scientifically.

Physical Assumptions

But the nuclear winter is based on many assumptions, some quite unrealistic, some merely uncertain. If any of these fail, then the nuclear winter fails to appear or becomes insignificant. And as mentioned earlier, if I introduce greenhouse effects, I can produce a nuclear summer instead of cooling. This result indicates that the theory of nuclear winter is not very robust.

I have pointed to the tenuous assumptions and to the neglect of greenhouse effects shortly after the TTAPS paper appeared, first in the *Wall Street Journal* (Singer 1984*b*), later in scientific journals where publication was much delayed (Singer 1984*c*, 1985*a*, 1988), and finally in *Scientific American* (Singer 1985*b*).

Sagan has disagreed heatedly, denying the relevance of greenhouse effects (Sagan 1984*b*, 1985). However, a recent review by Schneider and Thompson (1988) supports greenhouse effects, criticizes other physical assumptions of TTAPS, and arrives at minor cooling of the earth surface. The controversy is clear cut.

To make a respectable nuclear winter, TTAPS requires enough smoke, surviving for several weeks or months, and reasonably uniformly distributed. None of these conditions appears to be particularly likely (Penner 1986).

(i) To get much smoke requires the burning down of many cities. In cities, the loading of combustible material is much higher than in forests, prairies or tundras. But nuclear attack scenarios are directed against missile sites more than against cities.

Even if enough material is burned, much of the smoke created washes out in rainstorms. The scavenging of smoke particles is one of the major physical uncertainties in the theory; it determines the amount of smoke that survives.

(ii) Eventually all of the smoke must be washed out of the atmosphere. Smoke particles typically survive only a few days in the lower atmosphere below 5 km (15,000 feet), the lifetime depending on their physical properties and atmospheric moisture. The lifetimes can be much longer, on the order of weeks, for smoke particles that reach altitudes above the appreciable water vapor of the atmosphere. Hence, the altitude distribution of the smoke is important, but presents another major uncertainty. TTAPS favors high altitude smoke—naturally—but theory and observations point the other way. The analysis of the NAS report, when corrected, does not support high-altitude smoke, nor does the analysis of the Livermore Laboratory of the University of California (Penner *et al.* 1985), especially in the presence of crosswinds. Firestorms can certainly carry some smoke to high altitudes, close to the stratosphere. But what fraction? I have pointed out that much of the burning will really be smoldering, lasting for days; therefore, a goodly fraction of the smoke will end up at low altitudes and be quickly removed by rain (Singer 1984*a*, 1988).

(iii) The smoke clouds must start out as patches or plumes, and then gradually diffuse toward a more uniform pattern—provided the smoke survives long enough. While in the form of patches, the smoke will not be effective in cooling the surface. Once the smoke cloud is thick enough to cut off, say, 95% of sunlight, making it much thicker will not change the result by much. But patchiness means that periods of surface darkness will alternate with bright periods as the smoke patches drift with the wind.

It is worth noting that the TTAPS work has been done in one dimension—only up and down. Both latitude and longitude variations are ignored. This means that the diffusion of the smoke patches cannot be studied in their simple model; instead TTAPS assumes, quite unrealistically, that the smoke cloud is uniform to start with. Three-dimensional (3D) model calculations have now been done by a number of research groups; they find a more moderate cooling of the land surfaces than did TTAPS, largely because they take account of the influence of the oceans, which do not change temperature (Covey *et al.* 1984; Aleksandrov and Stenchikov 1983). But these calculations are *not* independent confirmations of the TTAPS work; they make the same physical assumptions, including an initial uniform smoke layer, and, of course, must end up with similar results. But even more advanced 3D models cannot realistically represent the atmosphere (MacCracken *et al.* 1985). The models are too coarse to show the diffusion of smoke and cannot yet calculate the rainout. Nevertheless, better models are being built, including some that operate on a small enough scale to show rainout in clouds.

Greenhouse Effects

We finally turn to the discussion of the greenhouse effects that are missing in the TTAPS study, and can drastically change the outcome of their calculations from cooling to warming. I have pointed to and performed model calculations on three separate and additive greenhouse effects (Singer 1988); there may yet be others that add to an overall heating result. (The numerical values for temperature given in this section are based on an extremely simple one-dimensional model and are purely illustrative. See Table 19-1.)

(1) If the earth had no atmosphere—or an atmosphere that is completely transparent, not only to the visible incoming solar radiation, but also to the outgoing infrared (heat) radiations

Table 19-1: Surface Temperature Ts Calculated for Different Model Parameters

	Albedo	Ts
(1): "Planetary" (no atmosphere, no smoke): T_P	0	5°C*
	(0.3	-19°C)
(2): Ambient (normal greenhouse, no smoke): T_A	0.3	30°C
(3): TTAPS (high-altitude smoke cloud)	0	-39°C**
(4): TTAPS (low-altitude smoke cloud)	0	35°C
(5): TTAPS + high IR opacity high-altitude smoke cloud	0	5°C
+ high IR opacity low-altitude smoke cloud	0	58°C
(6): TTAPS + cirrus	0	>35°C
+ reflecting cirrus	<0.2	>15°C

From: Singer 1988. See text for detailed explanation.
* An accurate calculation would give a value below 0°C.
** Only the extreme TTAPS case produces freezing temperatures, by neglecting any possible greenhouse effects.

from the earth's surface—then the average global surface temperature would be slightly below freezing. This result is derived from a simple energy balance in which the incoming visible energy, absorbed by the earth's surface, equals the outgoing IR radiation. This temperature is sometimes called the "planetary" temperature T_P; we will use it as a reference temperature. (See also Covey, this volume.)

(2) A real atmosphere is not transparent in the infrared, because of the IR-absorbing properties of water vapor (H_2O) and carbon dioxide (CO_2) that form important constituents. Their effect is to hold back some of the escaping IR radiation (in a part of the IR spectrum), emit it back to the surface, and thereby raise the surface temperature above T_p—hence the name "greenhouse" effect. A simple 1D atmospheric climate model shows the temperature increase as 53°C (95.4°F). If we take account of the reflecting properties (albedo) of clouds, then the increase above the planetary temperature is only 25°C (45°F). We call this the ambient temperature T_A; it corresponds to the present average global temperature.

(3) The TTAPS case can be represented by a smoke cloud so high in the atmosphere that the amounts of overlying H_2O and CO_2 are too small to give an appreciable greenhouse effect. With solar radiation virtually cut off, the surface temperature drops to 44°C (79°F) below T_P. Hence the full freeze effect of nuclear winter is a (25°+44°) = a 69°C drop below ambient T_A ("nuclear winter").

(4) But if the TTAPS smoke cloud is placed at low altitude, the normal greenhouse effect raises the temperature, to a full 30°C (54°F) above T_P, or 5°C (9°F) above T_A. One reason for this surprising result is that the albedo is zero; the black smoke cloud absorbs all incoming solar radiation.

(5) We now alter the physical properties of the TTAPS cloud in a significant but plausible way. We assume a size distribution of smoke particles which extends beyond the wavelength of visible light, so that the cloud itself becomes IR-absorbing. This IR absorption is greatly enhanced when we consider in detail the optical absorption properties of the combustion gases created in the fires: complex organic molecules that now reside in the smoke cloud. If we assume full IR absorption by the smoke cloud (plus combustion gases), we get surface temperatures ranging from T_P up to 53°C above T_P, which correspond from 25°C below to 28°C above ambient temperature T_A. We get either moderate surface cooling or moderate surface heating, depending on whether the cloud is high in the atmosphere or low. An "in-between" cloud altitude will yield intermediate temperature values. (Keep in mind also that a 3D calculation would bring in the moderating influence of the oceans and lower the size of all temperature excursions away from the ambient temperature T_A.)

(6) We now add another physical effect not considered by TTAPS or the NAS study. The thousands of nuclear explosions that precede and ignite the smoke-generating fires create nuclear fireballs that lift dust and water vapor high into the stratosphere. The amounts of dust and H_2O are given in the NAS study. Using those numbers, I can calculate the formation of a veil of cirrus in the lower stratosphere. The cirrus cloud particles are dirty ice crystals with well-known strong IR-absorption properties. The thin cirrus clouds, which may be barely visible, thus create a long-lasting *additional* greenhouse

effect. I have estimated (Singer 1988) that this will raise the earth's surface temperature from 17°C to 78°C (31°F to 140°F) above T_A, depending on the albedo value of the cirrus cloud. This astounding *theoretical* temperature rise would of course be moderated by the oceans, and through complicated feedbacks, by the increased evaporation of water from the ocean and land surfaces.

On second thought, the high temperature is not so surprising. After all, the planet Venus has a surface temperature of about 700° which is created by the extreme greenhouse effects of the Venus atmosphere—according to the original studies by Professor Sagan and his colleagues.

Conclusion

I have given here the scientific arguments why a surface heating (nuclear summer) is more likely than a nuclear winter. A small temperature change either way (at different points on the earth) may be the most likely outcome overall.

The analysis given here thus confirms my original doubts about the validity of a nuclear winter. Having made this point, however, I must immediately state that a general surface warming may have adverse ecological effects also, especially if accompanied by darkness. In fact, prolonged darkness, even with no surface temperature change, can be ecologically harmful—just by suppressing photosynthetic activity on the earth.

The discussion here presents the climatic and ecological aftereffects of a nuclear war in a new light. Above all, we must get the physics of the phenomenon straight and reduce uncertainties wherever possible. A whole host of experiments and calculations must now be carried out. The debate on nuclear winter has opened up a fascinating new area of research into earth climate processes. The investigations on nuclear winter (or summer) will enable us also to deal better with predicting the climate consequences of large-scale natural phenomena that are beyond our present control: volcanism and impacts of extraterrestrial bodies.

References

Aleksandrov, V.V. and G.L. Stenchikov. 1983. *Proc. Appl. Math.* Moscow: Computing Center of Academy of Sciences.

Alvarez, L. *et al.* July 1987. *Physics Today* 40 (No. 7): 24.

Ayres, Robert U. 1965. *Environmental Effects of Nuclear Weapons*. Report HI-518-RR, New York: Hudson Institute.

Ayres, Robert U. 1988. This volume.

Covey, C., S.H. Schneider and S.L. Thompson. 1984. *Nature* 308: 21–25.

Crutzen P.J. and J.W. Birks. 1982. *Ambio* 11: 114–125.

Disarmament 7. 1984. (No. 3). New York: United Nations: 33–62.

Ehrlich, P. *et al.* 1983. *Science* 222: 1292–1300.

Ehrlich, R. July 1986. *International Journal on World Peace III*, 3: 31–43.

Hanel, R.B. *et al.* 1972. *Icarus* 17: 423.

Katz, J. January 5, 1985. *New York Times*.

MacCracken, M.C., R.C. Malone *et al.* 1985. *Science* 230: 317–319.

National Academy of Sciences/National Research Council. 1985. *The Effects on the Atmosphere of a Major Nuclear Exchange*. Washington, D.C. 193 pp.

Penner, J.E. 1986. *Nature* 324:222–226.

Penner, J., L.C. Hasselman, and L.L. Edwards. January 1985. UCRL-90915. Livermore, California.

Sagan, C. 1983. *Foreign Affairs* 62: 257–292.

Sagan, C. December 29, 1984a. *New York Times*.

Sagan, C. February 16, 1984b. *Wall Street Journal*.

Sagan, C. January 1985. *Scientific American* 252 (No. 1): 9.

Schneider, S.H. and S.L. Thompson. 1988. *Nature* 333: 221–227.

SCOPE (Scientific Committee on Problems of the Environment). 1986. *Environmental Consequences of Nuclear War*, Vol. 1 (Physical). New York: John Wiley, 341 pp.

Singer, S.F. 1984a. *Disarmament* 7. New York: United Nations, (No. 3): 63–72.

Singer, S.F. February 3, 1984b. *Wall Street Journal*.

Singer, S.F. 1984c. *Nature* 310: 625.

Singer, S.F. 1985a. *Science* 227: 356.

Singer, S.F. April, 1985b. *Scientific American* 252 (No. 4): 8.

Singer, S.F. July 1986. "Comments on R. Ehrlich." *International Journal on World Peace III*, 3: 43–46.

Singer, S.F. 1988. *Journal Meteorology and Atmospheric Physics 38:* 228–239 (Springer Verlag).

Sparks, B. December 31, 1985. *National Review* 27: 28–38.

Turco, R.P., O.B. Toon, T.P. Ackerman, J.B. Pollack, and C. Sagan (TTAPS). 1983. *Science* 222: 1283–1292.

United States Department of Defense. March 1985. Report to Congress, *The Potential Effects of Nuclear War on the Climate*. 17 pp.

PART FIVE:

NATURAL PROCESSES

TWENTY

GLOBAL EFFECTS of METEORITE IMPACTS and VOLCANISM

Devendra Lal

The space age brought forth the realization that ours is a fragile planet, and that we have already altered it significantly. This awakening in the last 15 years has led to rather detailed investigations of the character of changes and their magnitude. This task is fraught with difficulties because the environmental complex, the atmosphere-hydrosphere system is a large one, with nested dynamic physio-chemical processes occurring over a wide range of space and time scales, extending to the size and age of the planet, respectively.

Today the problem of environmental changes is being tackled by the best of scientists. These studies use the most sophisticated techniques available, both for experimental analyses and theoretical modeling. We have learned that the atmosphere-ocean system is in a dynamic state, even on time scales of the order of decades or centuries; it is far from being in a near static or quasi-steady state as had been assumed earlier in many studies. Both the geochemical cycles and the exchange/storage of heat between the atmosphere and the oceans depends on a complex interplay between their elements. Thus geochemical cycles get perturbed due to changes in the global climate, and in fact this perturbation serves to provide a paleo-record of the global climate throughout the geological history of the earth.

Climate is a complex manifestation of exchange of matter and energy across the air-sea interface, and of mixing/transfer within the atmospheric and oceanic reservoirs. Yet, it is a large system and exhibits a high degree of stability, as is evident from the fact that primary climatic changes occur over long periods of time, in the range of 10^4 to 10^6 years. During the last million years, glaciation cycles of 200,000 years have been observed. The peak of the last glaciation occurred 18,000 years ago. A number of forcing causes have been identified for the observed climatic variations, and significantly, some of these adequately explain the observed periodicity in climate.

Thus the terrestrial climate system does respond to slow forcing, but the question before us is whether it has sensitive elements which can be perturbed significantly to lead to large enough oscillations in the system so that it does not return to the original state it had at the start of perturbation.

That the atmosphere-ocean system is stable can be easily judged from the climatic record through geological time. Surface temperatures have not been too different over most of the Precambrian and Phanerozoic eras. Microfossils and stromatolites provide evidence for existence of life on the earth during the past 3.5 billion years (b.y.). The oceans have been here for 3.8 b.y., and the climate system has been driven by solar energy much in the same way as today. To discuss further the variability in the climate system, one would have to decide on the time scales, which are best chosen in the light of geological records. Five time scales seem to be convenient: (1) Pleistocene era (01.7) million years (m.y.), during which time the continents have essentially remained in their present position; (2) (1–100) m.y. period for which we have adequate sedimentary data; (3) (0–200) m.y. period, marking the disruption of the early Mesozoic super-continent comprising most of the earth's continental masses; (4) (0–600) m.y., the Phanerozoic period, which includes the Paleozoic (for which period we have a complete record in climatic zonation); and (5) the Precambrian period, which includes the entire period of earth's history prior to the

Phanerozoic period. Climatic data for the Precambrian period are scarce and less precise.

Except for the early history of the earth, when dramatic changes occurred in the composition of the atmosphere and oceans, but for which our knowledge of climatic variability is inadequate, we are today in a fairly good position to comment on the nature of climatic variability and its possible causes. The first documentation of Ice Ages is at ~ 2300 m.y. ago. The subsequent Ice Age record is for the period just prior to the Phanerozoic (600–1000) m.y. B.P. when several glaciations occurred. These continued till the late Permian (about 240 m.y. B.P.). After that the continental ice sheets disappeared, beginning to wax at about 25 m.y. B.P. Continental glaciers have advanced and receded through most of the history of the earth.

There are three principal ways in which one can attempt to discuss the climatic changes, considering primarily the causative mechanisms. First of these are secular variations in the external forcing functions: solar flux variations due to changes in solar output or changes in the orbital parameters of the earth. The second of the changes are those due to rearrangements of land masses and seas, and changes in the orientations of the elevated features on the land, location of ocean gateways, etc. The 2 m.y. B.P. northern glaciation event may have been a result of the closing of the Isthmus of Panama (leading to deflection of previously westward travelling Caribbean waters into the Gulf Stream—causing increased moisture/snow fall at high latitudes). The third of the changes could be due to short-lived high-energy events which cause amplification of signals by positive feedback mechanisms. Two prominent examples of such events would be the impact of a large meteorite or the occurrence of a large explosive volcanic event.

The principal character of the first of the climatic changes is periodicity. It must be emphasized here that the climate system shows a good response to frequencies in the range 10^{-3}–10^{-7} cycles per year. The solar flux has increased steadily since the formation of the solar system, by about 20%; however, the orders-of-magnitude smaller variation in solar flux on the earth arising due to changes in the orbital parameters of the earth on time scales of 10^{4}–10^{5} years produce significant climatic changes, as established from careful studies of the sedimentary record.

The second and third types of climatic changes occur sporadically, and may or may not be rare. The marine record speaks very eloquently of sharp transitions or climatic steps, and the phrase "commotion in the ocean" aptly describes the nature of the transition. About a dozen prime examples of such changes are well discussed in the literature. The most prominent of these are the Permian-Triassic (230 m.y. B.P.), the Cretaceous-Tertiary (65 m.y. B.P.), and the Eocene-Oligocene (34 m.y. B.P.) biological extinction events, the mid-Miocene and 6 m.y. B.P. oxygen and carbon isotope shift events, respectively. There are records of a large number of events which may not belong to periodic causative mechanisms, but these are relatively less studied.

For a comprehensive discussion of climatic variations in geological history, reference is made to the publication "Climate in Earth History" (1983). Special reference is made to articles by W.H. Berger, J.C. Crowell, J.B. Pollack, S.M. Savin, and H.R. Thierstein in that publication and to the comprehensive discussions on climatic changes (Grove 1988).

In this chapter, I will dwell only on climatic events of the third type, to elucidate that local natural perturbations seem to be of importance. To keep the discussion within bounds, we will consider here two causative mechanisms only: (1) Impact on earth of a comet or a large meteorite; and (2) Large volcanic eruptions, continental and submarine.

One can consider, among this class, other events which can deposit sufficient amounts of energy in the atmosphere (such as solar flares or a nearby supernova event) or lead to dramatic changes in the earth's environment (e.g., passage of the solar system through a molecular or dust cloud). No doubt these events could have been very important in geologic history, but since the physics of their impact is quite similar, namely, the non-linear modification of the atmosphere-ocean system by a short-duration perturbation, I will detail only the two causative mechanisms discussed above. I will summarize the present state of our knowledge about the expected climatic influences due to such events. I also point out the tremendous potential of volcanism, especially submarine volcanism in bringing about global environmental changes. In fact, explanation of some of the continental and marine records may lie in events which bring about sudden and dramatic changes in oceanic structure. These suggestions must remain somewhat speculative at present, but seem quantifiable in the near future from the type of high-accuracy global synoptic data that are now becoming available because of earth satellites and the rapid increase in our modeling capabilities.

Nature of Influences of Meteorite Impacts and Volcanism

The principal influence of meteorite impact and/or volcanism is the reduction of solar insolation as a result of "dust" loading in the atmosphere. This would cause cooling and a reduction in photosynthesis. The atmospheric CO_2 content can increase significantly in the case of an explosive volcanic event. The same effect can also be produced, indirectly, in the case of a meteorite impact, due to oceanic mixing. These are the most extensively talked-of phenomena, and are relevant to the major historic volcanic events for which some climatological data are available. But we have no comparable data for the case of a meteorite impact. Recently, L.W. Alvarez and his colleagues made a bold hypothesis that the mass-extinctions at the Cretaceous-Tertiary boundary (K-T) were caused by the impact of a nominally 10 km diameter asteroid (Alvarez *et al.* 1980). There seems to be a general consensus today concerning the validity of this hypothesis, but contrary models have also been proposed.

I do not intend to discuss this hypothesis or the controversy from the point of view of reaching a conclusion about its correctness here, but will primarily be concerned with the physics/chemistry/biology of the meteorite impact event, and also the expectations for a volcanic event, causing global climatic changes. In fact, one of the alternate mechanisms proposed for the K-T extinction is volcanism. Therefore, I am gratified that the Alvarez hypothesis has attracted considerable attention so that detailed criticisms and evaluation, both pro and con, exist in the literature. The hypothesis has also triggered numerical modeling of the effects of dust-loading in the atmosphere, and of other related effects. Thus the most significant influence of Alvarez's meteorite impact theory is its impact on the scientific community calling for the urgent need to carry out detailed model calculations of the effects expected from dust blanketing of the atmosphere due to meteorite impact/volcanism, and the consequent effect on the atmosphere-ocean system.

Besides dust loading and possibly an increase in the atmospheric CO_2 content, a large meteorite/volcanic event appreciably disturbs the atmosphere-ocean systems, physically as well as chemically. These changes, discussed in some detail later in this chapter, need to be treated analytically to evaluate the expected global climatic changes.

Scales of Energies Involved in Geophysical Processes

Before going into any details of the models, it seems important to consider the approximate magnitude of energy which is associated with the two types of events, and that needed to perturb the atmosphere or the ocean, i.e., significantly to cause convective motions in the system. I have given, in Table 20-1, approximate energies involved in a variety of processes relevant to our discussions. It must be realized that the energy dissipated in an event may be significantly larger than, say, that required to remove the earth's atmosphere or to vertically stir up the oceans; but this may not be of any consequence, for example if it is all used up in the eruption of lava. On the other hand, a small amount of energy dissipated appropriately may cause significant effects in the global climate, for example if it serves as a trigger to perturb the atmosphere-ocean coupling.

A supernova explosion in the vicinity of the solar system had earlier been suggested as responsible for the sudden extinction of dinosaurs about 65 m.y. ago (Krassovskij and Sklovskij 1958). This is the same event with which mass extinction of small and large animals is associated, the K-T event. Although present evidence argues against this supernova hypothesis, it may be useful to note that if such an event occurs within a distance of 100 light years (the probability of which is estimated to be one event per 50–100 million years), the earth's atmosphere would receive a dose of 10^{28} ergs in the form of X-rays in about a week or less (Russell and Tucker 1971). This flux is two orders of magnitude larger than that in the ultraviolet and X-ray regions from the sun in a week. This energy is of the same order as the binding energy of the earth's atmosphere above

Table 20-1: Selected Terrestrial Processes and the Approximate Energies Involved

Process considered	*Energy*
1. Solar energy received by the Earth	3.5×10^{31} ergs/yr
2. Energy required to remove 1% of the earth's atmosphere	4.4×10^{30} ergs
3. Energy involved in the formation of deep waters in the oceans[a]	2×10^{29} ergs/yr
4. Total kinetic energy of the atmosphere	10^{28} ergs
5. Geothermal heat	10^{28} ergs/yr
6. Energy released from hydrothermal vents[a]	10^{28} ergs/yr
7. Kinetic energy of a comet of 10^{18}g and velocity of 45 km sec^{-1}	10^{31} ergs
8. Energy release in volcanic eruptions[b]	
(i) Tambora, 1815	8×10^{26} ergs
(ii) Krakatoa, 1883	10^{25} ergs
9. Earthquake of ninth magnitude	10^{25} ergs

Notes: a. The approximate values were calculated by the author; b. Based on sources from Williams and McBirney 1979.

60 km, or the total kinetic energy of the earth's atmosphere. (We may note here that the energy involved in a 1 Megaton explosion is 4.5×10^{22} ergs; the Tambora eruption energy release corresponds to 20,000 Megatons (MT) TNT-equivalent energy release.)

Clearly the key to perturbation of the atmosphere-ocean system is a catastrophic change in a sensitive part of the system or a sustained change in the entire system. The former is energy "economic." The figures in Table 20-1 do point to the possibility that meteorite impact or volcanism can produce a significant perturbation in the terrestrial climate system, considering the available energy.

Evaluation of Expected Climatic Influences of Meteorite Impact/Volcanism

In the foregoing I noted that meteorite impact and volcanism can produce similar effects as far as the basic processes are concerned. However, if the issues are examined in some depth, these similarities are possibly not so important since the global climatic effects are probably related to their other influences. Also as we shall shortly see, in both cases, the location of the event is of great importance. The primary difference depends on whether the meteorite impact or the volcanic eruption is on a continent or in the ocean; the exact location of the event would also be of consequence, sometimes in a rather important manner. For example, a high ash-yield volcanic eruption (particularly in the polar region) could decrease the albedo considerably and lead to massive erosion of the ice stocks. It was believed that the Krakatoa event may have been responsible, by this mechanism, for increasing the sea-level by several centimeters.

We will now consider some of the principal effects of the meteorite impact and volcanism.

Dust Loading in the Atmosphere

Dust loading of the atmosphere has been considered to be one of the important causes for global climatic change, either from the impact of a large meteorite or a large explosive eruption. Recent sophisticated hydrodynamic calculations, spurred by the Alvarez meteorite impact hypothesis for the K-T extinction (Alvarez 1980), show that in a major impact, dust would be emplaced worldwide within a few hours. The atmosphere would then become opaque to sunlight. Model calculations show that the duration of darkening would only be of the order of a few months. It must be noted here that if the duration of darkening had been as long as a few years, as was at first estimated without considering coagulation effects, this would have led to drastic extinctions among plants of the tropics because of their inability to remain dormant. However, nontropical plants could accommodate darkness periods of a few months.

The consequences of dust loading for instantaneous and slow global dispersion of the dust cloud have been modeled (Pollack *et al.* 1983). They find an important darkening effect and also cooling of the earth's surface. For dust loadings exceeding 0.1 gm/cm^2, the light level immediately after impact drops below that needed for human dark-adapted vision. Dust loading exceeding 0.01 gm/cm^2 yields a light level below the adopted threshold value for photosynthesis.

Because of such changes in the earth's radiation budget, the potential atmospheric and surface temperature changes are substantial. The oceans would cool by only a few degrees because of their large heat capacity, but the continents would cool by as much as 40°K, with maximum cooling occurring within 2 to 5 months of the impact.

These results suggest that marine extinctions at the K-T boundary resulted primarily from dust loading and consequential sunlight blocking—leading to a cessation of photosynthesis and disruption of the food-chain. The effect would have been less severe for terrestrial plants and animals because of the longer turn-over time scales of the food-reservoirs. Pollack and his colleagues suggest that blocking of sunlight for several months and the consequent low visibility may have reduced the effectiveness of land animals, particularly the large ones, to locate food. Further, they argue that the substantial decrease in surface temperatures over land may have killed organisms not adapted to the cold; small animals would show better adaptability under such circumstances. Other investigators (see Silver and Schultz 1982) have come to similar conclusions.

A large meteorite impact, or an intense solar ionization event (solar flare UV and charged particle radiations) can significantly alter the balance of ions and neutral constituents in the stratosphere. The shock waves surrounding a high velocity meteor can heat the surrounding air to tens of thousands of degrees and completely dissociate N_2 and O_2 molecules. Formation of NO is an important consequence which would lead to depletion of stratospheric ozone. Atmospheric ions created (also in solar flare events) can serve as condensation nuclei for the formation of cirrus clouds in the 10 to 12 km region. Thus solar activity and meteorite impact can both cause alterations in climate and environment. These influences have been examined, in the case of solar flare radiation by Hoffman and Rosen (1983), and in the case of a meteorite impact, specifically for the July 1908 Tunguska impact, by Turco *et al.* (1982).

The Tunguska event has been estimated to have produced up to 30 million tons (m.t.) of NO in the stratosphere (Turco *et al.* 1982). It has been shown convincingly by Prinn and Fegley (1987) that large amounts of NO would be formed by shock during the atmospheric entry of the meteorite. Their calculations yield about $7x10^{40}$ molecules NO from the entry of an ice-rich long-period comet with a mass of $1.25x10^{16}$ kg and velocity 65 km/sec. The NO would be converted to NO_2 and then nitrous and nitric acid rain with a pH = 0 -1.5-2 globally. This acid rain would traumatize the biosphere (Prinn and Fegley *op. cit*). There also appears to be evidence for the acid rain at the K-T boundary in the analyses of MacDougall of sea water strontium isotopes; the ratio of $^{87}Sr/^{86}Sr$ was appreciably higher at the K-T boundary for which the only logical explanation seems to be dissolution of large amounts of crustal strontium by the acid rain.

Turco *et al.* (1982), employing a one-dimensional photochemical model, estimated that the Tunguska event caused a 35–45% global ozone depletion in early 1909; 30% and 15% respectively in 1910 and 1911. In the decade following Tunguska, the northern hemisphere was observed to cool by ~ 0.3°K relative to the southern hemisphere. Based on radiation transport simulations of temperature changes associated with predicted O_3, NO_2, and dust perturbations, Turco *et al.* (1982) estimated a global cooling of 0.2–0.3°C. It is, however, not safe to attribute any cause-effect relationship based on this approximate agreement between observations and calculations, because of the intrinsically variable nature of the weather and complex positive and negative feedback processes not yet well understood.

The Tunguska event was caused by an object of estimated mass about 10^6 tons, smaller by a factor of 10^5 than Alvarez's hypothetical object producing the K-T event. Of course, the most interesting thing about the Tunguska event is that there exists a historical record of the aftereffects; one therefore hopes to learn about modeling such events from this example.

Although our understanding of the consequences of dust loading following the impact of a huge meteorite are largely theoretical, we stand on firmer grounds than for the case of volcanic eruptions. There exists considerable debate on the extent to which volcanic dust loading modifies global climate, but that there is a cooling effect is certain. Further, particles of dust serve as condensation nuclei; in subfreezing air saturated with water vapor, ice crystals are formed. Incidentally, this effect should also be incorporated in any model calculations of expected climatic changes. In volcanic eruptions, account also has to be taken of the addition of carbon dioxide, other gases and aerosols to the atmosphere.

The cooling effect of volcanic dust seems to be well documented (Williams and McBirney 1979; Lamb 1970), based on direct studies after several volcanic eruptions. Well-known examples of worldwide climatic effects are the eruptions of Laki (Iceland) and Asama (Japan) in 1783, Tambora (Indonesia) in 1815, and Krakatoa (Indonesia) in 1883. For relative energy releases in these events, reference is made to Newhall and Self (1982). The world temperature fell by about 1.1°C below the normal after the Tambora eruption; the year 1816 became known as the "poverty year" and the "year without a summer" (Stommel and Stommel 1983).

Changes in the Ocean Chemistry

Two types of changes are expected: those due to the introduction of chemical components into the atmosphere/ocean, and those due to modification of the environment, post-event. In the case of an oceanic impact, the largest effect would be expected in the case of a comet, which should release large quantities of H_2O, HCN, CH_3CN, CO_2, and possibly CO. The changes in the ocean chemistry would be very dramatic, as discussed by Hsu (1980), who has considered this problem with special reference to the K-T event.

I must stress here that an important change would be post-event, due to instantaneous and subsequent vertical mixing of the ocean, as discussed in the next section. This would upset the geochemical cycles significantly; hence evidence for such events should be found in the marine sedimentary record.

In the case of volcanic events we would expect major differences in the oceanic chemistry depending on whether the event is on the land or submarine. It is difficult to present any quantitative discussions here, but I will consider some of the implications of changes in ocean chemistry.

Changes in the Thermal Structure of the Oceans

The oceans are generally thermally stably stratified. In the upper Atlantic and some parts of the Pacific, surface waters are warm and saline, while the deep waters are cool and fresh. The cold deep waters originate from various sources in Antarctica, in the high salinity Norwegian Sea, and off Greenland. The general layered structure of the oceans is maintained by some form of continuous or intermittent meridional circulation, and the oceanic structure is therefore a manifestation of the climate. The principal features of the thermohaline circulation, i.e., poleward transport of heat received at tropical latitudes via major warm water currents, and equatorward (return) flow to the tropics at depths, are clearly climate-dependent. Paleoclimatic data speak eloquently of dramatic relative changes in the temperatures of the surface and deep waters during the past 70–80 m.y. for which data are available (Savin 1977).

The vertical structure of trace constituents, heat and salinity are a result of climate-balanced circulation. Any changes in the climate will change the vertical structure, or vice-versa. If the oceans are vertically mixed by some mechanism, the state of the atmosphere and the ocean itself would be dramatically altered. The average temperature of the mixed ocean would be lowered to ~ 4°C. The chemistry of the oceans would be modified significantly. The partial pressure of CO_2 in surface water would become 440μ atm, compared to the present average value of 330μ atm. For other species, the surface concentrations would change manifold.

The questions then before us are: What is required to change the vertical structure of oceans, with respect to its constituents? How much energy is required to bring this about? Can a meteorite impact or volcanism do it? The answer to these questions is that in fact it is easy to perturb the vertical structure on a global scale. This fact is generally not recognized and therefore I will dwell on this aspect briefly here.

Alternatively, the production of bottom water could be shut off if the characteristics of waters in the Greenland/Norwegian and Weddel Seas could be modified to lower salinity and higher temperature. This would lead to stagnation and the warming (due to geothermal heating) of deeper waters. This scenario may be the explanation of high bottom water temperatures in the Tertiary.

A hypothetical ocean constructed opposite to our present-day oceans would also be stably stratified, i.e., warm, saline waters at the bottom and cold, fresh waters at the top. Both the systems are stable, but this is a more stable configuration. In a present-day ocean type structure, the phenomena of "salt-fingering" occurs due to relatively rapid diffusion of heat compared to that of salt, by about two orders of magnitude. This was first pointed out by Henry Stommel (1983). In the alternate structure (i.e., the deeper layer is both warmer and saltier), salt fingering does not occur; mixing is then due to double-diffusion convection and is weaker.

A meteorite impact on the ocean or an explosive submarine eruption will cause disturbances locally in the ocean due to shock-waves, and over a wide area due to giant tidal waves. In the case of either a large meteorite impact or a large volcanic eruption, heating of the deep waters would also induce vertical mixing, in events comparable to or bigger than Tambora (See Table 20-1). For detailed discussions on this subject, reference is made to Emiliani *et al.* (1981), McKinnon (1982), O'Keefe and Ahrens (1982), and Croft (1982).

We have estimated the amount of energy required to vertically mix the oceans, instantaneously, i.e., without resorting to diffusion phenomena. Mixing the deep waters of the global ocean would require about 3×10^{29} ergs, of the same order as required in the formation of deep waters of the oceans (see Table 20-1). From the above discussion, we clearly see the possibility of a large modification of the earth's climate and environment, via changes in the vertical structure (thermal and compositional) of the oceans as a result of a "meteorite/volcanic" event. It is obvious that a detailed evaluation would have to be made considering the location of the event and the energy available. Based on these considerations, one can decide whether the effect would be local or global. In the case of a volcanic event, the sustained activity of a volcano can bring about climatic changes that are global in character. Reference is made here to the NAS publication "Beneficial Modifications of the Marine Environment" (Singer 1972), where some examples are discussed of major global climatic changes due to a small trigger from human modifications of the oceans, such as the possibility of pumping up cold water from the deep ocean.

Turning our attention to volcanic events, we are indeed aware of the climatic changes caused by volcanic explosions during the historical period (Rampino *et al.* 1979; see also Lamb 1970), but at the same time, none of these events have been able to trigger a longer-lasting global climatic change for periods of the order of several decades or more. The energy release of historic volcanoes lies in the range of (10^{20}–10^{27}) ergs (Williams and McBirney 1979). Geological evidence points to much more energetic events, based on the sizes of calderas, lava, and ash deposits. It is difficult (as well as not useful) to place energy limits on the largest volcanoes on the earth, but based on observations of ash and lava deposits, it can be stated safely that events two to three orders of magnitude larger than Tambora have been frequent in the geological history of the earth. Energies of the order of 10^{30} ergs are therefore often available in large volcanic events.

The frequency of impacts of meteorites of mass exceeding 10^{11} tons is estimated as about 1/100 m.y. (Grieve 1982; Shoemaker 1983). The event rate for meteorites of n-fold diameter varies as n^{-2}. That means for a meteorite ten times as small, corresponding to 10^3 times lower kinetic energy, the impact frequency would be 10^2 fold. We have no similar size-frequency relation for volcanic events, but based on our limited knowledge, it appears that the frequencies of occurrence of meteorites and volcanic events involving energy release of more than 10^{28} ergs is larger but similar, within an order of magnitude.

The importance of volcanic events in producing global climatic changes has been highlighted for the entire duration of the Pleistocene by Bray (1977). He has pointed out that a high correlation exists between massive eruptions and the cooling episodes for the past 2 m.y. and has concluded that these volcanic events might have in fact triggered each of the Ice Ages. Verification of links between volcanism and glaciation in the geological period of course needs much more extensive study. In a thought-provoking exposition, Keith (1982) has considered how major volcanism may lead to a stagnant ocean, (principally due to a greenhouse effect due to volcanic CO_2, decreased snow cover and polar ice, ocean warming and expansion, etc.), for which isotopic evidence ($^{13}C/^{12}C$ ratios) exists in limestones. The sign of climatic change is reverse to what we have considered so far. Nevertheless, the interesting feature here is the emphasis on changes in the ocean structure, and its effect on fertility.

Uncertainties in Modeling Large Meteorite/Volcanic Events

At the present state of our knowledge, it would be premature to claim any reasonable understanding of effects of large meteorite/volcanic events, let alone quantify the effects as a function of energy. But it is interesting to consider the effects of such events seriously in the geological and the modern perspective. Model calculations such as those reported for the K-T meteor and the Tunguska event are the first steps in our understanding the effects of such events: (1) processes, energies and frequencies of energetic meteor impact and volcanism, and (2) the response of the atmosphere-ocean system to pulsed energetic perturbations.

I will now consider some critical issues where in-depth knowledge would be necessary to improve on predictions. The acquisition of this knowledge is bound to be a slow process. Take for example item (1). We are dealing with rare events and it is unlikely that in the coming decades one would have an opportunity to study an event. Such events cannot be simulated experimentally; one has to rely entirely on modeling and paleorecords. The response of the atmosphere-ocean system is a highly complicated, non-linear problem, complicated by space and time scales. This investigation involves building a very rich database and the employment of the most sophisticated computational techniques and fast computers.

Our present-day knowledge of submarine volcanism and hydrothermal activity is inadequate. In large impacts, unwarping of the asthenosphere might lead to the formation of a long-lived thermal plume in the mantle. The volcanism could extend much beyond the crater because of fracturing of the thin oceanic lithosphere. The scar of a meteorite or volcanic event would be expected to be obliterated, because of the filling of the crater by surrounding materials. It has been argued that the sparsely scattered plateaus could possibly be signatures of impacts of large meteorites.

Hydrothermal activity along the mid-oceanic ridges has been found to be quite variable even on time scales of months. It is a fairly high-energy process (see Table 20-1); it is not unlikely that it may have contributed significantly to global climatic and environmental changes.

For reasons stated above, we have no documented case of the impact of a meteorite on the oceanic lithosphere. Seventy-five percent of impacts must occur on the oceans. Because of the high plausibility of the K-T event being due to a meteorite impact, in particular because of the associated global iridium anomaly (Alvarez *et al.* 1982; Kyte *et al.* 1980), the question has been asked: Where did the meteor impact? Oceanic impact has been considered because there is no known terrestrial crater; Emiliani *et al.* (1981) have, on the basis of the pattern of extinction of higher plant species, suggested the northern Pacific Bering Sea as the impact area. Whipple (unpublished) and Seyfert (1983) have suggested that Iceland may in fact be the result of an asteroid impact.

The mechanics of impact of a large meteor on the ocean have been modeled by Croft (1982), Gault and Sonnet (1982), and O'Keefe and Ahrens (1982). The velocity of a large meteor (10 km) is not significantly impeded by the atmosphere and the ocean. Depending on velocity, only 5–15% of the energy is taken up in the atmosphere and the ocean. An ejecta 10 to 100 times the bolide mass can be injected into the stratosphere. The vaporized, melted, and sub-mm solid ejecta would transfer up to 40% of their energy to the atmosphere/ocean water giving rise to a short lethal heating pulse. No calculations have as yet been made to study the propagation of tsunami and shock waves within the oceans, a process which would be expected to contribute significantly to vertical mixing in the oceans in regions near the impact site.

There are several interesting aspects of an ocean impact which have not yet been considered. About 10^{20} gm of saline water would be injected into the atmosphere. A hot spot would be created in the ocean at the impact site. If the impact takes place in a carbonate-rich zone, appreciable amounts of CO_2 would be added to the atmosphere. The evolution of the atmosphere-hydrosphere system and the consequent climatic changes are difficult to comprehend and would require a detailed evaluation of positive and negative feedback processes.

The subtleties of the atmosphere-ocean system, particularly as regards its influence on climate, probably present us with more uncertainties and surprises. The year 1982 has provided a very interesting example of how small perturbations can make dramatic global changes. Two terrestrial events combined to produce a measurable atmospheric/oceanic circulation perturbation; the El Chichon eruptions in March/April 1982 and the subsequent extreme El Niño conditions in the equatorial Pacific. The El Chichon eruptions have been judged to be similar to the 1963 Agung eruption, in several respects. The bulk volume of airfall deposits is estimated to be 0.3 km^3 and the amount of fine ash about 10^{11}g. The eruptions are supposed to have produced about 10^{13}g volcanogenic sulfuric acid (these estimates are based on Rampino and Self 1983). A few energy-balance-type climate models show appreciable global cooling effects in the year following the eruptions.

The oceanic changes have been very dramatic, as can be judged from changes in atmospheric pCO_2. The rate of CO_2 growth in the atmosphere decreased appreciably during 1982 by about 50%. This change can be understood as a result of two influences, which go in tandem, according to Gammon and Komhyr (1983). The strong positive sea surface temperature (SST) anomalies (3–5°C) in the equatorial Pacific led to the failure of the normal upwelling of cool CO_2-rich waters. This effect was enhanced due to the large negative SST anomalies (1–3°C), particularly in the important intermediate and deep water formation regions, thereby increasing the normal ocean sink for CO_2. Lal and Revelle (1983) have pointed out that the atmosphere-ocean processes, as observed during the anomalous period of 1982, are probably responsible for the large, rapid variations in atmospheric pCO_2 during the last glacial period, within time periods of the order of 100 years; based on studies of polar ice cores (Stauffer *et al.* 1983).

Atmospheric CO_2 concentrations have been found to be strongly correlated with air temperatures (Stauffer *et al.* 1983). The result supports the expected positive feedback character for changes of pCO_2 in the atmosphere. Thus, small changes in the atmospheric pCO_2 due to changes in the ocean circulation in the top 1 km depth, particularly at high latitudes, may be greatly amplified to lead within a few decades to large increases or decreases in both atmospheric pCO_2 and temperature.

Large meteorite impacts/volcanic events can easily produce rapid non-linear changes in climate. It must be noted here that although there are no published estimates of energy release for the El Chichon eruption, one can estimate an approximate value of 10^{24} ergs, making it a relatively small event. At this point, I would like to point out that sustained low-energy volcanic explosions can also be very effective in producing global climatic changes.

The tremendous sensitivity of ocean circulation changes to climate can be easily understood. About half the solar radiation absorbed by the earth-atmosphere system is finally absorbed in the surface layers of the tropical oceans. Both the atmosphere and the oceans transport energy poleward, the driving force being the general atmospheric circulation. The heat capacity of the oceans is three orders of magnitude larger than that of the atmosphere. In view of their large thermal and mechanical inertia, oceans play the role of a fly-wheel in the climatic system. However, any changes in the ocean currents can have a dramatic influence on weather, and can induce climatic changes. Variability in the ocean currents and in the equatorial sea surface temperatures and ocean flows can have rather profound effects to the extent they exert positive feedback effects.

Thus we have seen that oceans influence climate via their exchange of energy, momentum, and matter with the atmosphere at the air-sea interface. This exchange is governed by the large-scale atmospheric circulation which in turn is controlled by the large-scale oceanic circulation. I have identified, in this chapter, several elements of the atmospheric/oceanic circulation that are non-linear in character, and whose variation can trigger global climate and environmental changes. Our knowledge about these elements is still in its infancy, however. But based on observations of low-energy atmospheric perturbations, such as those due to volcanic eruptions, pCO_2 changes, we can be sure that certain high-energy atmosphere/ocean perturbations can be effective in triggering global climatic and environmental changes.

Conclusions

This chapter highlights the possible dramatic consequences to global climate and environment of two types of naturally occurring events, one extra-terrestrial and the other terrestrial. Geological data on craters bear testimony that large meteorite impacts and volcanic eruptions have been instrumental in causing major changes in the surface features of planets. Additionally, it has been pointed out that meteoritic bombardment during the early history of the moon, Mars, and other inner planets, may have been responsible for producing the present crustal dichotomy that we see on these planets (Lowman 1978).

In this chapter I have reviewed the present state of our knowledge about the possible effects of "large" meteorite and volcanic events to modifications of global climate and environment. Their most direct and common influence is to blanket the solar radiation due to dust loading of the atmosphere. Model calculations (Turco *et al.* 1982) show that Tunguska, estimated to have an initial kinetic energy of ca. 3×10^{25} ergs (meteor mass, ca. 3×10^{6} tons), produced dramatic environmental changes, leading to nearly total ozone depletion, and significant cooling effects in the northern hemisphere in the decade following the event. At present, discussions of larger meteorite events refer to a hypothetical meteorite impact to explain the K-T event, 65 million years ago. Alvarez *et al.* (1980), who proposed this scenario, estimate a meteor mass of 10^{11} tons and kinetic energy ca. 10^{31} ergs. The K-T boundary is associated with a massive biological extinction episode, including the demise of very large animals, the dinosaurs. Being a geological boundary it also necessarily includes an epoch of predominant global climatic changes, but the manner in which the meteor event directly produced these changes is not known at present. (This statement is of course subject to the proviso that the K-T event was indeed caused by a meteor impact.)

I would like to mention here that among the three other well-documented prominent biological extinctions, Eocene-Oligocene (34 m.y.), Permian-Triassic (230 m.y.) and Late Cambrian (500 m.y.), the iridium anomalies have been found only in the case of the first two (Silver and Schultz 1982); the latter events if they represent climate trigger events, would have then to be attributed to volcanism. It must be noted here that Magaritz *et al.* (1988) found, based on their studies of ratios of carbon isotopes in marine carbonate rocks at the Permian/Triassic boundary that the extinction (clearly marked by these ratios) did not occur in a single event: rather it happened as a result of complex changes spanning perhaps a million years.

The present emphasis on the investigations of the K-T impact relate on the one hand to building models to explain the causes of the rapid climatic changes, changes in the chemistry of the oceans and the atmosphere, and the causes of the intense trauma in the marine biosphere, and on the other, to attempts to explain the entire geophysical and geochemical evidence by working with a large volcanic event, as an alternative to impact. Presence of iridium and many meteoritic elements, and shocked quartz (Bohor *et al.* 1987) on a global scale are the principal evidences for a meteorite impact causing the K-T event. Scientists favoring the volcano hypothesis claim that appreciable mantle iridium gets volatilized in a large volcanic event and could reach the stratosphere from where it would be globally dispersed. And apparently, some evidence also exists for formation of shocked quartz in a large volcanic event. For an excellent scientific case for the meteorite impact theory, reference is made to Alvarez (1986), and for the volcano hypothesis, to Officer and Drake (1985), and Courtillot *et al.* (1986).

In a recent paper, Wolbach *et al.* (1985) have provided a fairly convincing evidence for wildfires associated with the giant meteorite impact. The event contributed to a global scale dispersion of 0.021 +/- 0.006 g C (soot), which the authors believe may have caused considerable optical attenuation and thereby the extinction.

Another interesting development in the studies of the consequences of meteorite impact is the new view that seems to be gaining some respectability, namely that some asteroid impacts may trigger geomagnetic reversals (Raup, 1985; Muller and Morris, 1986; Schwarzschild, 1987).

I must emphasize that I have so far concerned myself with the global influences of large rare events—meteorite impact and volcanism. At such low frequencies, we must remain alert to other rare astronomical events, such as giant solar flares, passage of the solar system through giant molecular or dust clouds, etc. A great deal of speculation has been made about such astronomical influences on terrestrial climate.

Clube and Napier (1982) have hypothesized that major impact episodes may be related to the passage of the solar system through the spiral arms of our galaxy. The impacts may have caused magnetic field reversals, plate movements, Ice Ages, and mass extinctions. Thus, Clube and Napier (1982) suggest that our geophysical past may have been closely related to interaction of the earth with the galaxy.

Clearly, there is no consensus today on what caused the K-T extinction but a greater number of scientists are impressed with the original meteorite impact hypothesis of Alvarez *et al.* (1980). In any case, we must now recognize that definite climatic changes can be induced by large energy fluxes at the surface of the Earth.

Acknowledgements

I am extremely grateful to Professor C.S. Cox for critical reading of the manuscript and for many valuable suggestions. Thanks are also due to Dr. F. Kyte for discussions and comments on the text.

References

Alvarez, L., W. Alvarez, F. Asaro, and H.V. Michel. 1980. "Extra-terrestrial cause for the Cretaceous-Tertiary extinction." *Science* 208J: 1095–1108.

Alvarez, L., L.W. Alvarez, F. Asaro, and H.V. Michel. 1982. "Current status of impact theory for the terminal Cretaceous extinction." *Geological Society of America*, Special Paper 190: 305–315.

Alvarez, W. 1986. "Toward a theory of impact crisis." *EOS* 67: 649–658.

Bohor, B.F., Modreski, P.J., and Foord, E.E. 1987. "Shocked quartz in the Cretaceous-Tertiary boundary Clays: Evidence for a global distribution." *Science* 236: 705–709.

Bray, J.R. 1977. "Pleistocene volcanism and glacial initiation." *Science* 197: 251–254.

"Climate in Earth History": *Studies in Geophysics*. 1982. Sponsored by Geophysics Research Board and National Research Council. Washington, D.C.: National Academy Press.

Clube, S.V.M. and W.M. Napier. 1982. "Spiral arms, comets, and terrestrial catastrophism." *Quart. J. Roy. Astrn. Soc.* 23: 45–66.

Clube, S.V.M. and W.M. Napier. 1982. "The role of episodic bombardment in geophysics, earth planet." *Sci. Lett.* 57: 251–262.

Courtillot, V., J. Besse, D. Vandamme, R. Montingny, J. Jaeger and H. Cappetta. 1986. "Deccan flood basalts at the Cretaceous/Tertiary boundary." *Earth and Planet Sci. Lett.* 80:361–374.

Croft, S.K. 1982. "A first-order estimate of shock heating and vaporization in oceanic impacts." *Geol. Soc. Am.*, Special Paper 190: 143–152.

Emiliani, C., E.B. Kraus, and E.M. Shoemaker. 1981. "Sudden death at the end of the Mesozoic earth planet." *Sci. Lett.* 55: 317–334.

Gammon, R.H. and W.D. Komhyr. 1983. "Response of the global atmospheric CO_2 distribution to the atmospheric/ocean circulation perturbation in 1982." In *IUGG 1983 Hamburg Programme and Abstracts* p. 828.

Gault, D.E. and C.P. Sonett. 1982. "Laboratory simulation of pelagic asteroidal impact: Atmospheric injection, benthic topography, and the surface wave radiation field." *Geol. Soc. Am.* Special Paper 190: 69–92.

Grieve, R.A. 1982. "The record of impact on earth: Implications for a major Cretaceous/Tertiary impact event." *Geol. Soc. America*. Special Bulletin 190:25–37.

Grove, J.M. 1988. *The Little Ice Age*. London: Methuen, p. 498.

Hofmann, D.J. and J.M. Rosen. 1983. "Condensation nuclei events at 30 km and possible influences of solar cosmic rays." *Nature* 302: 511–514.

Hsu, K.J. 1980. "Terrestrial catastrophe caused by cometary impact at the end of Cretaceous." *Nature* 285: 201–203.

Keith, M.L. 1982. "Violent volcanism, stagnant oceans and some inferences regarding petroleum, strata-bound ores and mass extinctions," *Geochim. Acta*. 46: 2621–2637.

Krassovskij, V.I., and Sklovskij. 1958. "Variation of the intensity of cosmic radiation during earth's geological history and their possible influence on life's evolution." *Nuovo Cimento Supplemento*, v. 8, No. 2, Serie X: 440–443.

Kyte, F.T., Z. Zhou, and J.T. Wasson. 1980. "Siderophile-enriched sediments from the Cretaceous-Tertiary boundary." *Nature* 288: 651–656.

Lal, D. and R. Revelle. 1983. "Atmospheric pCO_2 changes recorded in lake sediments." *Nature* 308:344–346.

Lamb, H.H. 1970. "Volcanic dust in the atmosphere; with a chronology and assessment of its meteorological significance." *Proc. Royal Sco. London* 266: 425–533.

Lowman, P.D. 1978. "Crustal evolution in the silicate planets." *Naturwissenshaften* 65: 177–124.

MacDonald, G.A. 1972. *Volcanoes*. Englewood Cliffs, New Jersey: Prentice Hall, Inc.

MacDougall, J.D. 1988 "Seawater strontium isotropes, acid rain, and the Cretaceous-Tertiary boundary." *Science* 239:485–487.

McKinnon, W.B. 1982. "Impact into the earth's ocean floor: Preliminary experiments, a planetary model, and possibilities for detection." *Geol. Soc. America*, Special Paper 190: 129–142.

Magaritz, M., R. Bar, A. Baud, and W.T. Holser. 1988. "The carbon-isotope shift at the Permian/Traissic boundary in the southern Alps is gradual." *Nature* 331:337–339.

Muller, R.A. and D.E. Morris. 1986. "Geomagnetic reversals from impacts on earth." *Geophys. Res. Lett.* 13: 1117–1180.

Newhall, C.G. and S. Self. 1982. "The volcanic explosivity index (VEI): An estimate of explosive magnitude for historical volcanism." *Jr. Geophys. Res.* 87: 1231–1238.

Officer, C.B. and C.L. Drake. 1985. "Terminal Cretacsous environmental events." *Science* 227: 1161–1167.

O'Keefe, J.D., and T.J. Ahrens. 1982. "The interaction of Cretaceous/Tertiary extinction bolide with the atmosphere, ocean and solid earth." *Geol. Soc. America* Special Paper 190: 103–120.

Pollack, J.B., O.B. Toon, T.P. Ackerman, C.P. McKay, and R.P. Turco. 1983. "Environmental effects of an impact-generated dust cloud: Implications for the Cretaceous-Tertiary extinctions." *Science* 219: 287–299.

Prinn, R.G. and B. Fegeley, Jr. 1987. "Bolide impacts, acid rain, and biospheric traumas at the Cretaceous boundary." *Earth and Planet Sci. Lett.* 83: 115.

Rampino, M.R. and S. Self. 1983. "El Chichon, eruption volatiles and atmospheric effects." In *IUGG 1983 IAMAP Abstracts and Programme*: 595.

Rampino, M.R., S. Self, and R.W. Fairbridge. 1979. "Can rapid climatic change cause volcanic eruptions?" *Science* 206: 826–829.

Raup, D.M. 1985. "Magnetic reversals and mass extinctions." *Nature* 314:341–343.

Russell, D. and W. Tucker. 1971. "Supernovae and the extinction of the dinosaurs." *Nature* 229: 553–554.

Savin, S.M. 1977. "The history of the earth's surface temperature during the past 100 million years." *Ann. Rev. Earth Planet. Sci.* 9: 319–355.

Seyfert, C.K. 1983. "Did the Cretaceous-Tertiary asteroid cause the beginning of separation of Greenland from Europe?" *Geol. Soc. Am.* 96th Annual Meeting, Indianapolis. Abstracts with programs. v. 15, no. 6, Sept. 1983, Abstract No. 28327, p. 684.

Shoemaker, E.M. 1983. "Asteroid and comet bombardment of the earth." *Ann. Rev. Earth Planet. Sci.* 11: 461–494.

Silver, L.T. and H. Schultz, eds. 1982. "Geological implications of impacts of large asteroids and comets on the Earth." Special Paper

190, Boulder CO: The Geological Society of America.

Singer, S. F., ed. 1970. *Global Effects of Environmental Pollution*. New York: Springer Verlag Inc.

Singer, S.F. (chairman). 1972. "Beneficial modifications of the marine environment." A symposium sponsored by the National Research Council, March 11, 1968. Washington, D.C. National Academy of Sciences.

Stauffer, B., H. Hofer, H. Oeschger, J. Schwander, and U. Siegenthaler. 1983. "Atmospheric CO_2 concentration during the last glaciation." Symp. on Ice and Climate Modeling, Evanston, Illinois, USA, June 27-July 1, 1983. *Annals of Glaciology* 5:160–164, 1984.

Stommel, H. and E. Stommel. 1983. *Volcano Weather*. Newport, Rhode Island: Seven Seas Press.

Schwarzschild, B. 1987. "Do asteroid impacts trigger geomagnetic reversals?" *Physics Today* Feb., 1987: 14.

Turco, R.P., O.B. Toon, C. Park, R.C. Whitten, J.B. Pollack, and P. Noerdlinger. 1982. "An analysis of the physical, chemical, optical and historical impacts of the 1908 Tunguska Meteor Fall." *Icarus* 50: 1–52.

Turner, J.S. 1981. "Small scale mixing processes." In *Evolution of Physical Oceanography*. B.A. Warren and C. Wunsch (eds) pp. 236–265. Cambridge, MA: The MIT Press.

Williams, H. and A.R. McBirney. 1979. *Volcanology*. San Francisco: Freeman, Cooper and Co.

Wolbach, W.S., Lewis, R.S., and Anders, E. 1985. "Cretaceous extinctions: Evidence for wildfires and search for meteoritic material." *Science* 230: 167–170.

Urey, H.C. 1973. "Cometary collisions and geological periods." *Nature* 242: 32–33.

TWENTY ONE

COMMENTARY on DEVENDRA LAL'S PAPER

Andrew A. Lacis and Sergej Lebedeff

Volcanic eruptions can inject light-scattering particles high into the stratosphere, thereby reducing the solar radiation transmitted to the ground. It is commonly thought that this reduction in solar heating will cool the surface temperature. However, comparison of the global temperature record with known volcanic eruptions does not show a convincing correlation—a conclusion reached previously by a number of investigators. We examine some examples where apparent agreement is found and conclude that in order more clearly to characterize the volcanic climate signature, it is necessary to know the optical thickness of the volcanic aerosol including its composition, particle size, and geographic distribution. The particle size distribution is particularly critical since it can determine whether the volcanic aerosol will heat or cool the surface temperature.

In qualitative terms it seems plausible that highly energetic events such as impacts by small asteroids or the eruptions of very large volcanoes (Lal 1989), aside from the obvious local devastation, may also have a significant impact on the global climate of the earth. However, when it comes to preparing a quantitative description for the expected climatic consequences, we encounter difficulties. First of all, the high energy catastrophes cited by Lal are orders of magnitude larger than our observational experience to date, and it is risky to expect linear extrapolations from present-day climate to remain valid. Other major sources of uncertainty relate to the mechanics for the global dispersal of aerosols following cataclysmic explosions and to the actual amount, composition, and size distribution of such aerosols injected into the stratosphere.

In general terms, we can specify the essential characteristics that an aerosol-producing event should have in order to significantly impact the global climate. These are: (1) A large fraction of the globe must be covered with sufficient optical thickness in order to have a noticeable effect on the radiation budget of the earth. (2) Because of the large heat capacity of the ocean, the aerosol residence time in the atmosphere should be longer than several months. This implies that the aerosol must be placed into the stratosphere since the tropospheric rain-out time is relatively short. (3) The aerosols must be substantially submicron in size. This is necessary to maintain a long residence time in the stratosphere and also to have the required radiative properties to produce a cooling effect on the surface temperature. Depending on particle composition, aerosols which are only moderately larger than a micron can have sufficiently large cross-sections for thermal radiation to produce a greenhouse warming of the surface temperature. (See, e.g. Singer, this volume.)

We know from historical records (e.g., Stothers and Rampino 1983a) that large volcanic eruptions have indeed produced globally extensive haze covers coinciding with, or followed by, noticeably poor harvests or exceptionally cold winters. Corroborating records of these events are also found in ice core data in the form of acidity peaks due to enhanced deposition of sulfuric acid aerosols following major eruptions (Stothers and Rampino 1983b). It appears that in all of these cases the global distribution of the volcanic aerosol was probably achieved because the aerosol-producing material was injected into the stratosphere and dispersed in gaseous form, following which photochemical reactions gradually produced the characteristic long-lasting sulfuric acid haze. The problem with particulate aerosols (ash) is that they generally consist of larger particles which fall out relatively quickly. There is the further problem that in order to produce a globally significant amount of ash particles small enough to stay aloft for extended periods, the number density of such particles in the immediate vicinity of the volcano would be so large that coagulation would become an important loss mechanism.

Besides ejecting particulate matter, an asteroid impact is likely to vaporize a large amount of solid material and water which may or may not become globally dispersed before condensing out as aerosol particles. Unfortunately, the necessary physics for properly modeling the condensation process of vaporized rock and water vapor, particle coagulation and the high-energy dynamic interactions with the atmosphere, is both difficult and complex, and not even the most sophisticated of present climate models are prepared to address this most critical part of the problem. Thus we have few guidelines to evaluate critically the speculations about the nature of cataclysmic explosions and their impact on the global climate.

To help illustrate the difficulty in estimating climate change perturbations, it is instructive to compare the global temperature record with known volcanic eruptions and meteoric impacts. The Tunguska meteor of June 30, 1908, which leveled some 8,000 km^2 of Siberian forest, is undoubtedly the most violent meteor impact in recent history. Figure 21-1 shows the monthly mean temperature trends (with the mean seasonal cycle subtracted out) in the northern latitudes for the decades preceding and following the Tunguska event. The solid line is an 11-month running mean average to define more clearly the temperature trend. The filled arrows depict volcanic eruptions during this period as tabulated by Newhall and Self (1982). Cursory inspection of Figure 21-1 shows no clearly defined relationship between temperature trends and volcanic eruptions. Increases in temperature seem to be just about as likely to follow an eruption as temperature decreases. If anything, the temperature trend appears to have a semi-periodic component with little correlation to volcanic occurrences. For any climatic impact due to the Tunguska meteor, the temperature record is somewhat ambiguous. The temperature does decrease following the Tunguska impact, but the decreasing trend had begun well before the impact and could just as easily

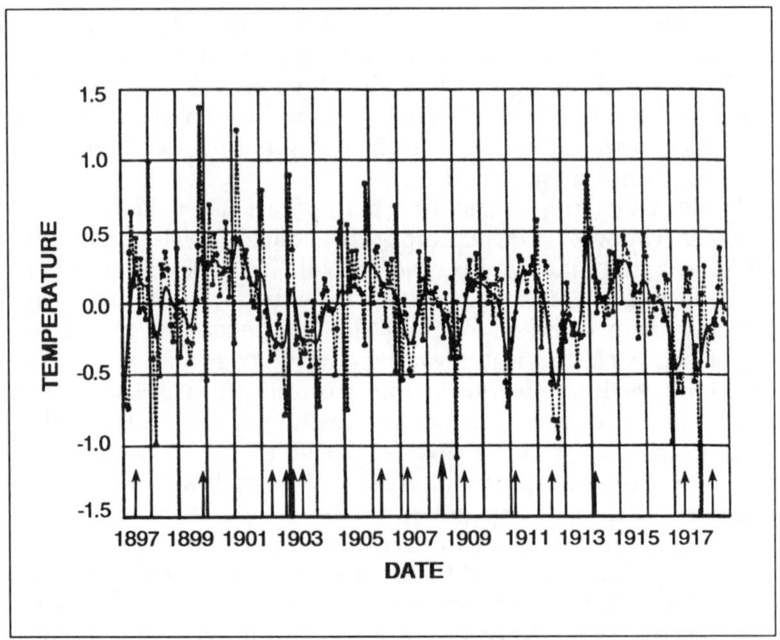

Figure 21-1: Northern latitude (23.4°N–90°N) temperature trends preceding and following the June 30, 1908 Tunguska meteor fall. Monthly mean temperatures are plotted after subtraction of the mean seasonal cycle. The solid line is an 11 month running mean average. The solid arrows depict volcanic eruptions tabulated by Newhall and Self (1982).

be part of the semi-periodic temperature oscillations that occur throughout the temperature record.

Turco *et al.* (1982) analyzed the circumstances of the Tunguska meteor in considerable detail. In their analysis, they estimated that the falling meteor could have generated up to 30 megatons of NO, and that the photochemical consequences of this large NO injection into the stratosphere could have depleted the ozone by 35–45% in the northern hemisphere. As corroborating evidence, they cite contemporary atmospheric transmission measurements which show a spectral dependence of optical extinction in the Chappuis band of ozone to be consistent with significant ozone depletion. They attribute a decade-long cooling trend of 0.3°C in the northern hemisphere as being due to the reduction in the ozone greenhouse effect.

Figure 21-1, however, shows no apparent temperature decrease in the decade following the Tunguska impact. Turco *et al.* apparently base their temperature change estimate on a warming trend in the southern hemisphere during this time. Also, a search of the Greenland ice core samples for nitrate deposits (Rasmussen *et al.* 1984) shows no evidence of enhanced nitrates in the years following the Tunguska event. Rasmussen *et al.* concluded that the nitrate production calculated by Turco *et al.* is 1–2 orders of magnitude too high. Thus, in the one major meteor event for which we have some observations, there appears to be no conclusive determination of its impact on the global climate.

In the case of volcanoes, global temperature records have been analyzed by many investigators attempting to extract an empirical climatic response to volcanic eruptions (e.g., Oliver 1976; Taylor *et al.* 1980; Self *et al.* 1981). The basic results of their analysis indicate that there is indeed some evidence for climatic cooling of a short duration following volcanic eruptions, but that there is also a large statistical uncertainty. One conclusion that may be drawn is that volcanoes are not identical in their impact on climate, and that therefore, statistical analysis of eruptions and temperature trends can provide only qualitative information about the volcanic climate signature. Clearly, we must go a step further and examine the radiative properties of volcanic ejecta (optical thickness, composition, particle size distribution) and compare these results against the temperature record.

Figure 21-2 shows the optical depth measured at Mauna Loa Observatory after removal of the background value. The results are plotted inversely to give the appearance of atmospheric transmission, which for the small optical thickness involved, is proportional to 1 minus optical depth tau. The solid line shows a GCM simulation of volcanic aerosol transport computed with the GISS tracer model for the following eruptions: Agung (1963), Awu (1966), Fernandina (1968), and Fuego (1974). Global optical thicknesses for these volcanoes were obtained by normalizing the tracer model results to the observed optical thickness variations measured at Mauna Loa.

Figure 21-3 shows the 20-year global temperature record coinciding with the Mauna Loa optical depth measurements.

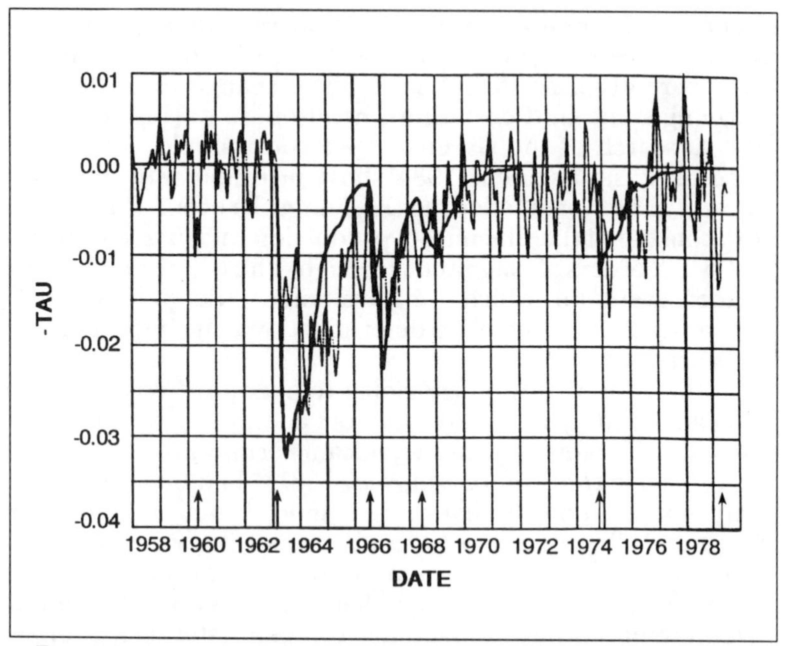

Figure 21-2: Optical depth (tau) measurements at Mauna Loa Observatory after subtraction of background average. The values are plotted inversely to give the appearance of transmission (1-tau). The solid line is the normalized fit of tracer model simulations to Mauna Loa observations. The arrows indicate the principal volcanic eruptions.

Again, the seasonal cycle has been removed, the solid line depicts an 11-month running mean to clarify the temperature trend, and the arrows identify the volcanic eruptions during this period as tabulated by Newhall and Self (1982). In comparing Figures 21-2 and 21-3, we find a considerable degree of correspondence between the atmospheric transmission measured at Mauna Loa and the global temperature record, particularly in the case of the Agung eruption. These results tend to corroborate the analysis of Hansen *et al.* (1978) in that significant cooling of the global temperature occurred after the Agung eruption. It is also clear that a number of the eruptions tabulated by Newhall and Self, which do not show up in the Mauna Loa transmission measurements, also do not appear to correlate with temperature decreases. Although the agreement

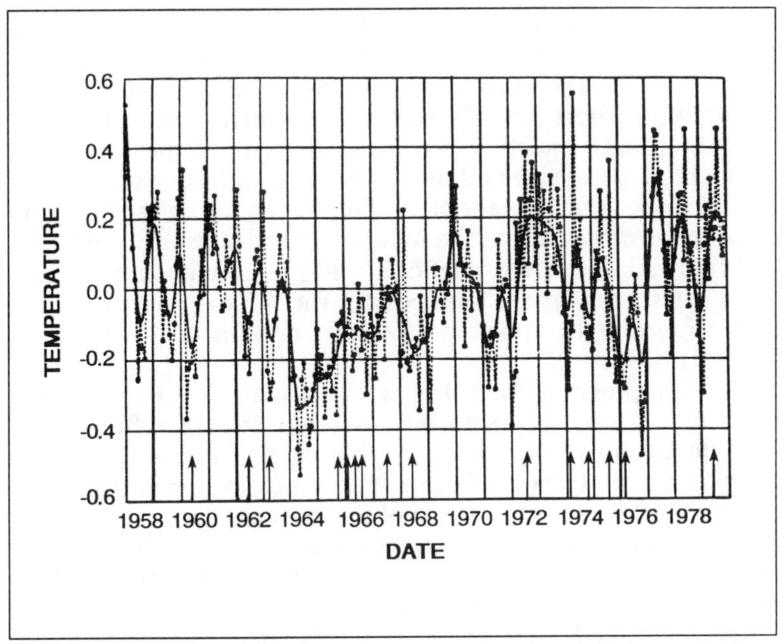

Figure 21-3: Global temperature record. The dots indicate monthly mean temperatures after subtraction of the mean seasonal cycle. The solid line is an 11 month running mean average. The arrows depict volcanic eruptions as tabulated by Newhall and Self (1982).

between changes in atmospheric transmission due to stratospheric aerosols and the global temperature decreases is not perfect, there is now a more physical basis for estimating the climatic effect of volcanoes.

As noted by Hansen *et al.* (1978), the particle size distribution of the volcanic aerosol is also a critical parameter in determining the climatic effect. This is because two opposing radiative effects are operating. On the one hand, the volcanic aerosols scatter and reflect incoming solar radiation and thus tend to cool the surface temperature. On the other hand, sulfuric acid and silicates have strong absorption cross-sections in the 10 μm region; this, for particle size distributions only moderately larger than 1μm, can produce a greenhouse warming effect strong enough to counteract the solar albedo effect and cause a net warming of the surface temperature.

As a case in point, Hofman and Rosen (1983) measured the effective size distribution of stratospheric aerosols a month after the El Chichon eruption to be about 1.5 μm. For this size distribution, the thermal greenhouse contribution predominates and causes surface warming to occur. Some 6 months later, the particle size had decreased to an effective radius less than 0.5 μm due to fallout of the larger particles. For this smaller size distribution, the solar albedo effect is stronger than the thermal greenhouse effect, leading to a cooling of the surface temperature. Thus, besides the optical thickness, it is also important to know the size distribution of the volcanic aerosols.

The long-term climate impact of sustained volcanic eruptions is not easy to assess because the radiative effects of a volcano last only a year or two, whereas the response time for the ocean mixed layer is decades and much longer for the deep water. It is conceivable that volcanoes may act to trigger an impending ice age or to perhaps delay its ending. In any case sustained forcing over long periods of time is needed to overcome the long time constant of the ocean.

References

Hansen, J.E., W.C. Wang and A.A. Lacis. 1978. "Mount Agung eruption provides test of a global climatic perturbation." *Science* 199: 1065–1068.

Hofman, D.J. and J.M. Rosen. 1983. "Sulphuric acid droplet formation and growth in the stratosphere after the 1982 eruption of El Chichon." *Science* 222: 325–327.

Lal, D. 1989. *Global effects of meteorite impacts and volcanism*. This volume.

Newhall, C.G., and S. Self. 1982. "The Volcanic Explosivity Index (VEI): An estimate of explosive magnitude of historic eruptions." *J. Geophys. Res.* 87: 1231–1238.

Oliver, R.C. 1976. "On the response of hemispheric mean temperature to stratospheric dust: An empirical approach." *J. Appl. Meteorol.* 15: 933–950.

Rasmussen, K.L., H.B. Clausen and T. Risbo. 1984. "Nitrate in the Greenland ice sheet in the years following the 1908 Tunguska event." *Icarus* 58: 101–108.

Self, S., M.R. Rampino and J.J. Barbera. 1981. "The possible effects of large 19th and 20th century volcanic eruptions on zonal and hemispheric surface temperatures." *J. Volcanology and Geotherma. Res.* 11: 41–60.

Singer, S.F. 1989. *Nuclear winter or nuclear summer?* This volume.

Stothers, R.B. and M.R. Rampino. 1983a. "Volcanic eruptions in Mediterranean before A.D. 630 from written and archaeological sources." *J. Geophys. Res.* 88: 6357–6371.

Stothers, R.B. and M.R. Rampino. 1983b. "Historic volcanism, European dry fogs, and Greenland acid precipitation, 1500 B.C. to A.D. 1500." *Science* 222: 411–413.

Taylor, B.L., T. Gal-Chen and S.H. Schneider. 1980. "Volcanic eruptions and long-term temperature records: An empirical search for cause and effect." *Quart. J. Roy. Met. Soc.* 106: 175–199.

Turco, R.P., O.B. Toon, C. Park, R.C. Whitten, J.B. Pollack, and P. Noerdlinger. 1982. "An analysis of the physical, chemical, optical, and historical impacts of the 1908 Tunguska Meteor Fall." *Icarus* 50: 1–52.

TWENTY TWO

FURTHER COMMENTARY on DEVENDRA LAL'S PAPER

Hugh W. Ellsaesser

I am encouraged to hear the reservations of Lacis and Lebedeff regarding the climatic effects routinely attributed to volcanic eruptions. I had cause to review again the evidence of this field (Ellsaesser 1983) and concluded that published climatogenic effects attributed to volcanoes primarily represent cases of mistaken identity.

There is no question but that volcanic eruptions and subnormal temperatures are statistically correlated—apparently on all time scales (Rampino *et al.* 1979). However, all who have looked closely have found discordant data which appeared to defy the presumed cause-and-effect relationships. Mitchell (1961) noted cooling even after non-explosive eruptions. Many have noted, as did Lacis and Lebedeff, that the cooling was generally underway before the eruptions. In the case of Tambora (1815) the available temperature records show that cooling even on a hemispheric scale was underway for about 50 years before the eruption; the entire period from 1799 to 1817 was substantially cooler than any period since 1605 to 1610. The New England crop failures of the notorious "Year Without a Summer" or "1816 and Froze to Death" were due to three isolated severe storms bringing killing frosts in each of the three summer months—June, July, and August. Weather between the storms was normal to unseasonably warm (Milham 1924). Much the same occurred in England. I, for one, am at a loss to explain such weather on the basis of a persistent dust layer in the stratosphere from a volcanic eruption occurring in 1815.

Lacis and Lebedeff said in part: "We can with reasonable confidence identify the temperature decrease due to Agung..." It should be noted that Starr and Oort (1973) created quite a stir when they called attention to an apparent 0.6°C cooling of the northern hemisphere 1000–500 mb layer between May 1958 and April 1963. How can anyone be *confident* that any cooling subsequent to the Agung eruption of March 1963 was not due to the same cause as that responsible for the *unexplained* cooling occurring over the *5 years preceding the eruption*? Angell and Korshover (1983) after having examined the data several times stated:

> There is evidence for an 0.3°C decrease in northern hemisphere surface temperature following the Agung eruption of 1963, but there was no obvious influence on southern hemisphere temperatures, cooling occurring before the eruption and continuing after the eruption.

How does one reconcile this as a cause-and-effect relationship when the Agung dust cloud was reported to be an order of magnitude thicker in the southern hemisphere than in the northern (Dyer and Hicks 1968)?

Self *et al.* (1981) and Rampino and Self (1982) noted two serious discrepancies in the cause-and-effect relationship. They found that the series of six major eruptions from 1881–1889, and five in 1902–1903 produced no greater response in the temperature record than did individual isolated eruptions. They also pointed out that Tambora (1815), Krakatoa (1883), and Agung (1963), with relative eruption magnitudes of 150:20:1 and sulfate aerosol magnitudes of 7.5:3:1, appeared to produce comparable temperature responses, reflecting no differences in the intensities of the presumed perturbing factors.

Recent studies by Kelly and Sear (1984), Sear *et al.* (1987) and Bradley (1988) found the maximum hemispheric cooling within two months after major volcanic eruptions. This contrasts strongly with most previous studies which claimed to find the maximum cooling one to three years after the eruption.

One of our most powerful means for distinguishing that which we know from that which we merely believe is the test of consistency. It is my considered opinion that the published studies on the climatic effects of volcanic dust veils in the stratosphere do not pass this test.

References

Angell, J.K. and J. Korshover. 1983. "Global temperature variations in the troposphere and stratosphere, 1958–1982." *Mon. Weather Rev.* 112: 901–921.

Bradley, R.S. 1988. "The explosive volcanic eruption signal in northern hemisohere continental temperature records." *Climatic Change* 12: 221–243.

Dyer, A.J. and B.B. Hicks. 1968. Global spread of volcanic dust from the Bali eruption of 1963." *Quart. J. Roy. Met. Soc.* 94: 545–554.

Ellsaesser, H.W. 1983. "Isolating the climatogenic effects of volcanoes." Paper presented at Meeting of Experts on Anthropogenic Climatic Change, July 4–10, 1983, Leningrad, USSR, UCRL-89161. Lawrence Livermore National Laboratory, Livermore, CA 94550.

Kelly, P.M. and C.B. Sear, 1984. "Climatic impact of explosive volcanic eruptions." *Nature* 311: 740–743.

Milham, W.I. 1924. "The year 1816—The causes of abnormalities." *Mon. Weather Rev.* 52: 563–570.

Mitchell, J.M., Jr. 1961. "Recent secular changes of global temperature." *Ann. N.Y. Acad. Sci.* 95: 235–250.

Rampino, M.R. and S. Self. 1982. "Historic eruptions of Tambora (1815), Krakatoa (1883) and Agung (1963), their stratospheric aerosols and climatic impact." *Quaternary Research* 18: 127–143.

Rampino, M.R., S. Self and R.W. Fairbridge. 1979. "Can rapid climate change cause volcanic eruptions?" *Science* 206: 826–829.

Sear, C.B., P.M. Kelly., P.D. Jones and C.M. Goodess, 1987. "Global surface-temperature responses to major volcanic eruptions." *Nature* 330: 365–367.

Self, S., M.R. Rampino and J.J. Barbera. 1981. "The possible effects of large 19th and 20th century volcanic eruptions on zonal and hemispheric surface temperatures." *J. Volcanology and Geothermal Research* 11: 41–60.

Starr, V.P. and A.H. Oort. 1973. "Five-year climatic trend for the northern hemisphere." *Nature* 242: 310–313.

TWENTY THREE

ASTEROIDS, VOLCANOES, and CLIMATE

H.G. Goodell

Fortunately, the likelihood of an asteroid or comet striking the earth is small. Shoemaker *et al.* (1979) calculated a mean impact rate of 2.5×10^{-9} per comet per perihelion passage for comets. It appears that the cumulative number (N) of celestial bodies between 1 and 40 km is distributed by diameter (d) such that $N \sim d^{-2}$ (Shoemaker and Wolfe 1982). This gives a possible population of about 10 near-earth-orbit-crossing or earth-approaching asteroids of 10 km or larger. The probability of collision by such a body is therefore about 10^{-8}/y (Wetherill and Shoemaker 1982). This probability is consistent with the geologic record of dated fossil asteroid craters (Grieve 1980).

At present only one earth-crossing asteroid of about 10 km is known (2212 Hephaistos, 8.7 km) although 3 others are earth-approaching and may eventually become earth-crossing (1036 Ganymed, 39.6 km; 433 Eros, 19.6 km; 1866 Sisyphus, 11.4 km). Based on meteoritic composition statistics (Mason 1962), an asteroid of 10 km is likely to be an aerolite (stone), with a density of about 3.0 g/cc, and have a velocity of about 20 km s^{-1} (Shoemaker 1983). Its kinetic energy would be almost equivalent to 10^9 million tons (Mt) of TNT. A comet of the same diameter would have a lower density but could have more than twice the velocity (short period 28.9 km s^{-1}, long period 56.6 km s^{-1}; Weissman 1982). The cumulative impact probability is greater for the SP comets because of their short periodicity and the low inclination of their orbits. Any such impact which converted energy of so much magnitude instantaneously would be cataclysmic (McKay and Thomas 1982; Lewis *et al.* 1982).

But is it necessary to postulate such an unlikely high energy impact to develop a scenario which imposes catastrophe on humanity? Gerstl and Zardecki (1982) calculate that between 1 and 4 x 10^{16} g of aerosols distributed globally in the stratosphere (about 10 ppm or 10^{-2} g cm^{-2}) would reduce photosynthetically active radiation (PAR) at ground level to 10^{-3} normal, thus initiating widespread extinctions. This aerosol mass could be ejected into the stratosphere by the impact of an asteroid between 0.4 and 3 km in diameter, events which have impact probability of about 10^{-5} and 10^{-7} y^{-1}, and energies of 10^5 and 10^7 Mt, respectively. Nevertheless, *widespread* extinction is still cataclysmic. Reduction of PAR to only 10^{-1} for 1 or 2 years would be catastrophic for a world population which now lives from agricultural production on a year-to-year basis. For example, total world reserve grain stocks had fallen to 12.7% of annual consumption by September 1983 (USDA 1983). A stratospheric dust loading of 2-3 x 10^{15} g would reduce PAR to about 10^{-1} (Gerstl and Zardecki 1982), a light intensity which would be like twilight. This requires scaling down the asteroid diameter slightly, which reduces its energy to 10^4 Mt. Its survival on atmospheric entry now becomes less likely. A switch to a siderite (iron) would reduce the probability of impact because of their scarcity relative to aerolites (Mason 1962). However, we are now at the energy range of volcanic events.

Simpkin *et al.* (1981) catalogue 125 volcanic eruptions in the last 480 years which had a volcanic explosivity index (VEI) of IV or greater. The previous 1500 years produced only 28 of these. It is perhaps not fair to compare the two time intervals given the extent of civilization during the first 1500 years A.D.; but it can be argued that most of the major (Class IV) events have probably been reported or found *post facto*, although many may have been underclassified (Newhall and Self 1982).

The explosive power of Krakatoa (1883) has been variously estimated at between 33 and 100 Mt, with conservative estimates of the mass of aerosols injected into the stratosphere of about 10^{14} g (Gerstl and Zardecki 1982). Krakatoa had a VEI of VI. Using Simpkin *et al.* (1982) scaling, a VEI of V would place 10^{13} g, and a VEI of IV 10^{12} g, into the stratosphere. St. Helens (VEI V) placed an estimated 1.3×10^{13} g or aerosol mass in partial confirmation of the scaling (Gerstl and Zardecki 1982). Tambora (1815) with a VEI of VII is estimated to have emplaced nearly 10^{15} g (10^3–10^4 Mt) into the stratosphere, and Toba (75,000 BP, Indonesia) to have emplaced about 10^{16} g (using 1.6×10^{-3} as the fraction of total mass erupted reaching stratosphere). Probabilities of volcanic eruptions are difficult to assess because of the uneven periodicity of centuries of high and low volcanism. The probability of a VEI of VII or greater is about 1.3×10^{-5} y^{-1} (Toba-Tambora). There have been 3 VEI VI, 16 VEI V, and 106 VEI IV events in the last 500 years giving the following approximate probabilities:

VEI VI	6.0×10^{-3} y^{-1}
VEI V	3.0×10^{-2} y^{-1}
VEI IV	2.0×10^{-1} y^{-1}

One is tempted to assign a probability of 10^{-4} to a VEI VII event. On the other hand, if the 1,500 years preceding the 16th century were included, the probabilities of the IV, V, and VI events would decline by a factor of about 10.

Lamb (1970) summarizes the effects of Tambora, which resulted in total darkness for 3 days within a radius of 500 km, a 1°C worldwide drop in temperature, and New England snows in June (Ludlum 1966). However, Tambora was imbedded in 13 years of intensive volcanic events commencing with Soufriere (1812), VEI IV, and punctuated with 5 VEI IV, 2 VEI V, and 1 VEI VII events, ending with Isonotski (1825), VEI IV. This is an average of about 100 Mt y^{-1} of eruptive power per year. Krakatoa, which caused only 4 to 5 hours of darkness 160 km away and a world temperature drop of 0.5°C, was imbedded in 3 decades of large volcanism commencing with Grimsvotn (1872), VEI IV, and containing 11 VEI IV, 1 VEI V, and 3 VEI VI events, ending with Santa Maria (1902), (VEI VI), an average of 30 Mt y^{-1}. Moreover, record cold winters followed in the United States (MacDonald 1972), including ice flows in the Mississippi at New Orleans (1899) for only the second time in its history (Ludlum 1966).

It seems well established that explosive volcanism has an adverse effect on climate, based on geological evidence (Kennett and Thunell 1975; Bryson and Goodman 1980; Hammer *et al.* 1980) historical volcanism (Lamb 1970; Bray 1978), empirical observation (Hansen *et al.* 1978; Mendonca *et al.* 1978), and theoretical modeling (Baldwin *et al.* 1976). What is not clear is what is the minimum volcanic activity which will severely impact world food production? More than just the VEI is now involved:

(1) The geographic position of the volcanic event: Most of the world's terrestrial volcanoes are located between 10°S and 70°N latitude. The tropical volcanoes and those north of 50° are much more capable of affecting climate because of a lowered tropopause at high latitudes and strong stratospheric circulation at low latitudes (Bryson and Goodman 1981).

(2) The time of year: The half-life of volcanic aerosols in the atmosphere is about one year. The first six months would therefore be the most critical in terms of drop in temperature and reduction in PAR. In the northern hemisphere a late winter or early spring volcanic event would have the greatest potential effect on agriculture.

(3) The existing aerosol loadings: Hofman and Rosen (1980) have pointed to greatly increased H_2SO_4 aerosols in the stratosphere and suggest that anthropogenic sources are increasing these concentrations by as much as 9% per year. Both previous volcanic and anthropogenic stratospheric aerosols would be background for any new volcanic event, whose effects would be cumulative. However, anthropogenic aerosols in the troposphere may actually cause warming (Idso and Brazel 1978). Volcanism of a frequency of 1 to 2 years will substantially increase existing stratospheric aerosol loadings. For example, the period between 1586 and 1600 experienced 6 VEI IV events, with an average spacing of 2.3 years; 1660–1673 4 VEI IV and 2 VEI V events with an average spacing of 2.5 years (Simpkin *et al.* 1981). Bray (1978) describes synchronous low temperatures and northern hemisphere glacial advances during these periods.

(4) The nature of volcanic explosive products: while fine ash in the stratosphere interdicts direct sunlight, the effect is partially compensated for by an increase in scattered radiation. On the other hand, SO_2 converted to H_2SO_4 aerosol absorbs insolation and heats the stratosphere, with attendant cooling of the troposphere. Simpkin *et al.* (1981) report 22 volcanic events in 1783 of which 3 had a VEI of IV. One of these, Lakagigak (Iceland), placed an estimated 10^7 tons of SO_2 in the atmosphere resulting in the loss of crops, livestock, and 24% of the population in Iceland in the following years (Sigurdsson 1983). The winter of 1783–1784 was exceedingly harsh in the United States, with ice flows in the Mississippi at New Orleans over 2 feet thick (Ludlum 1966). The importance of the SO_2 aerosols to the Little Ice Age has recently been underscored by Strothers and Rampino (1983).

Following the Katmai eruption (1912), the northern hemisphere experienced little volcanism until 1945. The temperature rose about 0.4°C. During this period hybrid maize became the most important grain crop in the United States; worldwide it equals wheat and rice as a cereal crop. Much of the corn is used as animal feed. Except for the last 80 years, food production since the 16th century has been dominated by C3 (Calvin Cycle) plants, such as tubers, legumes, wheat, and barley, which have some tolerance to cold but whose photosynthetic productivity drops rapidly only below 50% full sunlight intensity. Maize (and sorghum, millets) are C4 photosynthesizers and are particularly sensitive to chilling and low light levels.

In 1750, at the height of the Little Ice Age, the world had about 800,000,000 people; in 1850 about 1.3 billion. In the year 2000, the world population will be about 6.3 billion. Brown (1981) and Jensen (1978) have pointed to the increasing insecurity of world food supply and the narrowing margin between food production and increased population. In the 21st century it may be necessary to switch animal feed (grain and fish meal) to people. As this occurs, we will be particularly vulnerable to volcanic climatic influences which may be only partially offset by the rise in CO_2.

References

Baldwin, B. J.B. Pollack, A. Summers, O.B. Toon, C. Sagan, and W. Van Camp. 1976. "Stratigraphic aerosols and climatic change." *Nature* 263: 551–555.

Bray, J.R. 1978. "Volcanic eruptions and climate during the past 500 years." In *Change and variability*, Pittock, A.B., L.A. Frakes, D. Jenssen, J.A. Peterson, and J.W. Zielman, eds., Cambridge. Cambridge University Press: 256–262.

Brown, L.R. 1981. "World population growth, soil erosion, and food security." *Science* 214: 995–1002.

Bryson, R.A. and B.M. Goodman. 1980. "Volcanic activity and climatic changes." *Science* 207: 1041–1043.

Bryson, R.A. and B.M. Goodman. 1981. "The climate effect of explosive volcanic activity: analysis of the historical data." Symposium and Workshop in "Mount St. Helens Eruption: Its Atmospheric Effects and Potential Climatic Impact." Washington, D.C., manuscript, 9 pages.

Gerstl, S.A.W. and A. Zardecki. 1982. "Reduction of photosynthetically active radiation under extreme stratospheric aerosol loads." In *Geological Implications of Impacts of Large Asteroids and Comets on the Earth*, Silver, L.T. and P.H. Schultz, eds., GSA Special Paper 190, p. 201–210.

Grieve, R.A. 1982. "The record of impact on earth: Implication for a major Cretaceous/Tertiary impact event." *GSA* 190, p. 25.

Hammer, C.V., H.B. Clausen and W. Dangaard. 1980. "Greenland ice sheet evidence of post-glacial volcanism and its climatic impact." *Nature* 288: 230–235.

Hansen, J.E., W-C Wang and A.A. Lacis. 1978. "Mt. Agung eruption provides test of a global climatic problem." *Science* 199: 1065–1067.

Hofman, D.J. and J.M. Rosen. 1980. "Stratospheric sulfuric acid layer: Evidence for an anthropogenic component." *Science* 208: 1368–1370.

Idso, S.B. and A.J. Brazel. 1978. "Climatological effects of atmospheric particulate pollution" *Nature* 274: 781–782.

Jensen, N.F. 1978. "Limits to growth in world food production." *Science* 201: 317–324.

Kennett, J.P. and R.C. Thunell. 1975. "Global increase in quaternary explosive volcanism." *Science* 187: 497–504.

Lamb, H.H. 1970. "Volcanic dust in the atmosphere: With a chronology and assessment of meteorologic significance." *Philosophical Transactions of the Royal Society of London* 266: 425–533.

Lewis, J.S., G.H. Watkins, H. Hartman and R.G. Prinn. 1982. "Chemical consequences of major impact events on earth." *GSA* 190, p. 215–222.

Ludlum, D.M. 1966. "Early American winters 1604–1820." Boston: American Meteorological Society, 283 pages.

Macdonald, G.A. 1972. *Volcanoes*. Englewood Cliffs, NJ: Prentice-Hall, Inc., 510 pages.

McKay, C.P. and G.E. Thomas. 1982. "Formation of noctilucent clouds by extraterrestrial impact." *GSA* 190, p. 211–214.

Mason, B. 1962. *Meteorites*. New York: Wiley, 165 pages.

Mendonca, B.G., K.J. Hanson, and J.J. Deluisi. 1978. "Volcanically related secular trends in atmospheric transmission at Mauna Loa Observatory, Hawaii." *Science* 202: 513–515.

Newhall, C.G. and S. Self. 1982. "The volcanic explosivity index (VEI): An estimate of explosive magnitude for historial volcanism." *Jour. Geophysical Research* 87: 1231–1238.

Shoemaker, E.M., J.G. Williams, E.F. Helen, and R.F. Wolfe. 1979. "Earth crossing asteroids: Orbital classes, collision rates with earth and origin." In *Asteroids*, Gehrels, T., ed., Tucson: University of Arizona Press.

Shoemaker, E.M. and R.F. Wolfe. 1982. "Cratering time scales for the Galilean satellites." In *The Satellites of Jupiter*, D. Morrison, ed., Tucson: University of Arizona Press, p. 277–339.

Shoemaker, E.M. 1983. "Asteroid and comet bombardment of the earth." *Ann. Rev. Earth and Planet. Sci.* 11: 461–494.

Sigurdsson, H. 1983. "Volcanic pollution and climate: The 1783 Laki eruption." *EOS* 63: 601.

Simpkin, T., L. Seibert, L. McCelland, D. Bridge, C. Newall, J.H. Latter. 1981. *Volcanoes of the World*. Stroudsburg, PA: Hutchinson Ross Publishing Co., 232 pages.

Strothers, R.B. and M.R. Rampino. 1983. "Historic volcanism, European dry fogs, and Greenland acid precipitation, 1500 B.C. to A.D. 1500." *Science* 222: 411–413.

U.S.D.A. 1983. "World agricultural supply and demand estimates." United States Department of Agriculture, WASDE-155, Washington, D.C., 26 pages.

Weissman, P.R. 1982. "Terrestrial impact rates for long and short period comets." *GSA* 190, p. 15–24.

Wetherill, G.W. and E.M. Shoemaker. 1982. "Collision of astronomically observable bodies with the earth." *GSA* 190, p. 1–13.

TWENTY FOUR

GAIA: A NEW LOOK at GLOBAL ECOLOGY and EVOLUTION

Penelope J. Boston

The past history of life on Earth and the future fate of living things is a subject of intense interest both to science and to people on a personal level. After all, we too are living organisms and subject to the conditions of our planet. The scientific picture of the origin and evolution of life which has emerged since Darwin has included primarily the idea that the physical environment is a given to which organisms must adapt more or less successfully or become extinct. This would work quite well on a planet whose physical conditions remain constant. However, the concepts which have developed over the same period of time in astronomy, geology, and atmospheric science have painted a much more dynamic picture of the long history of our planet's development. These ideas have gradually filtered down to the level of biology and paleontology and are having impact upon the classical interpretations of ecology and evolution.

Perhaps the most comprehensive, radical and controversial idea in this general area put forward since the concept of evolution itself is the Gaia Hypothesis. It was conceived by J.E. Lovelock, a chemist and inventor, who came upon his ideas while working on life detection experiments for the Viking missions to Mars. The essence of Lovelock's theory is that life on Earth exerts controlling influences on the conditions which occur in the non-living parts of the planet. Life optimizes conditions for its own perpetuation. In fact, this control and the processes of adaptation and evolution are so closely coupled that essentially one process is at work. This postulated mega process indivisibly includes the biological, chemical, and physical events which occur on Earth and has resulted in the creation and evolution of a single entity, the entire planet, known as Gaia.

The idea of Gaia, the earth as a living thing, has a seductive appeal. It is poetic. It is compatible with much of the cultural heritage that Western civilization has acquired from the classical Greeks. Furthermore, it has maternal and nurturing overtones which appeal to both the romantically inclined and the environmentally desperate among us. It is comforting to think of Earth as "the Great Mother" who, by implication, will protect and nourish her children, namely ourselves. Disquiet creeps in when one realizes that if Gaia is the Mother, her first-born are the bacteria, not us. Her huge brood includes countless species of organisms whose conflicting needs cannot be reconciled. This irreconcilability has led to the natural extinction of the vast majority of all species that have ever lived. If this is motherhood, it begins to lose much of its comfort value for us.

While the romantic and poetic aspects of Gaia have attracted many people, they have been an equal detriment to the scientific consideration of the ideas of Gaia. Possibly because of faulty education, many scientists are instantly suspicious of anything which is poetic and react in an automatically and unthinkingly negative way. Much of the debate about Gaia has had more the flavor of a confrontation of opposing political viewpoints than of sound scientific criticism. In my judgment, the jury should be out on Gaia for a long time to come. Allying oneself in either the pro or con camp is premature. Strong opinions at this early stage of the development of Gaian ideas are probably more a reflection of one's basic personality structure than a result of deliberate and considered views on the actual epistemological and scientific merits of Gaia theory.

The context in which Gaia is debated can range from a discussion of its use as an ethical code, a guide for human conduct apropos stewardship of the Earth, to a consideration of its merits as a strict set of scientific mechanisms capable of explaining some of the complex phenomena which occur on Earth. When arguing the case for or against Gaia, it *must* be extremely clear at which of these levels the discussion is taking place or no progress can be made. The goal of a scientific treatment is to explain the workings of the planet, its development over time and its possible development in the future. What the Earth actually does is the target of interest. The goal of an ethical treatment of Gaia concerns how we can keep living on the planet without irretrievably destroying features and conditions which are beneficial and important to us. How we *should* behave with respect to the earth is the realm of interest there.

The following discussion provides a primer to the concept of Gaia and some of the strengths and pitfalls of considering it scientifically.

What are the Properties of Gaia?

Statement of the Gaia hypothesis: The living and non-living components of Earth operate together as a single entity. Life acts to control the effects of both external impacts on the planet and internal changes to keep the range of physical conditions within the acceptable window for the continuation of life. Because life exerts this control, the planet itself can be viewed in a certain sense as a living organism.

Properties of Complex Systems

Many of Gaia's putative properties may not be properties of life alone, but of complex multi-component systems whose components possess some of the features we associate with life. Such systems may have properties which are peculiar to the overall system and some properties which are the same as those possessed by the component parts. I invoke here the analogy of the multi-cellular system who goes to work, drives a car, reads Shakespeare and debates evolutionary theory. None of the component parts of this system, from cells to organs, are capable of these activities. Only the unitary system can perform and experience them.

The principles which operate between individual organisms and species in the competition for resources (selection, in classical evolution) need not dictate the properties of larger complex systems of which these individuals and species are merely components. Ordinary evolutionary theory is probably completely adequate to explain these species-level interactions. The principle of emergent properties, that is, that the sum of the parts is significantly less than the whole, could be easily imagined to operate at the Gaian level. It could be argued that Lovelock's theory, if valid, is to evolution what Einsteinian physics is to Newtonian physics, that is, an adjunct and expansion which analyzes phenomena at a different level. Relativity does not invalidate Newtonian physics within the framework of its operation, it merely transcends it.

Prediction of emergent principles are not to be found in any study of the individual disciplines of natural science. They will perhaps emerge from a study of the mathematics of systems, a branch of science known as cybernetics. We do not currently have all of the necessary tools to understand large complex systems. Some progress is being made in this area by ecological modelers (e.g. May 1974). For example, Tregonning and Roberts (1979) have demonstrated multi-component mathematical systems which evolve towards homeostasis without possessing any specific properties unique to life or ecosystems.

The question of what imparts relative stability or homeostasis to a complex system like the global biosphere is arguable on many levels. One view is that increasing complexity contributes to increasing stability (Odum 1968). A possible explanation for this involves a dilution effect. In a simple ecosystem with few components (i.e., numbers of species) and few connections between them, any change in one of the species or the addition or deletion of species can be quickly felt throughout the system. Each species is a relatively large percentage of the whole. In a system with numerous components and connections, changes in some of them make much less difference because there is so much of the system that will remain unaffected. An alternate view which grows out of the mathematical consideration of ecosystems seems to indicate that large numbers of species and connections are *more* vulnerable to impact and less stable than systems with relatively few components (May 1974). However, this result may be due to the artificial sparsity of interactions between components in these models.

Negative Feedback Loops

One of the key properties of Gaia identified by Lovelock is the negative feedback loop. This concept, borrowed from cybernetics, involves several steps. Some initial event or process in a system causes a response in some other part of the system. This response acts as a modulator which operates on the original event or process to reduce its intensity. The result is the control of the overall process within relatively narrow limits. Of course, to qualify as Gaian, such a loop must involve the biota as the modulating element.[1]

Negative feedback loops which rely solely on non-biological mechanisms may also serve to stabilize various of Earth's processes. For example, Walker *et al.* (1981) hypothesized a mechanism of global temperature control involving negative feedback between levels of carbon dioxide in the atmosphere and the resulting temperature effects on the weathering of silicate minerals. This qualifies as a true negative feedback loop but it is not an example of the planet's biology controlling the environment.

The demonstration of non-biological negative feedback cycles do not themselves constitute negative evidence for Gaia. Not all the features of living organisms are themselves living although they may serve life's needs. Physics applies to biology, too. For example, tiny particles of magnetic minerals found in certain marine bacteria respond according to the magnetic fields lines around Earth. Yet, the bacteria employ this basic physical behavior of their magnetite particles to orient themselves in the water column. No one insists that the magnetic particles themselves must have living properties.

Life-like Properties of Gaia

Scientifically, we have a very flimsy grasp of what constitutes "life." We have a deep intuitive grasp, however, and we must not be too quick to dismiss the idea of perceiving Gaia as an organism long before we may be able to demonstrate its entityhood in any scientifically acceptable terms. If Gaia exists, and can be said to be a cohesive entity, this fact will only become clear once we understand the cybernetic principles which allow individual organisms to be called single entities.

What exactly is life? Probably hundreds of definitions have been formulated in attempts to uniquely describe it. I offer here one partial definition to suit my present purpose. Life is the persistence of a collection of physical components in a specific series or oscillating series of patterns. Material and energy flow through the system, becoming parts of this persisting pattern and then flow out again to the environment while new matter and energy replaces them. Overall the system possesses lower entropy than the surroundings and resides in a state far from thermodynamic equilibrium. Life may also entail the ability to create new entities which possess this persistence of pattern. If one accepts this definition, the emphasis then shifts from simple enumeration of the properties of organisms to perpetuation of patterns and maintenance of a non-equilibrium state. We know that the chemistry and temperature of the earth are far from equilibrium. The question remains whether the earth has the persistence of pattern that organisms possess.

The analogy of Gaia-as-organism can be carried to extremes. One could imagine the physical structure of Earth as the skeleton of Gaia, the oceans and fresh water systems of the Earth as blood and the gaseous component of the planet as Gaia's breath. Does this crude analogy serve any purpose? Possibly it can serve as a heuristic tool as we try to map out the specific functions which a Gaian system must perform. For an organism or a Gaian system to operate it must have state memory of some sort and mechanisms for information transfer. It must be able to move material and energy from points of uptake to areas within the system requiring them. It must maintain the integrity of the boundary conditions which set it apart from the environment. It must be able to control the degree of internal entropy. It must be able to sense and control chemical and physical properties of its internal environment. The control of many of these functions comes about through negative feedback loops.

Development of Gaia

If Gaia can be considered an "organism" then it may be supposed to have an infancy, an adolescence, a maturity, and a senescence and death. This sequence of the development of an organism may not be so unique to life as we usually assume. Stars can be said to proceed through stages of this sort. It may be that these phases arise from the existence of massive, long-lived complex systems whether we understand them to be living or not.

There are three possible alternatives for the development of life on the earth:

1. Life was and is non-Gaian in perpetuity.
2. Life was initially non-Gaian but became Gaian at some point either gradually or abruptly.
3. Life was Gaian right from the beginning and will continue to be so until it ceases with the stellar death of the planet.

When could life on Earth be said to have achieved homeostasis or Gaian status? When the ocean was filled with microbes? When stromatolites[2] covered the continental shelves? When land plants colonized the continents? I perceive these as meaningless questions. Better to ask, "At what point in the development of this system do we find ourselves? In its infancy or its maturity?" Clearly the early development of a Gaian system may be very different from its mature state. Many organisms have larval stages whose lifestyle would be anathema to the adult mode of life.

Proponents of Gaia point to the fact that we currently see the earth in a mature, stable, successful state, but this situation is inevitable and does not constitute proof that Gaian mechanisms are at work. We look through an unimaginably long stretch of time composed of the result of innumerable successful changes. In addition to the highly selective view we have looking backwards in time, we ourselves are very short-lived entities and not well-prepared to perceive long-term trends in the system. Over the course of evolution, if an organism needs *x* and *x* is scarce, then if the organism can obtain or manufacture *x* or adapt to get along without it, it survives. If it cannot, it perishes. One is only left with the successful changes.

Many eons of sifting out unworkable processes and entities has refined the biology of the planet to a remarkable extent.

The mature biosphere is apparently so robust that in an interesting study by Raup (1982) of the mass extinction possibilities of major killing events, he found that lethal areas had to exceed half the Earth's surface before extinctions reached the levels typical for the *known* major extinctions of the geological record. If this is true, it means that you have to kick Gaia very hard to make a big impression.

What lies in the future for Gaia? One can only speculate that if Gaia progresses through stages like organisms do, senescence and eventual death will result. It may be that on any planet with life and the natural span of the global biology is fixed by internal as well as external factors. We know that our sun will eventually reach its own old age and go through tremendous changes which will result in the destruction of its planets. What we do not know is whether a Gaian system would age according to some clock of its own and perhaps reach termination long in advance of the actual demise of the planet by solar activity.

How can Gaia be Detected

One of the biggest operational difficulties with Gaian theory is agreeing on a set of distinctions which separate a Gaian from a non-Gaian biosphere. Developing a set of criteria for this is essential if we are to correctly interpret the results of experiments designed to look for Gaian mechanisms.

As discussed above, negative feedback loops emerge as the key feature of a Gaian system. Such loops appear to be the only clear distinction identified so far that unequivocally demonstrates some measure of control of life over the physical environment. To complete such a loop there must be four primary ingredients:

1) External event or internal input. ← 4) Reduction of impact of external event or input.

↓ ↑

2) Response of some portion of the biota. → 3B) Product or physical effect.

↓

3A) Classical adaption of the biota.

In practice, following this loop to closure is extremely difficult. For many phenomena, showing the first three steps is relatively simple, at least in principle. The difficult part is demonstrating that the product or physical effect then reduces the effects of the initial event or input. That is, that the loop goes from 2 to 3B to 4 rather than going from 2 to 3A, ending in the simple adaptation of organisms to their environment without any feedback involved.

If one could conclusively demonstrate that for a specific process, step 4 does *not* occur, would this be strong evidence against the existence of Gaia? Unfortunately, because of the logical nature of scientific proof, it is much easier to demonstrate the presence of Gaian processes than it is to demonstrate that the system is non-Gaian. Those attempting to disprove Gaia are facing a much tougher job than those trying to find evidence supporting it. Even just a few examples of closed, biological negative feedback loops would constitute strong evidence of Gaian operations on the planet. In contrast, the absence of negative feedback or its non-biological nature is much more difficult to establish and many more examples will be necessary to strengthen the non-Gaia case. There are many plausible explanations which can be brought to bear to explain away the absence of negative feedback for any given process. To wit, the overall Gaian system may not really *need* to control that particular function to maintain overall stability. Or it may not have responded yet. We have no real clues to the critical time periods during which Gaia would have to make a response to internal and external changes to maintain overall homeostasis. An even more slippery problem—if Gaian systems have a natural set of stages through which they pass, we could never be sure that changes in the system state were not part of one of those major shifts. Would microbe scientists living during the change from anoxic to oxygen-rich atmosphere during the early Proterozoic time interval[3] have been able to predict that things would never be the same again?

An example of a system currently under investigation as a possible Gaian negative feedback mechanism involves the emission of certain sulfur compounds from the ocean (Charlson *et al.* 1987; Lovelock 1982). Some species of marine algae emit a sulfur compound known as dimethyl sulfide (DMS).

They appear to do so in response to temperature, salinity, and possibly other environmental factors. When DMS is in the air it is rapidly converted to SO_2 and then to sulfate aerosols. These aerosols are very hygroscopic and make excellent nuclei for condensing cloud droplets. Unlike the land where there is much dust in the air, the oceanic air is otherwise poor in cloud condensation nuclei. There is, however, abundant water vapor, so clouds readily form in the presence of nuclei. The larger the number of particles per unit mass of water (the case with large numbers of condensation nuclei), the brighter the resulting clouds. These high albedo clouds reflect sunlight away thus cooling the ocean surface. The final critical step is the depressing effect that cool temperature has on DMS emission by the algae. Whether the algae produce less DMS or whether the species composition changes favoring naturally low emitters is unclear but not central to the functionality of the system. As the DMS emission is reduced, the cloud formation is suppressed allowing temperatures to increase again and, hence, DMS levels to rise once more. If each of the hypothesized links in this chain is eventually substantiated, then the cycle will fit all of the criteria to make it a truly Gaian negative feedback loop.

Trivial Criticism of Gaia

While there are legitimate criticisms that can be leveled at the Gaia Hypothesis, there are a few points of criticism which are trivial and dismissable. These have appeared several times in print and should be laid to rest.

One mistaken interpretation of Gaia concerns constancy. This has been seen by some as necessary for the demonstration of Gaian mechanisms. The notion of "constancy" is treacherous. The chemical composition of the atmosphere and the climate have *not* been constant through time; far from it. Nor has life been constant in form or function throughout its history. To truly fit Gaian criteria, Earth's condition must merely be far from equilibrium and able to support life in some form. Life makes pollutants. Other life learns to use, detoxify, or live with such pollutants.[4] Life may go on but the players constantly change. Constancy is a comforting anthropomorphic notion implying that nothing bad will happen to us but it is inconsistent with what we know about our planet and particularly what we know about the history of life.

Another rather naive criticism of Gaia points to the relatively small mass of the biosphere compared to the non-living components of earth and concludes that life cannot be important in many critical processes because there is not enough of it. Arguments about the relative mass contributions of life versus non-living processes on Earth are irrelevant. First, total flow through the system including turnover time is far more important than standing mass. Secondly, things often have importance out of proportion to their mass. In the bodies of plants and animals, extremely tiny quantities of hormones wreak massive changes on the whole system. In the atmosphere, the presence of trace gases which absorb solar energy can very efficiently exert tremendous effects on the thermal structure of the atmosphere and hence the climate of the Earth. Many of these compounds are present in the parts per billion or trillion range of concentration. In fact, substances which could act as global hormones should be actively sought as valuable clues to Gaian control processes.

Exorcising the Anthropomorphism

One of the least productive conclusions some have drawn from Gaia is that it could only work if the biosphere is somehow consciously guiding its own development. If we are to rationally consider the scientific merits and faults of the Gaian idea, we must overcome this dreadful idea of purpose which has crept into interpretations of what Gaia means. Images of species and organisms somehow sacrificing themselves for the overall good of the community or of life in general are ludicrous. The idea of purpose in this context is a preposterous and entirely unnecessary byproduct of sloppy thinking. It is not unlike the argument from design used in some theological attempts to prove the existence of God.[5] If one is walking down a beach and discovers a watch, one assumes that there must be a watchmaker. This kind of reasoning is no more applicable to Gaia than it was successful in proving the necessity of a deity to explain the existence of the universe. Control does not imply purpose. After all, no one seriously suggests that the highly controlled cells in a multi-cellular organism have in some way "conspired" for the common good. Simply, the myriad individual events of mutation, selection, and happenstance have resulted in a successful overall product. So also it may be with Earth.

We look at the earth and see a planet with a highly successful biology. In contrast, Mars and Venus appear lifeless. They may or may not have ever developed life, or it may have originated and failed. From our vantage point we cannot tell the difference. These are currently unanswerable questions. If purpose played a role in the development of life, why did life on the other terrestrial planets falter or fail to begin at all? Rather than ask why the earth has remained relatively stable and capable of supporting life, better to ask why *shouldn't* conditions remain stable? Remember, we reached this point because circumstances pruned out the unsuccessful or unstable processes and entities. The only answerable questions are *how* such a system has stabilized itself, what are the mechanisms? Only these questions are scientifically valid and amenable to research.

State inertia or homeostasis is probably a common property of large complex systems for several reasons having nothing to do with purpose or design. For one thing, large systems simply possess a lot of mass. Secondly, such systems have a huge amount of "baggage" in the form of connections between the multiplicity of their parts. Thirdly, once a system happens upon stable negative feedback, such feedback would tend to perpetuate itself unless disturbed beyond some outer limit. The potentially answerable questions revolve around how *much* forcing various parts of the system can tolerate before reaching a new condition of stability.

If Gaia is Real Will She Save Us From Ourselves?

There is no doubt in my mind that life will persist on this planet until some overwhelming event creates totally unsurvivable conditions. Either some factor like the gradually increasing solar constant will overwhelm the buffering in the earth's system (Gaian or otherwise) and result in a runaway greenhouse situation, or the Sun will eventually perform the devastating expansions of its old age, destroying at least the volatile parts of the inner planets. In the absence of these kinds of catastrophic events the persistence of life may be considered, in some sense, Gaian. This life, however, may be restricted once more to the microbial level of organization, depending upon the future course of the planet's history. The question that we *really* want to address now is: How far can natural and anthropogenic changes push the status quo before a new and possibly deleterious or disastrous (to us) state is reached? There are several interpretations of whether Gaia fits in positively or negatively with this anthropocentric question. In one view, the existence of Gaia would be helpful because it would tend to remain at the present overall condition in spite of environmental insults which we perpetrate or natural oscillations of climate or other conditions. In the second view, Gaia's existence would render us more vulnerable because she is a delicate living organism. If we stress her too far we may "kill" her, that is, destroy her ability to repair those same environmental insults and natural changes or, just as disastrous for us, push her to some new stable state or developmental plateau. Personally, I would not presume to guess at this point which interpretation is more plausible.

A creature can live on in a wide variety of states. It can be in the peak of health or in a state of malnourishment and underlying disease and live on nonetheless. It is hard to specify what actually constitutes Gaian health. However, taking a few clues from the trends apparently exhibited by the biosphere over time, one could point to several possible symptoms. Over time, species diversity seems to have increased (Simpson, 1969). As new areas have been colonized, e.g., the emergence of plants onto land, global biomass has increased. Perhaps trends like these could serve as indicators of Gaia's natural tendencies. Any process or human activity which seriously reduces species diversity or biomass could be looked upon as potentially harmful. As Lovelock points out in the conclusion of his first book on Gaia (1979), some of the earth's places may be more vulnerable than others. He cites wetlands, marine coasts, and tropical forests as examples because of their high biological productivity and species richness. As most of us know, we are already in trouble on all these fronts both in the developed and Third World nations of the Earth.

I think it is foolish to imagine that we can create an ever-increasing flow of materials and toxins and sustain an ever-burgeoning human population without consequences. The existence of Gaia would not *ultimately* be able to triumph in maintaining the status quo against our ravages. What she *may* be able to do for us is to stimulate development of a vastly better understanding of the way life and the planet works. As we rush around doing research to try to prove or disprove Gaia's existence we will inevitably advance our knowledge in spite of ourselves. Whether Gaia is a real, live organism or a lovely, elusive vision is almost unimportant. In either case, she may help *us* to save ourselves from ourselves. For this contribution, not for the ultimate truth or falsity of the Gaia hypothesis, does science owe Jim Lovelock an important place in its history.

Notes

1. Lovelock and collaborators have developed a simple model of their own which explores the process of negative feedback, one of the key properties of Gaian homeostasis. They call it Daisyworld (Lovelock, 1983 and 1986; Watson and Lovelock, 1983). The living organisms on this mythical planet are black daisies and white daisies. Their optimum growth temperature is the same. However, because the black daisies absorb more solar radiation than the white daisies, they will always be warmer for a given intensity of sunlight. Because they have radically different albedos, the two kinds of daisies change the planet's albedo as they grow differentially in response to temperature. This convenient arrangement allows direct feedback to the temperature control of the planet modulated simply by biological competition between the two differently colored daisies.
2. Stromatolites are fossils of complex microbial communities which formed often very large, layered structures. They appear to have once been the dominant form of life on Earth. There are many modern marine and freshwater stromatolites still living today.
3. The geological period from 2.6 to 1.6 billion years ago. During this span of time micro-organisms arose which released free oxygen as a byproduct of photosynthesis and gradually changed the atmosphere from an anoxic state to one containing a large amount of free oxygen. The modern atmosphere is 20% oxygen.
4. The most notable example of this is the rise of oxygen, a highly reactive gas, which organisms had to combat by new means of metabolic protection or by retreating to relict niches which remain oxygen-free today.
5. The argument from design is one of the five proofs of the existence of God formulated by St. Thomas Aquinas (Pegis 1945). The essence

of the argument is the necessity to imagine that a highly complex mechanism, e.g., a watch, a tree, or the universe must have a conscious designer who fashioned it. The only reasonable candidate for this designer is God, hence there must be a God. A complete discussion is beyond the scope of this chapter; however, this line of reasoning has been rather effectively countered by many scholars, most notably since Hume (Taylor 1963).

References

Charlson, R.J., J.E. Lovelock, M.O. Andrae and S.G. Warren. 1987. "Oceanic phytoplankton, atmospheric sulphur, cloud albedo and climate." *Nature* 326: 655–661.

Lovelock, J.E. 1979. *Gaia: A New Look At Life On Earth*. Oxford: Oxford University Press.

Lovelock J.E. 1982. "The production and fate of reduced volatile species from oxic environments." In E.D. Goldberg, ed., *Atmospheric Chemistry*. Berlin: Kahlem Konferenzen, Springer-Verlag.

Lovelock, J.E. 1983. "Daisy World: A cybernetic proof of the Gaia hypothesis." *Coevolution Qtrly*. 38: 66–72.

Lovelock, J.E. 1986. "Gaia: The world as living organism." *New Scientist*, December 18, 1986: 25–28.

May, R.M. 1974. "Afterthoughts." In R.M. May, ed., *Stability and Complexity in Model Ecosystems*. 2nd edition. Princeton, New Jersey: Princeton University Press.

Pegis, A.C. 1945. *Basic Writings of St. Thomas Aquinas*. New York: Random House Inc.

Raup, D.M. 1982. "Biogeographic extinction: A feasibility test." In L.T. Silver and P.H. Schultz, eds. *Geological Implications of Impacts of Large Asteroids and Comets on the Earth*. Geological Society of America Special Paper 190: 277–281.

Simpson, G.G. 1969. "The first three billion years of community evolution." *Brookhaven Symp. Biol.* 22: 162–177.

Taylor, R. 1963. "A contemporary form of the design argument." In *Metaphysics*. Englewood Cliffs, New Jersey: Prentice-Hall.

Tregonning, K. and A. Roberts. 1979. "Complex systems which evolve towards homeostasis." *Nature* 281: 563–564.

Walker, J.C.G., P.B. Hays and J.F. Kasting. 1981. "A negative feedback mechanism for the long-term stabilization of Earth's surface temperature." *J. Geophys. Res.* 86: 9776–9782.

Walker, J.C.G., C. Klein, M. Schidlowski, J.W. Schopf, D.J. Stevenson and M.R. Walter. 1983. "Environmental evolution of the Archaen-Early Proterozoic Earth." In J.W. Schopf, ed., *Earth's Earliest Biosphere: Its Origin and Evolution*. Princeton, New Jersey: Princeton University Press.

Watson, A.J. and J.E. Lovelock. 1983. "Biological homeostasis of the global environment: the parable of Daisyworld." *Tellus* 35B: 284–289.

POSTSCRIPT

S. Fred Singer

Hypotheses about global climate change are constantly coming to the fore to be tested by new and more detailed observations and by theoretical studies. This postscript, prepared in mid-1989, discusses three topics that relate to the subject matter of the present volume. A word of caution: The concepts presented here are still somewhat controversial and may not stand up after closer scrutiny.

Greenhouse Warming

We are all agreed that greenhouse gases are increasing in the atmosphere and that they absorb infrared radiation. The question is: Why don't we see a clear signal of warming in the global temperature record? It is generally accepted that temperatures increased by about 0.5°C between 1880 and 1940 (well before there was any appreciable amount of carbon dioxide released from fossil fuel burning); they decreased (!) between 1940 and 1965, and have been increasing since then.[1]

There are two ways to explain this puzzling temperature record (Singer, 1988):

1. The negative feedbacks may be large enough to make greenhouse warming negligible. (See for example the discussions by Covey, Ellsaesser, and Lacis, this volume). In particular, a small increase in cloudiness, caused by an increase in evaporation from the ocean, might increase albedo sufficiently to cancel most of the warming. This hypothesis is strengthened by the recent satellite observation that the net effect of clouds produces cooling. (Ramanathan *et al.,* 1989).

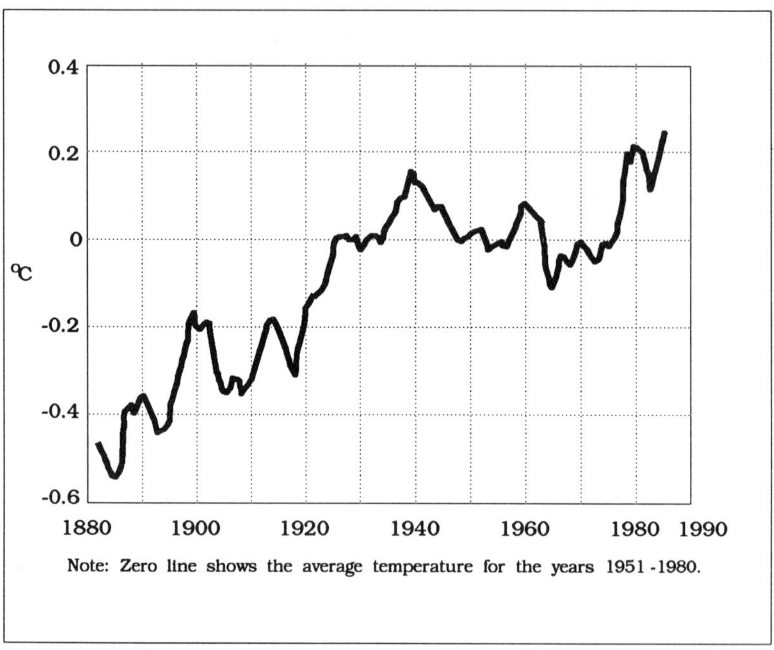

Figure P-1: Changes in the global surface air temperature since 1880. Source: Dr. James E. Hansen NASA - Goddard Institute for Space Studies.

2. Natural fluctuations in the climate may have been "swamping" the greenhouse warming, and could even be responsible for the mid-20th century cooling. According to this hypothesis, the full warming effect will soon become apparent. A particular version has been elaborated by P.R. Bell (1989) of the Oak Ridge Institute of Energy Analysis, who extends the work of Broecker (1975). Bell identifies four cyclical components: two strong ones with periods of 179 and 82.4 years from the oxygen 18/16 record in Greenland ice cores[2] of the past 800 years; and two weaker ones with periods of 18.6 and 22.3 years learned from western U.S. high plains drought records and eastern U.S. winter temperature records. The first three of the cyclic components are believed to be astronomical in nature and to produce their effects by increased tidal mixing; the last component Bell associates with solar luminosity changes.

Solar Cycle Modulation of the Weather

Connections between sunspot cycles and weather phenomena have often been suggested but have never stood the test of time. A sunspot-weather link that seems to be holding up is described in a series of reports in recent issues of *Science* (Kerr, 1989).

This most promising of sun-weather studies got its start in a 1987 paper in which Karin Labitzke of the Free University in Berlin pointed out that a clear-cut effect of the 11-year sunspot cycle on the atmosphere had been masked by its interaction with an internally generated modulator of the atmosphere, the Quasi-Biennial-Oscillation or QBO (*Science*, 23 October, 1987, p. 479). The QBO is a less than perfect periodic reversal of stratospheric winds over the equator every 13 months or so. Its effects reverberate through the stratosphere and down into the troposphere where weather occurs, but the magnitude of its effect outside the tropics is modulated by the solar cycle, Labitzke showed. It had been known since 1980 that the north polar stratosphere during winter tended to be colder during the QBO's west phase than its east phase, but Labitzke pointed out that at solar maximum its behavior was just the opposite of that general trend.

In a subsequent paper Labitzke and Harry van Loon of the National Center for Atmospheric Research in Boulder showed how sorting out east-phase from west-phase winters transformed a complete muddle, in which the opposing effects of the solar cycle in opposite phases wiped out any correlation, into dramatic correlations. Correlations of the solar cycle with temperature could be traced from the north pole stratosphere, into the troposphere, and to the surface, at least in large regions over the hemisphere. Labitzke and van Loon tested the significance of the correlations at the center of these regions using both a Monte Carlo technique and the bootstrap technique, methods intended for this kind of data. The correlations were significant at the 95% confidence level or better. Labitzke and van Loon (1988, 1989a, 1989b) have been extending their initial correlations.

The solar cycle correlates with sea level atmospheric pressure and surface air temperature during the Northern Hemisphere summer as well as during winter, but primarily over the oceans. The separation of the record into east- and west-phase years reveals correlations in the Southern Hemisphere too, but the correlations are especially marked in the east phase there rather than the west phase. The strength of the east-phase correlation is particularly efficient in the Antarctic stratospheric temperature, which is a crucial factor in the formation of the Antarctic ozone hole (*Science*, 28 October, p. 515).

Even if the correlations hold up, there would remain the question of mechanism. How could the feeble variations in solar output over a cycle be amplified to give the apparent changes in the weather, a required amplification of over a million? No one has any answers.

Volcanic Influences on Climate

The largest volcanic eruption in recorded history occurred in the year 536 AD. It produced a dry fog which darkened the sun for 18 months. The rays of the sun were so feeble that the crops did not ripen and there was famine recorded from the Mideast to China. The written records describing this volcanic aerosol are almost identical to the description given by Benjamin Franklin for the eruptions of 1783 AD in Iceland and Japan.

If large eruptions such as those of 536 AD and 1783 AD caused large climate change, then it is not hard to believe that smaller eruptions should cause smaller climate change. What are the climate effects of eruptions such as Krakatau and El Chichon which were 10 to 50 times smaller than the 536 AD eruption? Slight cooling directly beneath the aerosol have been observed as expected. However in 1984, Paul Handler, a physicist at the University of Illinois, published a paper in the Geophysical Research Letters which showed that a) low-latitude volcanic aerosols were the most probable cause of the El Niño[3] and its associated climate anomalies around the world, and b) high-latitude volcanic aerosols were the cause of the anti-El Niño. This separation of volcanic aerosols into two classes, high- and low-latitude, seems to have been the key to understanding the climate impact of volcanoes. The Handler (1984) paper showed very clearly that prior investigators by averaging over all volcanic aerosols, both high- and low-latitude, would obtain a null result.

The physical mechanism for the climate impact of volcanic aerosol is the decrease in radiation over narrow bands of latitude at the surface of the earth resulting from the establishment of the volcanic aerosol. In a paper (1989), Handler describes how the differential cooling of the land with respect to the oceans induces a higher sea level pressure over Eurasia and especially the monsoon lands. The increased sea level pressure over land decreases the normal onshore monsoon winds and results in less precipitation over India, Southeast Asia and all of Africa just prior to the peak temperatures of the El Niño. The higher sea level pressure over the monsoon countries implies that there is less airmass over the oceans and in particular the South Pacific Anticyclone. The winds out of the South Pacific Anticyclone control the wind over the eastern tropical Pacific Ocean (the El Niño region). Since under the aerosol the airmass now resides over the land regions rather than the oceanic anticyclones, the winds coming from the center of these regions are weaker. In the case of the South Pacific Anticyclone, the weaker winds result in less upwelling along the West Coast of South America and the Equator and thus the appearance of the characteristic warmer than normal waters of the El Niño events.

In a Monte Carlo simulation, Handler and Andsager (1989) showed that hypothesis of volcanic aerosols causing the El Niño events was above the 99.95 significance.

Any global model of climate must satisfy three criteria. 1) There must be a plausible physical model connecting cause and effect. 2) There must be a body of statistical evidence which fits the physical model. 3) Finally and most important, the model must predict the future. On November 13, 1985, the low-latitude volcano, Nevado del Ruiz, erupted in Colombia. Within three months the sea surface temperature in the eastern tropical Pacific Ocean began to turn warmer than normal and the 1986/87 El Niño was born. Handler states that his model also predicted 6–12 months in advance the following climate events in 1988 and 1989:

1. The anti-El Niño; 2. Drought in the USA; 3. Excellent rains in all of the monsoon countries of the world; 4. Excellent rains in South Africa and the Sahel; 5. The above average number of tropical storms in the North Atlantic Ocean; 6. The persistence of the anti-El Niño into 1989.

In spite of the fact that there are no alternative hypotheses with an equivalent level of explanatory and predictive power, Handler's model is not yet widely known and accepted.

Notes

1. See Figure P-1.
2. For a discussion of this technique, see, e.g. Oeschger (1990).
3. For a detailed discussion of El Niño and other ocean-climate interactions, see, e.g. Kraus (1990).

References

Bell, P.R., "Reprise and Extension of Broecker's Prediction: Factors in Hemispheric Temperature Records." To be published. 1989.

Broecker, W.S., *Science* 189, 460 (1975)

Handler, P., Possible Association of Stratospheric Aerosols and El Niño type events, *Geophysical Research Letters*, 11, 1121–1124, 1984.

Handler, P., Volcanoes, The effect of volcanic aerosols on global climate, *J. Volcanology and Geothermal Processes*, Vol 3/4 June 1989.

Handler, P. and K. Ansager. "Volcanic Aerosols, El Niño and the Southern Oscillation" *J. Clim.*, (in press, 1989)

Kerr, R.A.., *Science* 242, 1124 (1988).

Kraus, E., "The Role of the Ocean in Climate Fluctuations," *The Ocean in Human Affairs* (S. Fred Singer, ed.) Paragon, to be published 1990.

Labitzke, K., and H. van Loon, 1988: Association between the 11-year solar cycle, the QBO and the atmosphere. Part I: The troposphere and the stratosphere in the Northern Hemisphere in winter. *J. Atm. Terr. Phys.*, 50, 197–206

Labitzke, K., and H. van Loon, 1989a: Association between the 11-year solar cycle, the QBO, and the atmosphere. Part III: Aspects of the association. *J. Clim.*, 2, 554–565.

Labitzke, K., and H. van Loon, 1989b: The 11-year solar cycle in the stratosphere in the northern summer. *Annal. Geophys.*, 7, in press.

Oeschger, H., "Longterm Climate Stability," *The Ocean in Human Affairs* (S. Fred Singer, ed.) Paragon, to be published 1990.

Ramanathan, V. *et al.*, *Science* 243, 57 (1989).

Singer, S.F., "Fact & Fancy on Greenhouse Earth," *Wall Street Journal*, Aug 30, 1988.

CONTRIBUTORS

Gad Assaf: Chief Scientist, Solmat Systems, Yavne, Israel.

Robert U. Ayres: Professor of Engineering and Public Policy, Carnegie-Mellon University, Pittsburgh, Pennsylvania.

Penelope J. Boston: National Research Council Associate, NASA-Langley Research Center, Hampton, Virginia.

Ralph J. Cicerone: Director, Atmospheric Chemistry and Aeronomy Division, National Center for Atmospheric Research, Boulder, Colorado.

Curt Covey: Lawrence Livermore National Laboratory, Livermore, California

Hugh W. Ellsaesser: Guest Scientist, Lawrence Livermore National Laboratory, Livermore, California.

A.G. Everett: President, Everett and Associates, Consulting Scientists, Rockville, Maryland.

H.G. Goodell: Professor, Department of Environmental Sciences, University of Virginia, Charlottesville, Virginia.

William W. Kellogg: Senior Scientist, National Center for Atmospheric Research, Boulder, Colorado.

Andrew A. Lacis: Research Scientist, Goddard Institute for Space Studies, New York, New York.

Devendra Lal: Visiting Scientist, Institute of Geophysics and Planetary Physics, University of California at San Diego, La Jolla, California.

Helmut E. Landsberg: Former Professor of Meteorology, University of Maryland, College Park, Maryland.

Sergej Lebedeff: Research Scientist at the Goddard Institute for Space Studies, New York, New York.

Avraham Melamed: Engineer, Tushia, Ltd., Consulting Engineer, Tel Aviv, Israel.

Kenneth Mellanby: Director Emeritus, Monk's Wood Experimental Station, Huntingdon, England.

Michael G. Norton: Warren Springs Laboratory, Stevenage, U.K.

P. Kilho Park: Office of Chief Scientist, National Oceanographic & Atmospheric Administration, Washington, D.C.

F. Sherwood Rowland: Professor of Chemistry, University of California, Irvine, California.

Colette Serruya: Research Scientist, Israel Oceanographic and Limnological Research, Ltd., Haifa, Israel.

S. Fred Singer: Professor of Environmental Sciences, University of Virginia, Charlottesville, Virginia

Gerald Stanhill: Research Scientist, Volcani Institute of Soils and Water, Agricultural Research Organization, Ministry of Agriculture, Bet-Dagan, Israel.

LIST OF FIGURES

LIST OF TABLES

INDEX

D

E

F

G